The Heavens on Fire

The Great Leonid Meteor Storms

The years 1998, 1999, and 2000 offer the last chance people on Earth will have for a century to see the most spectacular of all meteor showers, the Leonids. In 1966, when they last filled the sky in great numbers, observers reported 40 every second. When this storm blazed in 1833, two widely separated observers described the sight as "the heavens on fire."

With the returning Leonids now reaching their peak of activity, *The Heavens on Fire* tells the story of meteors, and especially the Leonids, whose terrifying beauty established meteor science. Mark Littmann traces the history and mythology of meteors, profiles the fascinating figures whose discoveries advanced the field, and explores how meteors have changed the course of life on Earth. His book offers advice on how and where to make the best of the 1998, 1999, and 2000 return of the Leonids.

The Heavens on Fire

The Great Leonid Meteor Storms

MARK LITTMANN

CAMBRIDGE
UNIVERSITY PRESS

PUBLISHED BY THE PRESS SYNDICATE OF THE UNIVERSITY OF CAMBRIDGE
The Pitt Building, Trumpington Street, Cambridge CB2 1RP, United Kingdom

CAMBRIDGE UNIVERSITY PRESS
The Edinburgh Building, Cambridge CB2 2RU, UK   http://www.cup.cam.ac.uk
40 West 20th Street, New York, NY 10011–4211, USA   http://www.cup.org
10 Stamford Road, Oakleigh, Melbourne 3166, Australia

First published 1998

Printed in the United Kingdom at the University Press, Cambridge

Typeset in Adobe Utopia 9.45/14pt using QuarkXPress ™ [SE]

A catalogue record for this book is available from the British Library

ISBN 0 521 62405 3 hardback

For Peggy, Beth, and Owen
with love

CONTENTS

ACKNOWLEDGEMENTS

I would like to thank the scientists, historians, and writers who critiqued portions or all of the manuscript and/or answered questions I had about aspects of this story. They added accuracy and readability to the manuscript. The flaws that remain are mine.

Peggy Littmann, Knoxville
Bea Owens, Salt Lake City
Ann Rappoport, Philadelphia
Donald K. Yeomans, Jet Propulsion Laboratory
Ruth Freitag, Library of Congress
Neil Bone, British Astronomical Association
Stephen J. Edberg, Jet Propulsion Laboratory
Iwan P. Williams, Queen Mary and Westfield College, London
Christopher F. Chyba, Lunar and Planetary Laboratory, University of Arizona
Jacques Sauval, Royal Observatory of Belgium
Thomas J. Ahrens, California Institute of Technology
Dorrit Hoffleit, Yale University
Brian Marsden, Harvard-Smithsonian Center for Astrophysics
Paul Ashdown, University of Tennessee
David Hughes, University of Sheffield
Stanley Hey, Royal Radar Establishment
John Mason, British Astronomical Association
Larry Lebofsky, Lunar and Planetary Laboratory, University of Arizona
Carl Littmann, St. Louis
Jane Littmann, Columbia, South Carolina
Harry Y. McSween, Jr., University of Tennessee
John Noble Wilford, *New York Times*
Alan MacRobert, *Sky & Telescope*
David Morrison, NASA Ames Research Center

Donald Brownlee, University of Washington
Alan Fiala, United States Naval Observatory
Mark Davis, North American Meteor Network, Charleston, South Carolina
Robert Hawkes, Mt. Allison University, Sackville, New Brunswick
Paul Roggemans, International Meteor Organization, Mechelen, Belgium
Joe Rao, Yonkers, New York
Tom Gehrels, Lunar and Planetary Laboratory, University of Arizona
Ken Willcox, Astronomical League, Bartlesville, Oklahoma
Muriel Littmann, St. Louis
Two anonymous reviewers

For contributing vignettes and/or photography that enhance the value of this book, special thanks are due to:
Stephen J. Edberg, Jet Propulsion Laboratory
Neil Bone, British Astronomical Association
Fred L. Whipple, Harvard-Smithsonian Center for Astrophysics
Jacques Sauval, Royal Observatory of Belgium
Jim Young, Table Mountain Observatory, Jet Propulsion Laboratory
Jose Olivarez, Chabot Observatory and Science Center, Oakland
James Van Allen, University of Iowa
Jay Anderson, Environment Canada, Winnipeg
Dorrit Hoffleit, Yale University
Donald Brownlee, University of Washington
Peter D. Hingley, Royal Astronomical Society
Roy Goodman, American Philosophical Society

Rob Landis, Space Telescope Science
Institute
Stanley Hey, Royal Radar Establishment
Bernard Lovell, University of Manchester
Richard P. Binzel, Massachusetts Institute
of Technology
Eleanor Helin, Jet Propulsion Laboratory
James Cornell, Harvard-Smithsonian
Center for Astrophysics
Martha Leigh, Seventh-Day Adventist
Church, Knoxville
Peter A. Grognard, Embassy of Belgium,
Washington, D.C.
Kevin Grace, University of Cincinnati
Billie Broaddus, University of Cincinnati
Medical Heritage Center
Barbara Goyette, St. John's College
Ed Eckert, Lucent Technologies
Maria Schuchardt, Lunar and Planetary
Laboratory, University of Arizona
Lone Gross, Copenhagen University
Astronomical Observatory
Anthony F. Aveni, Colgate University
Dorothy Schaumberg, Mary Lea Shane
Archives of the Lick Observatory
David A. Kring, University of Arizona
Judith Ann Schiff, Yale University
Joyce Anderson, *Denver Post*

It would not have been possible for me to
obtain the documents I needed for this
project without the highly competent and
extraordinarily gracious help of the interli-
brary services librarians at the University
of Tennessee: Jim Hammons, Kathleen
Bailey, Tina Bentrup, Tracy Luna, Eric
Arnold, and Kelley Shelburne.

The reference librarians at the
University of Tennessee found information
for me that I thought was unfindable and
cheerfully steered me in the right direc-
tion. Thanks to Rita Smith, Teresa Berry,
Gayle Baker, Margaret Casado, Russ
Clement, Karmen Crowther, Lana Dixon,
Felicia Felder-Hoehne, Marie Garrett,
James Lloyd, Janette Prescod, Jane Row,
Linda Sammataro, Flora Shrode, Alan
Wallace, and Flossie Wise.

I needed help translating some languages in which important discoveries
about meteor science were first published.
That help came generously and skillfully
from Peter Höyng, Nicole Laroussi, and
Esther Littmann for German; Leon Stratikis
and David Littmann for French; and Beth
Secrist and Salvatore Vecchiuzzo for Italian.

University of Tennessee communica-
tions scholars Dana Williams and Tom
Carter indefatigably tracked down many
old newspapers and periodicals that
carried fascinating reports about early
sightings of the Leonids.

A picture is worth at least a thousand of
my words. The diagrams in this book were
created by Will Fontanez and Tom Wallin of
the University of Tennessee Cartographic
Services Laboratory. The artwork is by
Knoxville space artist Mark Maxwell.

Special thanks to Simon Mitton, direc-
tor of Science, Technology, and Medical
Publishing for Cambridge University Press,
for his confidence in this book and to the
production staff at Cambridge University
Press for their editing, design, and publica-
tion of this work: Jo Clegg, Miranda Fyfe,
Brian Watts, Rob Sawkins, Debbie Garrett,
Nicola Stearn, Kirsten Watrud and Cristin
Curry. The fine index is the work of Alexa
Selph of Atlanta.

I am deeply grateful to the Dudley
Observatory for a Dudley/Pollock Award
and to the University of Tennessee for two
grants that greatly assisted research for this
project.

CHAPTER 1

The night the stars fell

The most grand and brilliant celestial phenomenon
ever beheld and recorded by man.   RICHARD DEVENS (1876)[1]

DENISON OLMSTED's left eye cracked open a slit. It was still night. He shut his eye. Something had awakened him. A noise from outside – sort of a muffled moan? Or was it the brightness of the sky visible through the curtains? Just the light of the full moon high in the heavens. Yes. No.

Suddenly both his eyes shot open and he sat bolt upright in bed.[2] November 12 – no, now the 13th. It's new moon. The sky should be dark.

Then came a pounding on the door and the voice of his next-door neighbor, "Denison. Denison. You must see this. Look out your window."

He sprang from his bed – too fast – weaving dizzily, lightheadedly, the few steps to his window and pulled the curtain aside. He looked. A shiver of ice shot down his back.

In a moment he was heading out his door. He did not remember pulling his coat on over his nightclothes or wedging into his shoes. He stepped outside and looked up, his vision no longer framed and limited by his window – and a chill shuddered through him again. He saw not a meteor or two that would have made an ordinary night memorable. He saw dozens of shooting stars, fireballs, at every moment, in every direction. The sky was full of . . . fireworks. His first instinct was to duck, to fall to the ground and cover his head to protect himself from these falling objects.

The appearance of these meteors was striking and splendid beyond anything of the kind he had ever witnessed.

There were so many, and yet – it was so . . . organized. It appeared that the meteors were all spreading out from a single region – no, a single point – high in the sky, near the zenith. Where in the star field? The falling stars made it hard to get his bearings. But it was Leo. The meteors were all . . . all? – yes, as best he could see – all radiating from the constellation Leo, the Lion. From the curve of the sickle-shaped pattern of stars that is supposed to be the Lion's mane.

But it wasn't that all the meteors started right at that point in Leo and raced away in all directions. Leo stood out because there were few meteors visible there. Only short streaks, sometimes bright. Or just momentary glowing dots, like fixed stars swelling in brightness and then fading away. No, Leo was the center of this pageant because all the meteors he could see in every corner of the sky were streaking away from Leo. With so many of them, he could see it clearly. Even if a meteor trail started almost halfway across the sky from Leo, its trail was always away from Leo. The starry show was radiating out of Leo.

Olmsted wondered how widely this spectacle could be seen and what the reaction to it was. In that instant, he sensed that he, a college teacher of astronomy, was privileged to be seeing an astronomical happening of historic magnitude. He had trained most of his life for such an opportunity. Would he be worthy? He redoubled his efforts to look closely and remember everything. He concentrated his thoughts on seeing the event scientifically.

Date. November 13, 1833.

Time. He didn't have his pocket watch. His neighbor did. It was 5:15 a.m.

Conditions. Excellent. No clouds. No Moon. Mild for November. No wind. Orion, Sirius, and Procyon in the southwest. Balancing that brightness were Venus and Saturn in the southeast. He could see them through the blizzard of falling stars primarily because they were among a tiny minority of stars that were not in motion.

He tried to focus on one meteor at a time, then another. It was difficult to do. Bright flashes caught by the corners of his eyes continuously distracted him. Each bright meteor left behind a vivid streak of light, ending in what sometimes seemed to be a puff of smoke, sometimes an explosion, sometimes . . . just vanishing.

And then he noticed – it was eerie – there was no noise from the heavens, no sound of explosion. He strained to listen more carefully. He did hear a noise, rising and falling with the bursts and lulls of the shower. It was . . . a moan. It was . . . people. A collective noise of shock and awe and – fear. People awakened as he had been from a sound sleep. Standing beneath a sight they had never seen – or been warned about – or imagined. Perhaps like no starfall ever seen. He wondered for just an instant whether this vision might yet be a dream.

Concentrate. Some meteors were brighter than others. Some brighter than Mars or Jupiter or even Venus, the planets that outshine the brightest of the true nighttime stars. His neighbor had seen one nearly as bright as the full moon, and seemingly that large in the sky.

When he first stepped outside and looked at the meteors, it appeared that they were all raining down from a point nearly straight up – the zenith – where Leo, the Lion, stood. Now, half an hour later, it occurred to Olmsted and his neighbors observing near him that the point of apparent radiation of the meteors had shifted. The meteors were still radiating from

Leo, in fact, from near the star Gamma Leonis, but that radiating point had moved westward. The radiant of the meteor shower was moving westward in synchronism with the stars. The westward motion of the stars was created by the Earth spinning on its axis from west to east, once around in a day, causing the stars and Sun and Moon and planets to rise and set. Clearly, these shooting stars were not traveling with the Earth. They must come from space, far beyond our planet.

Light was creeping up the eastern horizon. Dawn had come. The meteors were fewer and fewer now, but still they fell. Was the shower declining or was it just getting harder to see the shooting stars because of the twilight? He could still see bright ones. The fall of meteors must be continuing unseen into the morning light.

Not much time left. Notice everything. You never know what might be important.

The weather. On the evening of the 11th, it had rained copiously, giving way on the 12th to gusty winds from the west. The skies cleared by evening and he had seen . . . a few falling stars . . . before bedtime. And now, the weather was perfect. Any connection? Time for that later.

Had anything like this deluge of meteors ever happened before? It seemed to him he had read about something like this. He thanked the neighbor who had awakened him and said good night – good morning, rather – to the other citizens and rushed inside to begin writing while the memory was still fresh. He might just have time to dash off a brief report and offer it to the newspaper, ending with a request for accounts from other observers.

Now where had he read about a previous extraordinary meteor storm? A travel book. Yes, Humboldt. He owned a copy. Yes, there it was, near the beginning of his travels in South America. Humboldt and Bonpland saw a meteoric spectacle from Cumaná in 1799. Huh. Look at that. The meteors fell in the early morning hours of the 12th of November. That was worth noting.

Olmsted delivered his impressions to the *New Haven Daily Herald* and his article was published that same day, including his hope to hear from other observers.

The response overwhelmed him. Other newspapers across the young nation had picked up his report and Denison Olmsted, 42 years old, professor of mathematics and natural philosophy at Yale College, found himself the clearinghouse for information and interpretation of the stars that fell on November 13, 1833. It was to be the defining moment of his career.

The ecstasy of the sight

From Alexandria, Virginia:
A more magnificent and splendid spectacle was never presented.[3]

From Baltimore, Maryland:
It seemed to rain fire.[4]

From Charleston, South Carolina:
Those who were up before the dawn yesterday witnessed a most glorious
sight – one glance at which "were worth ten years of common life." [5]

While Denison Olmsted was noticing that the meteors were streaming
out of Leo and that this point of radiance was sliding westward with the
star field as the Earth turned, observers all over the young United States
were recording their impressions, overwhelmed by the sight of meteors
beyond counting.

From Boston, Massachusetts:
Meteor succeeded meteor in such rapid succession that it was impossible
to count them; at times the sky seemed full of them, and the earth was
illuminated as with a morning light.
... Those who were so fortunate as to witness the scene describe it as
brilliant beyond conception ... [6]

From Natchez, Mississippi:
From 3 to 5 o'clock, the scene was truly magnificent – thousands upon
thousands [of meteors] were darting about in all directions without an
instant's cessation. It was so light that upon first awaking many thought
that the city was on fire.[7]

From Bowling Green, Missouri:
The most perfect master of language would fail of conveying to others a
full picture of this extraordinary and uncommon appearance ... [8]

It even got to be a matter of competition between cities as to which was
blessed with the most radiant celestial performance. New York claimed
victory over Philadelphia.

The celestial exhibition of yesterday morning is noticed in the
Philadelphia papers, but it is evident from their accounts that it fell far
short both in the number of meteors and the brilliancy of their light of the
splendors visible in our city. A correspondent of the [Philadelphia]
National Gazette estimates their number at two thousand one hundred
and sixty in the compass of two hours and a half. More than that number
were visible here within every ten minutes of that period, and it was as
difficult to count them as to number the raindrops." [9]

This disparity was to play a major role in the subsequent controversy
about the cause of the meteors.

Newspapers throughout the country were proud of their rational, unsuperstitious coverage of the meteor barrage. They began the day the stars fell by publishing the accounts of local observers and continued on and off for the next six weeks, as the papers reprinted each others' stories and offered commentary from local scientists and from Olmsted.

The press, South and North, reported the terrified response of illiterate slaves to the unexpected and unparalleled sight of the meteoric avalanche, but the tone was not smug hauteur or amused condescension. Instead, the white plantation owners and newspapermen generally responded with respectful understanding and even agreement.

From Hartford, Connecticut:

The negroes [in] the South who saw the phenomenon describe it as *"snowing fire"*: they generally thought the Judgment day had come.[10]

From Combahee, South Carolina, a planter wrote:

I was suddenly awakened by the most distressing cries that ever fell on my ears. Shrieks of horror and cries for mercy I could hear from most of the negroes of three plantations, amounting in all to about six or eight hundred. While earnestly listening for the cause, I heard a faint voice near the door calling my name. I arose and, taking my sword, stood at the door. At this moment, I heard the same voice still beseeching me to rise, and saying "O my God, the world is on fire!" I then opened the door, and it is difficult to say which excited me most – the awfulness of the scene or the distressed cries of the negroes. Upwards of one hundred lay prostrate on the ground – some speechless and some with the bitterest cries, but most with their hands raised, imploring God to save the world and them. The scene was truly awful; for never did rain fall much thicker than the meteors fell towards the earth; east, west, north, and south, it was the same.[11]

It was not just the blacks screaming. From Raleigh, North Carolina:

The scene was truly awful and indescribably sublime; . . . it carried to the bosoms of many terror and consternation. Some imagined the world was coming to an end and began to pray; and a gentleman from the country states that such was the alarm produced in the neighborhood where he was [that] the welkin every where around him resounded with cries of distress.[12]

Observers frequently found validity in the slaves' reactions. The author of the Raleigh newspaper story, struggling to describe the shower, said that it looked, "to use to the striking expression of an untaught son of Africa, 'like it was snowing stars.'"[13]

Nestled among the accounts of terrified slaves were reports that the educated white population was not one bit less stunned and confused and scared. A Macon, Georgia newspaper minced no words: "We do not jest when we say that stubborn knees were bent and flinty hearts melted into deep contrition at the alarming prospect of 'the heavens on fire.'"[14]

The scene looked to many Christians like the end of days portrayed in the New Testament. An anonymous observer in Bowling Green, Missouri wrote in the *Salt River Journal*:

> Forcibly were we reminded of that remarkable passage in Revelations which speaks of the great red dragon . . . drawing the third part of the stars of heaven and casting them [down] to the earth. . . . That figure appeared to be fully painted on the broad canopy of the sky – spread over with sheets of light and thick with streams of rolling fire. There was scarcely a space in the firmament which was not filled at every instant with these falling stars.[15]

The meteors inspired bizarre behavior:

A reliable witness

It was the predawn hours of November 13, 1833 in Annapolis, Maryland. The Reverend Hector Humphreys was sleeping soundly. In only two years as principal (president) of St. John's College, he had saved the school from bankruptcy by prevailing on the Maryland state legislature, meeting a few blocks away, to provide an annual subsidy. Using this endorsement, he could raise funds to build new buildings and expand the student body.

Hector Humphreys. Courtesy St. John's College

His wife Marie startled him awake. "Fire!" Her voice was shaking – "Fire! – as she stumbled toward a window. The room was bright, lighted from outside. The sky was a shower of sparks. Yet they could see no building on fire; could smell no smoke; heard no alarm. They dressed hurriedly and rushed outside.

That same day Humphreys sat down to write what he had seen for the *Annapolis Republican*. When an article on the meteors by a Yale professor appeared, requesting other accounts, he sent his newspaper report to, what was his name? – Olmsted. After all, Yale was his alma mater.

> They all appeared to move outward from a common centre, at or near the zenith. At times, they completely filled the whole heavens with beautiful brilliant streams of light, extending to the horizon. I do not mean that all the trains actually extended from the zenith to the horizon, but that the lines of light were *so directed* that if extended backwards, they would all converge to a point in the zenith. Their appearance was often so incessant that all the stars of the firmament seemed to be darting from their places. Many persons thought a shower of fire was falling and became exceed-

A fellow near Georgetown, District of Columbia, who had robbed a hen roost and was carrying off his booty, is said to have been so much frightened at what he believed a threatened judgment that he ran back and was caught in the act of returning his plunder.[16]

An observer in Fredericksburg, Virginia found in the meteors a presage of the Civil War, still more than a generation away:

> The whole starry host of heaven seemed to be in a state of practical secession and revolt . . . which finds parallel only in the affairs of earth.[17]

Newspapers did not stop with recording public reaction. Reporters, editors, and readers made an effort to explain the meteor deluge scientifically. The most frequently expressed belief was that the shooting stars were caused by a change in the weather. As the *Huntsville* (Alabama) *Democrat* wrote in its weekly issue the day after the great meteor storm:

ingly alarmed. The light was so intense that some sleepers woke up thinking that their dwellings were in flames.

The phenomenon must have continued more or less vividly for four or five hours. Many intelligent people in this city saw them and agree that there was an *almost infinite number of meteors*. They fell *like flakes of snow*.

It is said that some of the meteors were seen to fall upon the earth and to rebound into the air. As no vestiges, however, have been discovered upon the ground, it may be presumed that this was an optical deception. No audible explosion attended any of the meteors. It was a perfectly *silent and simultaneous dance of the stars*. It is probable that the phenomenon was seen over a wide range of the country. A gentleman living several miles beyond the Severn River saw the meteors there in as great abundance as they occurred here. The steamboat Maryland was about to leave Cambridge, on the eastern shore of Chesapeake Bay, so the hands were up at an early hour, and all on board agree substantially with what was witnessed at Annapolis.

Notwithstanding the strong persuasion of several observers that the meteors fell upon the ground, I am convinced that their paths were in the upper and rarer strata of the atmosphere, since optical principles show that in darting away to the horizon, they would *appear* to descend and to strike into the earth. The usual theory of meteors – that they are caused by inflammable gases high in the atmosphere – does not appear to explain the phenomena. If we admit that the gases are generated and diffused sufficiently to kindle up the whole heavens with light, the combustion of them would not present those *innumerable distinct sparks* which shot from the region of the zenith with such *perfect uniformity of direction*. This uniform direction of motion was in fact the most remarkable point in the whole phenomenon.*

* Condensed and slightly paraphrased from Humphreys' letter to Denison Olmsted as it appeared in "Observations on the Meteors of November 13th, 1833," *American Journal of Science and Arts*, volume 25, number 2, January 1834, pages 371–373. Humphreys' italics for emphasis have been retained.

  Here and elsewhere in long quotations of 19th century English I have retained the phrasing but modernized the punctuation to avoid confusion.

A practical joke

Late in life, James Flanagan, a judge in Clark County, Kentucky, recalled the "falling of the stars" on November 13, 1833, "the memory of which shall remain with me as long as life lasts." It "scared everybody to prayers," he said.

"The people were struck with awe, and thrown into great consternation," Flanagan remembered, "and one of the effects of the remarkable occurrence was to awaken a pious feeling, causing a general religious revival throughout Christendom." At the Log Lick Church, the congregants took the falling stars "as a sign that the end of the world was near at hand." Even after dawn brought an end to the meteor shower, "the little church was crowded to overflowing day and night with an eager and earnest people, singing and asking pardon for their many sins. Old feuds were reconciled, enemies were made friends ... for they expected at any moment to hear the last trump sound and be called to an account of their doings here below."

A few citizens, however, "took no part in the religious mania, among them Thomas F. Danaldson and M. Fritz, who were noted as practical jokers."

One night when the congregation had gathered at the little church and were listening with fear and trembling to the awful warning that the end was near at hand, Danaldson and Fritz, having procured a long tin bugle and a ladder, climbed to the top of the church, and just as the preacher was exhorting his hearers to be ready as Gabriel's trumpet might sound at any moment, Danaldson blew a blast "both loud and shrill" on the tin bugle, and Fritz proclaimed from the house top in stentorian tones that he was the Angel Gabriel sent to proclaim the end of the world and summon all nations to arise and come to judgment.

Hearing all this, the audience inside fell in a promiscuous heap on the floor, some begging for a little more time and others begging for immediate mercy and pardon. They remained in the church until broad daylight, and when they had finally mustered courage to venture out, the two "Gabriels" were nowhere to be seen and the old world was standing just as it always had stood. For a long time this affair remained a mystery, except to a few who were on the inside of the joke.*

* James Flanagan: "Falling of the Stars. The Remarkable Phenomenon that Scared Everybody to Prayers in 1833. Danaldson's Long Trumpet, and What He Did with It. M. Fritz Proclaims Himself the Angel Gabriel." Transcribed by George F. Doyle. No publisher, no date; cataloged as "Clark County, Kentucky – Meteorites"; four typewritten pages held by the University of Kentucky library.

It is no doubt the effect of an impure state of the atmosphere – the weather for some days having been warm and damp, but suddenly changed to cool or frosty.

Other newspapers fell back on the most prominent scientific theory of shooting stars, which from the time of Aristotle had linked meteors to meteorology – part of the everchanging weather patterns of Earth. Thus meteors were not associated with the planets or the stars; meteors were not part of astronomy.

The *Florence* (Alabama) *Gazette* looked up shooting stars in Conrad Malte-Brun's *Universal Geography* textbook and explained that meteors were caused by hydrogen gas that had been "sulphurated." As this sulphurated hydrogen gas rose to high elevation, it mixed with oxygen and was ignited by a spark so that the oxygen and hydrogen formed water and the sulfur fell to the ground as a "fetid, glutinous matter of a whitish color bordering upon yellow."[18]

Another theory, a favorite among Americans because of Benjamin Franklin's prominence in experiments with electricity, was that the meteors were an electrical disturbance somewhere in the atmosphere – a special form of lightning.

Virtually every observer of the great starfall of November 13, 1833 agreed with Virgil H. Barber of Frederick, Maryland: "I observed the most brilliant phenomenon of nature I ever [saw]." Denison Olmsted was equally awestruck: "The appearance of these meteors was striking and splendid, beyond any thing of the kind [I have] ever witnessed."

But Olmsted did not stop with a recollection of grandeur. He decided to take another step – to try to understand what he had seen.

The meteoric shower
of November 13, 1833.
Courtesy Seventh-Day
Adventist Church

Trying to draw what words cannot convey

Joseph Harvey Waggoner was 13 years old and living in eastern Pennsylvania when he and his family were overwhelmed by the great meteor storm of November 13, 1833. Some people he knew "hid themselves behind the curtains or under the bedclothes . . . refusing to take a second glance." When the meteors were at their peak, Waggoner recalled, "They could no more be counted than one can count the fast falling flakes of snow in a hard storm." Despite the bitter cold, he and his family couldn't bring themselves to go inside while any falling star could be seen. "It was a sight never to be forgotten."

As an adult, Waggoner converted from Baptist to Seventh-Day Adventist and became a minister. But he never forgot the meteors of 1833. Late in life, while on a mission in Switzerland, he thought that nothing could better illustrate the expected signs of the Apocalypse prophesied in the New Testament than the meteors he had seen as a boy. In the Gospel According to St. Matthew (24:3), Jesus had been asked by his disciples "what shall be the sign of thy coming, and of the end of the world?" Jesus had replied in part (24:29), "Immediately after the tribulation of those days shall the sun be darkened, and the moon shall not give her light, and the stars shall fall from heaven, and the powers of the heavens shall be shaken." Similarly, the Book of Revelation (6:13) described the opening of the sixth seal: "And the stars of heaven fell unto the earth, even as a fig tree casteth her untimely figs, when she is shaken of a mighty wind."

Waggoner approached artist Karl Jauslin to draw the meteor fall to preserve for others what had through the years remained so vivid in his mind. Waggoner was thrilled with the results: the drawing gave "the best idea of the actual scene of all the representations that I ever saw." A wood block engraving was made and it appeared in the Seventh-Day Adventist weekly publication *The Signs of Our Times* in 1888, 54 years after Waggoner saw the meteors as a boy. The image was then used in the 1889 edition of the church's question-and-answer guide *Bible Readings for the Home Circle*. Sketches claiming to be realistic that were published right after the night the stars fell suggest that, despite the passage of more than half a century, Jauslin's depiction of Waggoner's memory may not have exaggerated the event.*

* This vignette is based on David W. Hughes: "The World's Most Famous Meteor Shower Picture," *Earth, Moon, and Planets*, volume 68, 1995, pages 311–322.

NOTES

1. R[ichard] M. Devens: *Our First Century: Being a Popular Descriptive Portraiture of the One Hundred Great and Memorable Events of Perpetual Interest in the History of Our Country* (Springfield, Massachusetts: C. A. Nichols, 1876). The memory of the 1833 meteor storm was still so vivid in 1876 that Devens chose it for inclusion when he set out to observe the centennial anniversary of the United States by writing one of the earliest coffee-table books, *Our First Century: Being a Popular Descriptive Portraiture of the One Hundred Great and Memorable Events of Perpetual Interest in the History of Our Country*. Event 36 was the "Sublime Meteoric Shower All Over the United States – 1833," with the first section summarized as "The Most Grand and Brilliant Celestial Phenomenon Ever Beheld and Recorded by Man."

2. "The flashes of light, though less intense than lightning, were so bright as to waken people in their beds." Denison Olmsted quoted in *Mechanics' Magazine*, volume 2, number 5, November 1833, pages 287–288.

3. *Alexandria* (Virginia) *Gazette*, November 14, 1833. Also quoted in *Richmond Enquirer*, November 19, 1833.

4. From a letter by Thomas Kenny, November 13, 1833, to the *Baltimore American*, quoted in the *Richmond Enquirer*, November 19, 1833.

5. *Charleston Mercury*, November 14, 1833.

6. *Boston Transcript*, November 13, 1833.

7. *Natchez Courier*, November 15, 1833.

8. Olmsted was not the only one receiving correspondence about the meteor display. Benjamin Silliman, founder and editor of the *American Journal of Science and Arts*, the leading American scientific publication, received a copy of the November 20, 1833 edition of the *Salt River Journal* from Bowling Green, Missouri, about 75 miles northwest of St. Louis. The newspaper contained an anonymous article consisting of a pastiche of meteor observations of erratic accuracy and scientific value. Silliman passed it along to his Yale colleague, requesting that Olmsted write a summary and interpretative article on the meteors for Silliman's Journal (as it was often called).

9. *New-York Commercial Advertiser*, November 14, 1833.

10. *Hartford Times* (weekly), December 2, 1833.

11. No author given. Slightly condensed from the *Charleston Mercury*, November 15, 1833. This account was reprinted in Thomas Milner (The Rev.): *The Gallery of Nature: A Pictorial and Descriptive Tour Through Creation, Illustrative of the Wonders of Astronomy, Physical Geography, and Geology* (London: Wm. S. Orr, 1849), page 140, and in R[ichard] M. Devens: *Our First Century: Being a Popular Descriptive Portraiture of the One Hundred Great and Memorable Events of Perpetual Interest in the History of Our Country* (Springfield, Massachusetts: C. A. Nichols, 1876), page 330.

12. *Raleigh Star*, November 15, 1833. Also quoted in *Richmond Enquirer*, November 19, 1833.

13. *Raleigh Star*, November 15, 1833. Also quoted in *Richmond Enquirer*, November 19, 1833.

14. *The Georgia Messenger* (Macon), November 14, 1833.
The idea of the heavens on fire was also chosen to describe the event by the unnamed author of "The Meteoric Shower" in *New England Magazine*, volume 6, January 1934, page 52 of pages 47–54. "And who could behold unmoved this fearful lighting up of the midnight sky – the heavens apparently on fire – millions of stars seeming to fall from their spheres, and the elements as if about to melt with fervent heat?"

15. *Salt River Journal*, November 20, 1833; reprinted in Denison Olmsted: "Observations on the Meteors of November 13th, 1833," *American Journal of Science and Arts*, volume 25, number 2, January 1834, pages 381–383 of 363–411.

16. Variously reported; this quotation from the *Hartford Times* (weekly), December 2, 1833.
In the same issue, the *Hartford Times* reported:
In Portsmouth, N.H., a lady of rather wiry nerves was called up from her slumbers to behold this wonder of the natural world. This sudden transition from her pleasant dreams to such strange realities at such an hour threw her into the most frantic delirium, and she fainted away before her friends could have an opportunity of explaining the cause of the light and dancing stars which met her vision. She was positive the world had come to an end and it was no use for any body to convince her to the contrary.

17. From Fredericksburg, Virginia on November 15, 1833, reported in *Richmond Enquirer*, November 19, 1833.

18. As quoted in the *Huntsville Democrat*, November 21, 1833.

CHAPTER 2

Sifting

The early morning storm of meteors seen in the eastern United States
on November 13, 1833 marked the birth of modern meteor astronomy.

DONALD K. YEOMANS (1981)[1]

DENISON OLMSTED gathered newspaper articles about the meteor
storm from throughout the United States and its western territories.
Correspondence poured in to – and out from – him. His Yale colleague
Benjamin Silliman shared accounts of the meteors that had been submitted for his *American Journal of Science and Arts*. Olmsted dug through
observational reports and tried to extract what was meaningful.[2]

It wasn't easy. There were contradictions. There were bizarre observations. What should he make of them?

A man reported that on the night after the meteor display, the tips of the
ears of the horse he was riding became luminous.

At sunrise in West Point, New York, as the meteors became invisible in
the sky, a woman milking cows saw something land with a "sposh" in front
of her. It resembled "boiled starch" and was about the volume of a coffee
cup. About 10 a.m. she thought to show it to others, but when they went to
look, "no vestige of it remained."[3]

Olmsted dutifully recorded these oddities just in case they might be
significant, but focused his attention on what he gauged to be the essential
science of the event.

Where had the deluge of meteors been seen? From the newspaper
reports and letters he had received, the shooting stars had showered the
entire United States. Beyond that? Alexander Twining, a Yale graduate now
studying engineering at the United States Military Academy in West Point,
New York, had done some checking. While in New York City, Twining had
visited the docks to talk with the captains and crews of 15 ships about what
they had seen in the early morning hours of November 13.

Five ships had been in the North Atlantic east of Canada on that eerie
night, New York-bound from Europe. The skies were overcast; they saw
nothing. Other ships, closer to Europe or to the equator, had clear skies but
saw nothing special – meteors in no unusual numbers.

On the ship *Junior* in the Gulf of Mexico, however, Captain Gideon
Parker and his crew quickly lost count of the meteors they saw under partly
cloudy skies from 3:00 to 4:30 a.m. The clouds gave Parker and his sailors a

chance to observe that no matter how low or close or bright a meteor seemed to be, all "passed behind the clouds – not one between the cloud and the observer." The meteors were more distant than the clouds. The meteors had to be very high in the atmosphere.

Other ships under fair skies – the *Tennessee* in the Gulf of Mexico north of Cuba; the *Hilah* and the *Phoenix* off Bermuda – saw displays comparable to what Olmsted saw in New Haven.

"Accounts received from London," Olmsted reported, "make no mention of the meteors, whence we infer that they were not seen there, and probably not in any part of Europe." Only North America had been privileged to see this greatest of meteor showers.

Meteors of November 13, 1833 over Niagara Falls – a contemporary illustration from *Mechanics' Magazine*, November 1833. In 1876, Richard M. Devens wrote in *Our First Century*: "At Niagara, no spectacle so terribly grand and sublime was ever before beheld by man as that of the firmament descending in fiery torrents over the dark and roaring cataract!"

How many?

Olmsted tried to estimate the number of shooting stars seen. Most people exaggerate their observations, he warned. He doubted claims that "millions" or even "thousands" of meteors were visible at a single moment.

In passing, Twining had offered a rough estimate of not less than 10,000 meteors during the hour he watched. An unnamed observer in Boston had attempted a systematic count, although by the time he began, he said, the meteors were declining in number. Nevertheless, in 15 minutes, from 5:45

to 6:00, he counted 650 meteors. He felt he had missed at least one-third of those visible in the sector of the sky he was watching. If so, his estimate would correspond to a rate of about 4,000 meteors an hour. The Boston observer, however, expressed his estimate in terms of meteors visible in the *entire* sky – 8,660 meteors in 15 minutes, equivalent to 34,640 an hour, not a meaningful number for comparative purposes. Olmsted thought the Boston estimate was "considerably too low" based on what he had seen.

Olmsted was not yet aware that physicist Joseph Henry in New Jersey had estimated 20 meteors a second[4] and that a New York newspaper reporter had written: "Within the scope that the eye could contain, more than twenty could be seen at a time shooting (save upward) in every direction."[5] Twenty meteors a second was the equivalent of 72,000 an hour.

Olmsted marveled at the total number of shooting stars that must have fallen. "On the supposition that the meteors seen at places remote from each other were not the same," he concluded, "the entire number that descended towards the earth must have been *indefinitely great.*"

How noisy?

Olmsted was confused as to whether or not the meteors emitted a noise. As he watched the falling stars, he had been struck by the silence of the display. The great majority of the observers mentioned no noises, especially those with good scientific educations. A few accounts recalled occasional popping or hissing or crackling. There was even one claim of a loud explosion from offshore at Charleston, South Carolina. Olmsted was skeptical. "It is well known," he wrote, "that persons unaccustomed to observations in the stillness of night are apt . . . to hear sounds which they associate with any remarkable phenomenon that happens to be present, although [the sounds are] wholly unconnected with it."

Olmsted then offered a good reason why the meteors most likely made no sound: "Meteors which are distinguished for their brightness and apparent magnitude and which would therefore be expected to afford sounds might still be too distant for such sounds to be audible or might be in a region of the atmosphere where the air is too much rarefied for the purposes of sound."

The radiant

Of all his observations, Olmsted was most fascinated by the image of the meteors shooting outward "from a fixed point in the heavens." He and

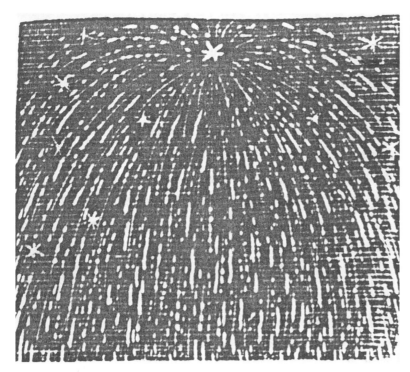

Meteors of November 13, 1833 depicted in the Boston *Evening Transcript* on November 16 – perhaps the earliest published illustration

Alexander Twining were the first to publish in a scientific journal their observation that meteors radiate from a specific position in the sky, the first to explain the phenomenon, and the first to give that point the name *radiant*.[6] Olmsted fired off letters to people whose meteor reports he had read to clarify this never-before-observed phenomenon and to gather their interpretations of it.

The more knowledgeable observers identified that point of origin as the constellation Leo, the Lion, near the star Gamma Leonis. This point migrated as the hours passed – westward with the stars, retaining its position in Leo as the Earth turned. No matter how far north or south they were of Olmsted's position in New Haven, observers remarked that virtually all the meteors radiated from Leo. There was no displacement of that point as one moved north or south.

So intense was the shower of falling stars that all but a few observers noticed that the meteor trails traced back to a common point of origin. Most observers, however, were too awestruck or unfamiliar with the constellations to pinpoint where the radiant was among the stars. Instead, these observers assigned the radiant to the zenith or very near it. From 4:00 to 5:00 a.m., when the shower was at its maximum, Leo was high in the sky, close to the zenith for observers in the United States. "Those who are unaccustomed to astronomical observations," Olmsted explained, "are apt to assign a wrong position to the zenith from the difficulty of looking directly upwards. The error frequently amounts to ten or fifteen degrees..."

All meteor showers have radiants

Of the many spectators of the great meteor fall who wrote to Olmsted or whose newspaper accounts were found by Olmsted, the most careful observation and most thoughtful analysis came from Alexander C. Twining, a Yale graduate studying civil engineering at the United States Military Academy at West Point, New York. Olmsted's article on the meteors for the *New Haven Daily Herald* was reprinted widely. On November 15, Twining saw in a local newspaper Olmsted's report, with its request for additional observations. Twining sat down and wrote to Olmsted immediately.

Dear Sir,

I presume that you will be glad to receive from various quarters observations upon the brilliant and wonderful phenomenon which appeared in the skies on the morning of Wednesday, the 13th. It was not my fortune to witness it from the beginning; but I observed it for more than an hour, from a few minutes past five o'clock, by the watch, till the morning light made it no longer visible. I shall describe only those things which passed under my own notice.

Of the numerous luminous bodies which were darting out on every side and at every altitude, the greater multitude were like stars suddenly lighted up in a state of rapid motion, shooting a certain distance and gone in a second, leaving where they had passed a luminous trace.

There was a point a few degrees south and east of the zenith which was evidently the directrix of all the apparent motions: every line and track of motion if continued backward would have passed, as nearly as the eye could discern, through that specific point. In the vicinity of that point, a few star-like bodies were observed possessing very little motion and leaving very little length of trace. Farther off the motions were more rapid and the traces longer; and most rapid of all and longest in their trace were those which originated but a few degrees above the horizon and descended down to it, like flaming sparks driven swiftly athwart the sky by a strong wind.

The position in the heavens from which the meteors seemed to emanate was between the stars in the breast and shoulders and those in the head of the Lion. This point cannot be supposed to have been a real part of space from which the luminous bodies actually proceeded, but the vanishing point of sight for motions which were truly or nearly parallel. If a multitude of bodies moving in parallel directions had entered the earth's atmosphere from that quarter of the heavens and become luminous by contact with the atmosphere and had been dissipated by motions through it, they must have presented the apparent motions, very nearly if not exactly, as those which I observed.

These bodies did not seem to bear affinity to those meteors which explode, throwing down masses to the surface; but to those shooting stars and fire balls which are often seen in the sky in the evening and which, I am now persuaded, might all be found capable of being referred as to their line of motion to a determinate point.

With these words, Twining became the first person to propose what we know today to be true: that *all* meteor showers have radiants.*

* Condensed from Twining's letter to Olmsted as it appeared in Denison Olmsted: "Observations on the Meteors of November 13th, 1833," *American Journal of Science and Arts*, volume 25, number 2, January 1834, pages 369–371 of pages 363–411.

A new field emerges

In his initial article for the *American Journal of Science and Arts*, published in January 1834, Olmsted had pretty much confined himself to reporting details of the great November 13, 1833 meteor shower. In his second paper, which appeared in the next issue of the *American Journal* in April, Olmsted gathered his courage and seized the initiative to go beyond reportage. He offered an analysis and interpretation of the meteors which sprang from Leo.

He began modestly enough with the hope of providing "materials for those who are better qualified than myself to give an explanation of these sublime but mysterious phenomena." He was genuinely intimidated by the responsibilities he was assuming. Having begun deferentially, Olmsted then asked himself 11 fundamental questions about the meteors and proceeded to offer his own conclusions. Perhaps he sensed that he was laying the foundation for a new field of scientific study.

The meteors of November 13, 1833, Olmsted argued, "had their origin beyond the limits of our atmosphere." If they were part of the Earth, like the atmosphere, they and their point of radiance would have moved west to east with the Earth, carried along by the rotation of our planet. Instead, the radiant of the meteors moved east to west in synchronism with the stars, "independent of the earth's rotation and consequently at a great distance from it, and beyond the limits of the atmosphere."

Olmsted concluded that the meteors emanated from a height of at least 2,238 miles (3,603 kilometers) above the Earth. Knowledgeable observers all reported that the meteors radiated from the background of Leo, near the star Gamma Leonis. W. E. Aikin, M.D., professor of chemistry and natural philosophy at Mount St. Mary's College in Emmitsburg, Maryland; John R. Riddell, a science lecturer at Ohio Reformed Medical College in Worthington, Ohio; and Olmsted in New Haven, Connecticut saw the meteors emanating from the same point at the same time. For all three of them, widely separated, to see the radiant of the meteors against the same stellar background, the radiant must be very high. Olmsted tried triangulating pairs of sightings to give the height of the radiant. The distances between the different pairs of observations – New Haven to Emmitsburg, Emmitsburg to Worthington, Worthington to New Haven – gave widely varying minimum altitudes, from 1,657 to 3,082 miles (2,668 to 4,962 kilometers). Olmsted assigned the variance to inexactness "in observations made loosely with the naked eye." None of the observers had (or, in the case of Aikin, used) instruments at hand to make precise measurements.

Olmsted used an average of the altitude calculations – 2,238 miles (3,603 kilometers) – as the minimum height of the *source* of the meteors above the surface of the Earth (not the height at which the meteors were burning up). The key for Olmsted was not the number, which he knew to be imprecise, but more evidence that the meteors came from outer space, not from the Earth.

Recalling that the meteors radiated from a common point and that they all traversed arcs of great circles in the sky, Olmsted then deduced that the meteors must be traveling together on parallel paths, because when you stand between parallel lines (such as railroad tracks) and look toward the horizon from which they come, the parallel lines seem to diverge from a single point. He added a diagram to help his readers visualize this phenomenon.

Meteor speeds

Olmsted next computed the speed of the meteors to be about 4 miles per second, 14,400 miles an hour (23,200 kilometers an hour) – about one-tenth their actual speed.[7] Olmsted missed on this point because he thought the meteors fell to Earth by gravity alone, falling with no initial velocity of their own relative to Earth.[8] Four miles per second seemed to a person of the early 19th century a most remarkable speed, and Olmsted marveled at it – 10 times the velocity of a just-fired cannonball, 19 times the velocity of sound at sea level. What would be the consequence of a solid object hitting the Earth's atmosphere at "this prodigious velocity"? The body, Olmsted reasoned, must generate "a powerful condensation of the air before it, thus retarding its progress and producing a great evolution of heat."

Olmsted considered that the Earth's atmosphere began at an altitude of

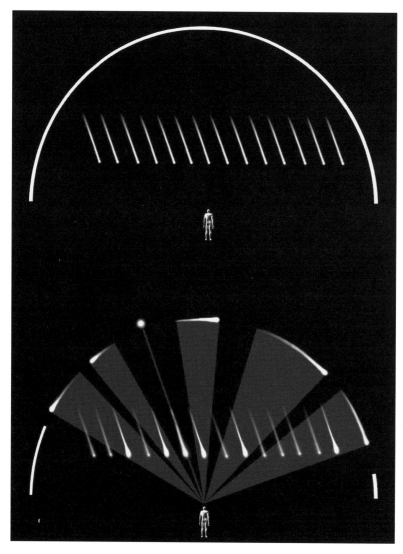

Olmsted drew a diagram like this to explain how meteors could be traveling together on parallel paths (*top*) and yet seem to an observer (*bottom*) to be radiating away from a spot in the sky. Notice that a meteor traveling directly toward the observer is seen as a momentary bright point rather than a streak. Artwork by Mark Maxwell.

about 50 miles (80 kilometers). It was a good approximation. At the beginning of the Space Age, astronaut wings were awarded to test pilots of the American X-15 rocket plane for flights above 50 miles, where there was too little air to allow the craft to maneuver by using its elevators, rudder, and ailerons. Instead, as the atmosphere thinned into the near vacuum of space, the X-15 pitched, yawed, and rolled like a spacecraft by emitting squirts of compressed gas.

Even though the air was very rarefied at an altitude of 50 miles, Olmsted

Using railroad tracks to explain meteor paths: the origin of an analogy

Joseph Henry

Perhaps the most distinguished American scientist to see the 1833 meteors was Joseph Henry, whose pioneering work on electromagnetism was unappreciated at the time. By 1833, he had built an electric motor and invented the telegraph.

Before dawn on November 13, Henry was awakened by two of his students at the College of New Jersey (now Princeton University) who were astounded by the meteors. With good cause. "They were so numerous," Henry said, "that 20 might be counted almost at the same instant . . ."

Henry noticed, as did Olmsted and most sophisticated observers, that the meteors all radiated from a single point in the heavens near the zenith, although Henry did not identify the constellation from which the meteors seemed to rain – nor did he observe that the position of this radiant seemed to move westward with the stars as the Earth turned.

What he did understand immediately was that the meteors spraying in every direction as if from a single point in the sky meant that the meteor bodies themselves must have been traveling together in paths parallel to one another. In a letter to his brother James, he explained this illusion with what may have been the first use of the railroad track analogy – still the favorite way to visualize the process:

> . . . although they [the meteors] fell nearly parallel like flakes of snow in a calm day, yet they appeared to diverge from each other like the two branches of a railroad which appear at the most distant point to be near each other and to widen as they approach the observer.*

* Dated November 14, 1833. For comments by Henry about the 1833 meteors, see *The Papers of Joseph Henry*, volume 2, edited by Nathan Reingold (Washington, D.C.: Smithsonian Institution, 1975), pages 116–121; 128–130; 133.

proposed, it would "abstract" motion from the meteor. At every stage of the meteor's descent, it would "meet with denser and denser air" until the heat generated by atmospheric friction would consume it. Olmsted thought that a meteor particle would be heated so greatly by friction that it would catch fire and burn like wood – the rapid combination of the meteor with oxygen. He was wrong in thinking that meteors glowed by conventional burning, but correct in recognizing that atmospheric friction could impart a high temperature to a speeding meteor.

He was right also that because meteors were not luminous in their original state out in space, their burnup had to be occurring during the few seconds while they were within the atmosphere. That was certainly the impression they gave as they raced across the sky – glowing, leaving luminous trails.

However, said Olmsted, the meteors in this shower did not seem to be of the same composition as meteorites which were known to have fallen from space onto the surface of Earth. Those meteorites were composed of iron and other heavy metals and compounds. They were too large and dense to have been fully consumed in their dives through the atmosphere. It would be 140 years until scientists could collect and analyze the smallest of meteor particles. There were some differences in compositions between meteor shower particles and meteorites, but primarily the November meteors were less dense, more crumbly, and much smaller than the objects that reached the ground as meteorites. It was a good thing they burned up, Olmsted noted, because the meteors in the shower were so numerous that had they hit the surface of the Earth, "we should have had . . . appalling proofs of the fact in the destruction and ruin that would have marked the places where they fell. Yet," said Olmsted, "no evidence has yet appeared of a single meteoric stone having been found, and it is even somewhat doubtful whether any palpable substance reached the ground which could fairly be considered as a deposit from the meteors."

Height of the meteors

Olmsted realized that the brightest meteors might give him the means to calculate the altitude at which they burned up. If observers at two widely separated sites saw the same meteor, they would see the streak against a slightly different background of stars. That displacement in degrees combined with the distance between the observers would, by trigonometry, allow a scientist to compute the distance to the meteor.

Both Olmsted in New Haven and Daniel Tomlinson in Brookfield, about 25 miles northwest of New Haven, had seen a dazzling fireball at about 5:45 a.m. Both recorded that it was traveling in the same direction, with a comparable trail. But they had seen it against a slightly different background of stars. If indeed both had seen the same meteor, its height could be calculated. Olmsted calculated that the meteor's burnup occurred at an altitude of about 30 miles (50 kilometers) – somewhat too low according to modern findings.[9]

A burnup altitude of 30 miles for the meteor had it own scary implications however. The brightest meteors left glowing trails 10 degrees or more in length. At a height of 30 miles, that meant that this trail was more than 5 miles (8 kilometers) long and the speed with which the trail drifted in the sky indicated (correctly) high-altitude winds of 300 miles an hour (500 kilometers an hour).[10] Such winds had never been observed on Earth, even in the most violent storms. Olmsted could not believe wind speeds

How the altitude of meteors is determined. Two observers (or cameras), typically 25 miles (40 kilometers) apart, see the same fireball (gauged by time and direction) but they see it against a different background of stars. The distance between the observers and the angle between the stellar backgrounds are known, which allows the height of the meteor to be calculated by trigonometry. Diagram by Will Fontanez and Tom Wallin, University of Tennessee Cartographic Services Laboratory.

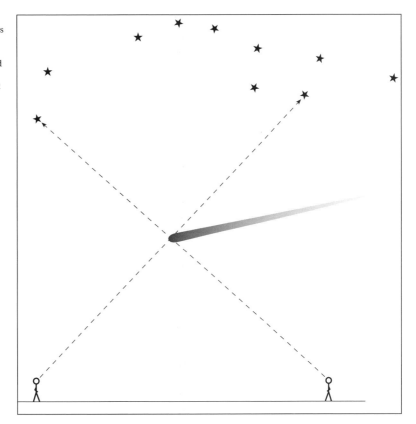

that great, which caused him to doubt the height he calculated for meteor burnup. "I feel constrained, therefore," he wrote, "contrary to my early impressions, to believe that the large meteors frequently descended to the region of the clouds." The lower the meteor burnup, the shorter the trail and the slower the wind velocity. He felt more comfortable with the lower altitude because the density of the air was greater, more quickly slowing the meteor by friction and offering more air for the combustion of the intruder. Olmsted was wrong. He should have trusted his observations.

Meteors in orbit

Olmsted deduced that the source of the meteor shower must be a sort of cloud of individual particles. A solid body that broke up as it neared Earth and thereby caused the meteor shower would have been too large, he figured (correctly), to have escaped detection as it approached our planet.

Olmsted then made a startlingly accurate guess – "that the cloud which produced the fiery shower consisted of nebulous matter analogous to that which composes the tails of comets."

> We do not know, indeed, precisely what is the constitution of the material of which [comet tails] are composed; but we know that it is very *light*, since it exerts no appreciable force of attraction on the planets, moving even among the satellites of Jupiter without disturbing their motions, although its own motions, in such cases, are greatly disturbed, thus proving its materiality; and we know that it is exceedingly *transparent*, since the smallest stars are visible through it.

The final question Olmsted asked himself was about the path of the particles that furnished the meteor shower. Was this family of particles a satellite – a sort of "terrestrial comet" – in orbit around the Earth? Or "was it a collection of *nebulous matter* which the earth encountered in its annual progress?" Or was it a comet crossing the Earth's path in its revolution around the Sun?

Olmsted ruled out a satellite of Earth. Many observers, spread widely across the United States from the Mississippi to the Atlantic, had noticed that the meteor shower radiated from Leo, the Lion, and they all agreed very closely about the position of the radiant. That meant that the meteors were originating high above the Earth and its atmosphere, at least 2,238 miles (3,603 kilometers), Olmsted calculated. Now an object in orbit around the Earth 2,238 miles above its surface would, Olmsted calculated, have to travel fast to stay in orbit, so fast that it would complete its journey around the Earth in 2 hours 45 minutes. Thus, in the course of the meteor storm, seen for several hours, the radiant of the meteors would have made at least one complete circuit of the sky. It didn't. It remained fixed to one position in the head of Leo.

Olmsted also veered away from the idea that the Earth had plunged through some kind of nebula "which was either stationary or wandering lawless through space" – that is, not part of the solar system.[11] The shower had lasted, according to reports, at least eight hours. During that period, the Earth had traversed more than half a million miles (900,000 kilometers) of space. Olmsted doubted that any nebula within the solar system could be that large.

Could the Earth, then, have encountered a comet? The particles in the meteor shower were tiny and spread out like the particles in a comet's tail.

Olmsted's homework and correspondence now paid dividends. The meteor shower, he announced, "returns at stated periods." It had appeared in Mocha, Arabia a year earlier, on November 13, 1832. It had been described independently by Alexander von Humboldt and Andrew Ellicott 34 years earlier in 1799.

Then Olmsted stumbled. He concluded that the cometlike body that the Earth encountered had to be traveling in the same direction as the Earth. If it had been stationary with respect to the Earth's orbital velocity, the Earth would have plowed through it at 18.5 miles a second (29.8 kilometers a second) and, Olmsted thought, the meteor display would have been very brief. If the cometlike body had been moving toward the Earth, the speed of collision would have been even higher and the display even briefer. Olmsted had conceived of a relatively short clump of debris rather than a stream of particles many millions of miles long.

Olmsted was still working carefully. The orbit of the cometlike body that provided the meteors must be an ellipse, he reasoned, because all other members of the solar system travel in ellipses.

Because the meteor shower had been seen a year earlier, Olmsted concluded (incorrectly) that the orbit of the meteor material must have an orbit of approximately one year or a fraction of a year that would allow the Earth to run into this debris once a year at or about November 13. If the orbit were less than a year, it must lie inside the Earth's orbit, closer to the Sun, and spray the Earth with meteors only when it was at aphelion, its farthest point from the Sun.

Olmsted doubted that the meteor debris had an orbital period of one year because then the path traveled by the particles must closely resemble the path of the Earth around the Sun. That would mean that the meteors-to-be would be traveling alongside the Earth for a long time and constantly colliding with it, causing shooting stars throughout that period. Yet the intense meteor shower lasted only hours.

But for the Earth to crash into the meteoric nebula every year on about the same day, as Olmsted thought it did, the orbit of the particles had to be some simple fraction of a year. That fraction, Olmsted decided, had to be one-half. It must take half a year for the meteoric material to make its circuit. A clump of debris in solar orbit could venture out as far as the Earth and still complete its journey in six months if its elliptical orbit then carried it close to the Sun. In that way, while the Earth revolved once around the Sun, the clump would make two revolutions and meet the Earth at the same point – the same date – in its orbit year after year.

Was there another possibility: a period for the meteor particles of ⅓ year? No, Olmsted realized. Even if the meteors traveled a highly elliptical orbit that grazed the Sun at closest passage, an orbit with a period of ⅓ year would not stretch out far enough to reach the Earth.

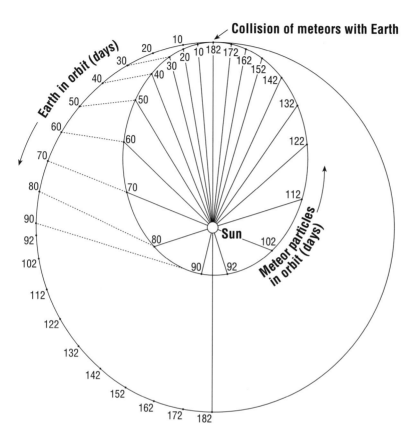

Collision of meteors with Earth

Earth in orbit (days)

Meteor particles in orbit (days)

Sun

Olmsted drew a diagram like this to illustrate his (incorrect) idea of a ½-year orbit of the November meteor particles. The particles, he thought, travel an elliptical orbit inside the orbit of Earth. As the Earth each year returns to its mid-November position, it runs into the meteor concentration, which in the meantime has completed two orbits of the Sun. In the caption for his diagram, Olmsted twice refers to the meteor particle swarm as a "comet." Diagram by Will Fontanez and Tom Wallin, University of Tennessee Cartographic Services Laboratory, based on Olmsted's April 1834 diagram.

A source for the meteors

But if a large, mostly transparent cloud of meteoric material passed so close to Earth, why wasn't it visible coming and going by sunlight reflecting from it? It must cover a substantial area in the heavens.

Perhaps it was seen, said Olmsted, at least as it receded from Earth. Olmsted imagined this nebulous body to be elongated in shape, like a comet with a tail. Olmsted recalled the testimony of James N. Palmer, a New Haven surveyor who came to see him several weeks after the shower with his rather fanciful recollection of the meteors. (Almost certainly, Palmer was a liar.) Supposedly, as dawn was approaching during the meteor storm, a patch of light appeared on the horizon in the east. "This light was so bright," Palmer told Olmsted, "that a member of his family got his pail to milk the cows, supposing it to be daybreak, but found it was only 4 o'clock." How curious – and unbelievable – that Palmer's relative would

Zodiacal light (a drawing)
Robert S. Ball: *The Story of
the Heavens*, 1888

be awakened by the patch of light in the east and not notice the phenomenal meteor shower. Nevertheless, Olmsted (and Twining) took Palmer's lies seriously and looked for confirming evidence, probably because Palmer claimed to have used scientific instruments while observing the meteors. Olmsted seized on an observation by Darius Lapham in Cincinnati who thought he saw the aurora near the horizon a little north of east during the meteor shower.

In its first very brief report on the meteors, the Boston *Columbian Centinel* described "a vapor in the atmosphere, visible round the horizon, which in the southeast assumed a very beautiful appearance during ten minutes, about half an hour before sunrise." All subsequent, more detailed reports in the *Centinel*, however, identified this light as meteors, not the aurora.

Olmsted himself had seen a cone-shaped glow extending upward from the sunset position on the horizon after the end of twilight on December 1 and 3, 1833, about two weeks after the great starfall. He identified it as zodiacal light. He saw the zodiacal light several more times that winter and spring as he prepared his report for Silliman's Journal.

Olmsted concluded that zodiacal light was a cometlike body – by which he meant a nebulous clump of meteoric particles. Thus, Olmsted proposed, zodiacal light was the source of the great meteor storm on November 13, 1833. Today, zodiacal light is recognized as the debris from comets and asteroids distributed in the plane of the solar system, so that the debris close to the Sun shines faintly by reflected sunlight and thus is sometimes visible in the east just before sunrise twilight begins or in the west just after sunset twilight ends.

The reports of Palmer, Lapham, and the Boston *Columbian Centinel* were not zodiacal light. Palmer's fraudulent glow was too early and too bright to be zodiacal light. Lapham's auroralike light was north of east, whereas a November 13 sunrise occurs south of east. And the *Centinel*'s vaporous glow, identified as shooting stars in subsequent issues of the newspaper, occurred about half an hour before sunrise, when the sky was too bright with dawn colors for the zodiacal light to be seen.

> From all the foregoing considerations [Olmsted wrote], I feel authorized finally to conclude, *That the Meteors of Nov. 13th consisted of portions of the extreme parts of a nebulous body which revolves around the sun in an orbit interior to that of the earth, but little inclined to the plane of the ecliptic, having its aphelion near to the earth's path, and having a periodic time of 182 days, nearly.*

Olmsted was wrong about the short orbit of the meteors, but right in his conception of the meteors as a diffuse swarm of particles in space, traveling around the Sun.

Olmsted's observation of the meteors of November 13, 1833 and his collection and sifting of reports had set meteor studies on a new course. Before Olmsted, meteors were part of meteorology, the study of the Earth's atmosphere. Between 1794 and 1803 scientists realized that rocks – meteorites – occasionally fell from the sky, but the source of meteorites was unknown and their relation to meteors was unclear.

What Olmsted surmised was that

- Meteor showers radiate from a specific point or small region of the sky.
- Meteors must be traveling together in a swarm on parallel paths, so that the radiant is simply the vanishing point where parallel lines seem to converge.
- Most meteors burn up at an altitude between 50 and 30 miles (80 to 50 kilometers (somewhat too low).
- Meteors are heated by the friction they experience when they strike the Earth's upper atmosphere at very high speeds (although Olmsted thought the particles burned rather than vaporized by friction).
- Meteors are particles traveling through space. They do not originate in the Earth's atmosphere or in Earth orbit.

If you stand in the middle of a straight road, with telephone poles or road markers on either side of you, all those parallel lines seem to converge to a point in the far distance. Similarly, meteors traveling together on parallel paths appear to radiate outward as they approach the observer as if they came from a distant point. Artwork by Mark Maxwell.

- Meteoric particles are similar to those in comet tails.
- Meteoric particles travel on cometlike orbits.
- Meteor showers are periodic, even annual.
- The zodiacal light is related to comets and meteors (although not as Olmsted imagined).

More than anyone else, Denison Olmsted had set in motion the transfer of meteors from meteorology to a field of their own within astronomy.

Denison Olmsted (1791–1859)

Good fortune poured down on Denison Olmsted on November 13, 1833. He wasn't the only person to see the great meteor shower, but he, more than any other, had the presence of mind and the tenacity to do something of scientific importance with it. Scientists of his time and ever after have praised his collection of observations and his efforts to sort out the facts.

Olmsted's life was filled with accomplishment, but very little good fortune. His father died when he was one. He was sent to live with an uncle on a farm until he was nine and his mother remarried. Then, so that he could attend a better school, he was sent away during the school year to live with and work for Judge (later Connecticut Governor) John Treadwell. But the school, typical for those days, taught no arithmetic, just reading and writing. Treadwell taught him elementary mathematics and then apprenticed him to his son as a store clerk. But Olmsted craved higher education. He studied with a tutor for two years and then, at age 18, went to Yale, graduating among the top in his class, and was immediately hired as a schoolmaster. Two years later, still hungry for education, he returned to Yale as a tutor and to receive his master's degree. In his master's oration, he proposed the establishment of teacher-training colleges to improve American public education. He may have been the first to urge this concept. He lectured and wrote extensively on this idea, and advocated long-term contracts for women teachers.

Denison Olmsted

As soon as he had his master's degree, the University of North Carolina hired him as professor of chemistry, which included mineralogy and geology. He stayed at North Carolina for seven years and conducted the first geological survey of any state.

In 1825, Yale brought him back as professor of mathematics and natural philosophy – although Olmsted's previous studies were in chemistry, geology, and theology. Yale saw in him a born teacher, a dedicated worker who could take advantage of opportunities, and a man of principle. Olmsted by now was married with a rapidly growing family. The financially strapped college paid him a meager salary. Olmsted did not complain. Always alert to opportunities, he spotted means of supplementing his income. Colleges and public schools lacked good introductory science books, so Olmsted wrote a series of general science and astronomy textbooks that were very successful. He lectured widely. He invented a stove which gave more even heat.

Misfortune struck in 1829 when his wife died, leaving him with five children, ages 5 to 10. He forged ahead. His students appreciated his constant updating and reworking of his lectures and the star parties he organized each fall to introduce students to the constellations. He remarried and had two more children. "His children revered him;" said the president of Yale, "they were trained by him to exercise their powers of thinking."

When the meteors fell in 1833, Olmsted was ready. When Halley's Comet was scheduled to return in 1835, he and tutor Elias Loomis used Yale's small telescope to become the first people in the Americas to find it in the sky. He urged Yale to build a major observatory with a large telescope, but there was no money. He was not depressed. He turned his efforts toward a more modest observatory for astronomy instruction and donated his own money to the cause. But the project fell through. Yale lacked the funds for even that. Olmsted was not discouraged. "I do not remember that I ever saw him irritated or uncourteous," said Theodore Woolsey, the president of Yale, in his memorial address on Olmsted. "He was devoid of the petty selfishnesses which mar many characters."

Woolsey characterized Olmsted as not greatly inventive or original, but possessing a lucid and discriminating mind, well organized, and a steady worker who did not oscillate between bursts of energy and fits of listlessness.

Four of Olmsted's five sons attended Yale. Francis Allyn graduated in 1839, sailed to the Sandwich Islands (Hawaii) to stave off consumption (tuberculosis), returned to Yale to earn a degree as a medical doctor in 1844 . . . and died that same year at age 25. John Howard delayed his education because of consumption, received his degree in 1845, and died in Florida the next year. Denison, the third son, showed promise as a scientist and remained healthy throughout his childhood and college education. He graduated in 1844, and died in 1846, of consumption. Alexander Fisher, the feeblest of the sons, also graduated Yale in 1844. He was racked by consumption and took a professorship in chemistry at the University of Alabama, a more favorable climate. He returned to New Haven in 1851 to do chemical research at Yale, published a chemistry textbook, and died in 1853, age 30.

*This vignette is based on [Theodore Dwight] Woolsey: "Discourse Commemorative of Professor Denison Olmsted, LL.D.," *New Englander*, volume 17, August 1859, pages 575–600; *Dictionary of American Biography*; *National Cyclopædia of American Biography*; and *Appleton's Cyclopædia of American Biography*.

Olmsted continued on without complaint, despite a growing pain in his stomach. He taught all his classes through the regular school year of 1858–1859 and was planning to teach summer term. "He died in the midst of his work," said the president of Yale. Woolsey went to see his old friend six days before his death. Olmsted's final request to the president of Yale was "to assure his colleagues that he loved them."*

NOTES

1. D[onald] K. Yeomans: "Comet Tempel-Tuttle and the Leonid Meteors," *Icarus*, volume 47, 1981, page 492 or pages 492–499.

2. Denison Olmsted's quotations and thinking in this chapter come from "Observations on the Meteors of November 13th, 1833," *American Journal of Science and Arts*, volume 25, number 2, January 1834, pages 363–411; continued under the same title in volume 26, number 1, April 1834 (although bound as "July 1834"), pages 132–174.

   During this period, the *American Journal of Science and Arts* was published in two "volumes" each year –

January and July – composed of four "issues" – April, July, October, and January. The April and July issues formed the July volume; the October and January issues formed the January volume. Thus part 1 of Olmsted's article in the January 1834 *volume* of the *American Journal of Science and Arts* was actually published in January, whereas part 2 of Olmsted's article, bound in the July 1834 *volume*, actually came out in the April issue.

3. Alexander Twining, a Yale-educated, tough-thinking engineer, was called to the spot "and concluded that if I heard of any analogous facts from other quarters,

I would consider this as entitled to notice, but not otherwise."

4. Joseph Henry: *The Papers of Joseph Henry*, volume 2, edited by Nathan Reingold (Washington, D.C.: Smithsonian Institution, 1975), page 116 of pages 116–121; 128–130; 133. Please also see the vignette in this chapter on Henry's creation of the railroad track analogy to explain meteor radiants.

5. *New York Commercial Advertiser*, November 13, 1833. The article was reprinted by the *National Intelligencer* (Washington, D.C.), November 16, 1833. The reporter goes on to say that "for the two hours intervening between four and six, more than a thousand per minute might have been counted . . . [T]hey fell as thick as the flakes in the early snows of December."

6. Denison Olmsted: "Observations on the Meteors of November 13th, 1833," volume 25, number 2, January 1834, page 400 of pages 363–411. Twining's letter that uses the term *radiant* is reprinted as part of this article on page 405.

7. Olmsted did not calculate meteor speeds as he might have, based on the burnout altitude he had derived combined with how many degrees of arc the meteor traversed in how much time. He did not adequately trust the altitude he had calculated for the meteors or how long they were visible. Instead, his proposal for the speed of the falling meteors was based strictly on gravitational theory: at what speed an object will fall to Earth from a certain altitude.

8. Alexander C. Twining calculated the velocity of that fireball as at least 14 miles a second (50,000 miles per hour; 81,000 kilometers per hour) – about one-third the actual speed of the November meteors. He favored 20 miles a second (32 kilometers a second), but included in his estimate the report of James N. Palmer, a fraud, who claimed that he put a watch to his ear to time how long meteors (not their trails) were visible. Alexander C. Twining: "Investigations respecting the Meteors of Nov. 13th, 1833. – Remarks upon Prof. Olmsted's theory respecting the cause," *American Journal of Science and Arts*, volume 26, number 2, July 1834, pages 320–352.

9. Twining calculated the altitude of that fireball and concluded that it was first seen at an altitude of 80 miles and exploded at 29½ miles (129 to 47 kilometers). Alexander C. Twining: "Investigations respecting the Meteors of Nov. 13th, 1833. – Remarks upon Prof. Olmsted's theory respecting the cause," *American Journal of Science and Arts*, volume 26, number 2, July 1834, pages 342–343 of 320–352.

10. Twining calculated the upper air wind velocity to be 180–240 miles per hour (290–390 kilometers per hour). Alexander C. Twining: "Investigations respecting the Meteors of Nov. 13th, 1833. – Remarks upon Prof. Olmsted's theory respecting the cause," *American Journal of Science and Arts*, volume 26, number 2, July 1834, page 344 of pages 320–352.

11. Denison Olmsted: "Observations on the Meteors of November 13th, 1833," *American Journal of Science and Arts*, volume 26, number 1, April 1834, page 163 of pages 132–174.

Plains Indians record the November meteors

In the Great Plains of the United States and southern Canada, the Sioux, Blackfoot, and Kiowa Indians kept track of each passing year by painting a symbol on an animal skin or, later, on fabric or paper to recall a distinctive event by which that year would ever after be known. This was the tradition of the winter count, in which the Plains Indians kept track of time by how many snow seasons a person had lived through. They did not number the years.

The winter count might depict an Indian with an arrow in him, the year to be known as the "Came-and-killed-five-Oglalas winter" (1838–1839). Other winter counts depicted "First-hunted-horses winter" (1811–1812) and "Smallpox-used-them-up winter" (1779–1780). Each year the Sioux, Blackfeet, and Kiowas added a new picture, often spiraling outward from the center so that there was no doubt about the sequence of the years. The leaders would show the painted skin or fabric to the tribe to recall tribal history – what happened in each named year. Everyone in the tribe knew what each symbol meant and it served as a device to remember other important events in that year.

For the year that included the winter of 1833–1834, the decisive event was the meteor deluge the night of November 12–13. The great majority of Plains Indian tribes named that winter after that stunning sight. Battiste Good, a Brulé Sioux chief whose Indian name was Wa-po-ctan-xi (Brown Hat), kept the most famous of the winter counts. He painted a tepee with stars all around it and referred to 1833–1834 as the "Storm-of-stars winter." Lone Dog (Shunka-ishnala), a Miniconjou Sioux, painted a torrent of bright stars falling past the Moon, explaining that 1833–1834 was set apart because in that year "The stars fell."*

* Garrick Mallery: *Picture-Writing of the American Indians* (Washington, D.C.: Smithsonian Institution, 1893), pages 280, 320, and 723; and Von Del Chamberlain: "North American Plains Indian Calendars," *Archaeoastronomy*, number 6, pages S1–S54, supplement to *Journal for the History of Astronomy*, volume 15, 1984. The Kiowas added calendar pictographs twice a year or even monthly.

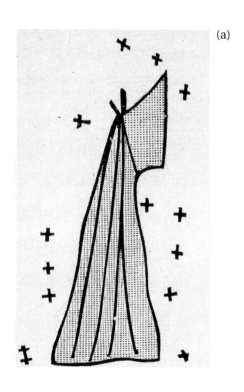

(a)

Winter counts for 1833–34:
(a) Battiste Good (Brulé Sioux): "Storm-of-stars winter; (b) Lone Dog (Miniconjou Sioux): "The stars fell"; (c) The Swan (Miniconjou Sioux): "Sioux witnessed magnificent meteoric showers; much terrified"; (d) Cloud Shield (Oglala Sioux): "It rained stars."
Garrick Mallery: *Picture-Writing of the American Indians*, 1893

(b)

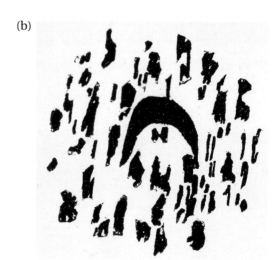

(c)

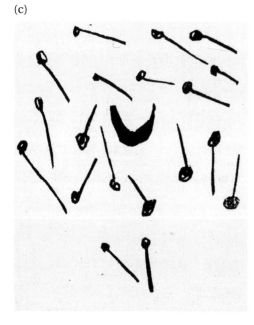

(d)

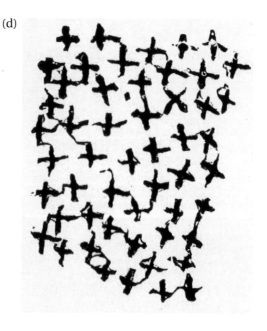

CHAPTER 3

Struggling to understand meteors

The Moon is an unfriendly neighbor that greets the Earth with stones.

GEORG LICHTENBERG (1797)
on the idea that meteors and meteorites come from the Moon[1]

M ETEORS HAD been part of meteorology for more than two millennia. Of all Aristotle's mistaken teachings in the realm of astronomy, his opinion about meteors persisted as accepted doctrine longest. Aristotle's view of shooting stars prevailed from his publication of *Meteorology* about 340 B.C. until scientists witnessed and analyzed the great November meteor shower of 1833. That stretch of nearly 2,200 years marks perhaps the longest run of any scientific error.

The problem was that Aristotle placed meteors in the realm of meteorology. *Meteor* is a Greek word that means things which happen in the air. Aristotle included meteors in meteorology along with clouds, wind, rain, snow, sleet, hail, lightning, thunder, halos around the Sun and Moon, rainbows, auroras, earthquakes, comets, and the Milky Way, thinking that meteors *originated* in the realm of earth and air rather than just becoming luminous there.

Meteor myths around the world

Meteor mythology dates back more than 3,000 years. The Mesopotamians generally thought that meteors were evil omens, signs of death and disaster.

Five hundred years ago in the Middle East, Moslems considered meteors to be the artillery in the war between angels and devils. Shooting stars were celestial fire that the angels rained down upon the evil ones.

In Brunswick, formerly a state in central Germany, meteors were dangerous fire dragons, but they could be appeased. If you see a shooting star, you should duck under the eaves of a house and shout, "Fire Dragon, share with me." The fire dragon might then drop you a ham or a side of bacon. Better yet, he might drop money. That was how some people suddenly got rich.*

* John G. Burke: *Cosmic Debris: Meteorites in History* (Berkeley: University of California Press, 1986), pages 217–218.

Aristotle classified meteors and comets with clouds, wind, and rain because of the transitory nature of these happenings, appropriate to the pattern of decay and death of things on Earth. The heavens, however, were the realm of permanence. Meteors, clouds, lightning, and such phenomena must be close to Earth, said Aristotle, because of their relatively fast apparent motion. Faraway objects do not appear to move fast.

In his treatise *Meteorology*, Aristotle taught that shooting stars begin as vapors created as the Sun warms the Earth. These vaporous "exhalations" from Earth rise because of their heat high into the air. There they catch fire in either of two ways. The rapid motion of a stream of vapor creates heat in the air so that it ignites and burns like a long strand of straw, creating the impression of a moving flame. Or the vapor may find itself rising in a column of air until the air cools and condenses so that the hot exhalation is ejected "by being squeezed out like slippery fruit seeds pinched between one's fingers."[2]

Aristotle separated ordinary shooting stars from fireballs. A fireball, he said, was called a "goat" if it seemed to scatter sparks. If it looked like a fast-moving flame yet cast no sparks, it was called a "torch." "But if the whole length of the exhalation is scattered in small parts and in many directions . . . , we get what are called shooting stars."[3]

Aristotle's doctrine echoed down the centuries that followed, a touch of folklore added here, a pinch of superstition added there. Virgil (70–19 B.C.), in his first book of *Georgics*, made meteors into weather forecasters:

> And oft, before tempestuous winds arise,
> The seeming stars fall headlong from the skies,
> And, shooting through the darkness, gild the night
> With sweeping glories and long trains of light.[4]

Meteors and term limits

Thirty-two hundred years ago, meteors provided the first term limits for governmental officials. In the city of Sparta in Greece about 1200 B.C., there was a special night once every eight years during which the priests surveyed the sky. If they saw a shooting star, that meant that the king had sinned and that he must be deposed.*

What were the king's chances of remaining on the throne? Even when no meteor shower is in progress, an observer will see about eight sporadic meteors an hour, meaning that unless the skies in ancient Sparta were cloudy all night or unless the priests were very well paid, a change in ruler was certain.

* David Hughes: "Meteor Myths," *Astronomy Now*, volume 3, November 1989, pages 43–46.

Seneca (4 B.C.–65 A.D.) doubted Aristotle's doctrine that comets were a phenomenon in the Earth's atmosphere, but he agreed that meteors were exhalations from Earth kindled high in the sky. Seneca visualized the exhalations as tiny particles ignited by the rays of the Sun just as powder or dust thrown into a lighted candle can burst into flame.

Seneca shrewdly observed that shooting stars must be occurring in the day as well as the night, but the brightness of the sky hides them from view – all but those of extraordinary luminosity.

And Seneca realized that shooting stars are not really stars: "[I]t is the stupidest thing to suppose the stars actually fall," he said, because, if that were true, there would be no stars left in the sky. Every night many stars seem to fall, yet each star is still found in its usual place and with its usual brightness. "It follows," he said, "that the fires are produced below the stars."

Seneca agreed with Virgil and sailing lore that "Sailors think it is a sign of storm when many stars fly across the sky." This sea lore, said Seneca, is more evidence that shooting stars belong to the realm of the atmosphere, where winds and storms are spawned."[5]

Meteor myths

Shooting stars figured rarely but vividly in mythology and superstition. Jacob Grimm recorded some of the explanations and warnings about meteors in his *Teutonic Mythology*.*

Stars, you know, are very fussy about the luster of the starlight they emit. They must clean themselves from time to time. This they do by a cosmic sneeze – that we see as a shower of shooting stars.

Whoever sees a falling star should say a prayer. If you make a wish while a meteor is falling, your wish will come true.

But meteors are sudden and unexpected. They look indeed like falling stars and this impression has led to some baleful interpretations.

The appearance of a meteor means that a poor girl is losing her money.

Or, as a folk song lamented:

Over the Rhine three stars did fly,
Three daughters of a widow die.

Another myth said that at the moment of your birth, an old woman in the heavens starts to spin a thread. From that thread your star hangs in the firmament. Sooner or later that thread breaks, the star falls, and you die.

* Jacob Grimm: *Deutsche Mythologie* (Berlin: Ferd. Dümmlers, 1876), page 602; and Jacob Grimm: *Teutonic Mythology*, translated by James Steven Stallybrass (Gloucester, Massachusetts: Peter Smith, 1976), volume 4, pages 1506 and 1801.

Pliny the Elder (c. 23–79 A.D.) thought that fiery shooting stars fall into clouds where they are extinguished with a hissing noise "just as when red-hot iron is plunged into water." Sometimes the wounded cloud gives off flashes of lightning and claps of thunder.[6]

Aristotle loses ground

The first of Aristotle's meteorological notions to crumble was his teaching that comets were part of meteorology – the study of effects in the air. In 1577 in Denmark, Tycho Brahe, the greatest of pre-telescopic observers, saw a bright comet. By comparing its position against the background stars with reports from astronomers elsewhere in Europe, Tycho calculated that the comet was at least four times the distance of the Moon from Earth. Hence comets must be considered part of astronomy, not meteorology.

Still, shooting stars remained within meteorology, their brief and rapid motion aligning them with lightning. And, unlike comets, meteors owed their visibility to their interaction with the air. Where Aristotle went wrong was in thinking that shooting stars had an earthly origin.

For 2,100 years and more after Aristotle, virtually all writers parroted his conception that shooting stars began as vapors caused by solar heating rising from the surface of the Earth to the high reaches of the atmosphere where they caught fire or their fiery nature was released.

William Fulkes was the first to write about shooting stars in English in his 1563 book with the book-length title *A Goodly Gallerye with a Most Pleasaunt Prospect, into the Garden of Naturall Contemplation, to Behold the Naturall Causes of All Kynde of Meteors, as Wel Fyery and Ayery, as Watry and Earthly, of Which Sort Be Blasing Sterres, Shooting Sterres, Flames in the Ayre &c. Thõder, Lightning, Earthquakes, &c. Rayne Dewe, Snowe, Cloudes, Springes &c Stones, Metalles, Earthes &c. to the Glory of God, and the Profit of His Creatures.*[7]

There was nothing new scientifically in Fulkes. "As science it is, from the point of view of today's knowledge, approximately 100 per cent wrong," his modern editor wrote.[8] All of Fulkes' ideas were derived from Aristotle and his Roman echoes Seneca and Pliny the Elder. But Fulkes brought these concepts to a wider audience: those who could read English but not Latin. He also reflected current educated opinion in Europe about 1,900 years after Aristotle. A generation earlier, Copernicus had begun the new assault on the idea that the Earth was the center of the universe. In the realm of shooting stars, however, no such revolutionary voice had yet arisen.

When it appears that "starres fal down," Fulkes wrote colorfully, "it is nothing els but the *Exhalation* that is thinne [then] kindled in many partes, sparkling as when sawe dust [sawdust] or coole dust [coal dust] is cast into the fyre."[9]

Meteors as astronomical objects

Edmond Halley

A century and a half passed. Galileo made his first telescopic observations in 1609. Newton published the law of gravity in 1687. But shooting stars were still classified with rain and wind.

On the subject of shooting stars, Isaac Newton thought that volatile gases rise to the top of the atmosphere where they mix with others, and there "taking fire cause Lightning and Thunder and fiery Meteors."[10] Not much had changed in the understanding of shooting stars in more than 2,000 years since Aristotle.

The first person to attempt an overhaul of the system was Edmond Halley in 1714. It was nine years after Halley had concluded that comets are members of our solar system by noticing that comets seen in 1531, 1607, and 1682 all traveled the same path. Either they were three different comets following one another at 76-year intervals on virtually identical courses, or those three comets were actually one comet returning to the Sun every 76 years. In 1705, Halley used Newton's mathematics and his theory of universal gravitation to calculate an elliptical orbit for this comet and to predict that it would be back in 1758. On Christmas night 1758, the comet returned to view – and ever since has been known as Halley's Comet.

Thus comets were astronomical – not of this Earth. Moreover, comets were members of the solar system, orbiting the Sun like the planets do and obeying the same law of gravity.

Time now – in 1714 – to analyze meteors. Halley used observations of a meteor seen in 1708 by different observers at different sites against different backgrounds of stars to estimate that this meteor and others perform their exhibitions at heights between 40 and 50 miles.[11] In 1686, Halley had used a barometer to measure atmospheric pressure on hilltops, mountain peaks, valleys, and seacoasts. He observed that pressure declined with altitude, so that, he calculated, at about 45 miles, there was essentially no atmosphere left.[12]

Based on records of a fireball widely observed over Italy on March 21, 1676, Halley calculated that it had to be traveling at least 160 miles a minute, at least 9,600 miles an hour. How could rising vapors reach that velocity? That was much faster than a cannonball. That was faster than the Earth was spinning on its axis. That speed suggested to Halley that meteors had something to do with the greater speed of the Earth in orbit around the Sun. He proposed that meteors are a collection of particles formed in space and that the Earth collides with these particles as it orbits the Sun.

But five years later Halley retreated, based on an experiment he had witnessed. By 1716 Halley had seen a demonstration performed by John Whiteside at Oxford in which gunpowder was placed in a vacuum bottle and heated so that vapors from the gunpowder rose to the top of the vessel and glowed. Vapors that could rise from the ground into a realm of vacuum and there give off light seemed to vindicate Aristotle.[13]

In 1719, Halley reported on a "wonderful luminous meteor" flying southwest over England and descending from an altitude of 73½ to 69 miles. It was traveling at more than 18,000 miles an hour – "a swiftness wholly incredible," he marvels. But now Halley explains the meteor as rising vapors that had condensed at high altitude into a long strand of flammable material. The apparent velocity was not due to actual motion of an object, said Halley, but to the rapid burning of the strand of condensed vapor like a fuse.[14]

Halley had been the first to calculate the height and speed of meteors and measure their descent. He was also the first to suggest an extraterrestrial origin for meteors, but he didn't feel that the observational evidence was conclusive. On this point he returned to the Aristotelian fold.

A new effort

Meteors continued to fall and entice scientists to ponder them. Reports of a bright fireball streaking northwestward over England on November 26, 1758 lured a Scottish physician to pause in his medical work to consider this oddity. John Pringle had started his career by introducing the terms *septic* and *antiseptic* into medicine. He conceived and worked for the idea that military hospitals should be sanctuaries exempt from enemy attacks. He served as physician to the royal family. He wrote *Observations on Diseases of the Army* (1752), which provided practical rules for military hygiene and expressed the notion that contagious diseases were spread by animalcules.

On December 20, 1759, Pringle laid before the Royal Society of London his analysis of the fireball seen 13 months earlier.[15] Reports from witnesses along the meteor's 400-mile-long path from central England to western Scotland allowed him to estimate that the object became visible at an altitude of 90 to 100 miles and descended to a height of between 26 and 32 miles. Pringle thought the fireball then skipped out into space, bouncing upward as it struck the denser part of our atmosphere. "The meteor," he explained, "might be reflected by the air in the same manner as a cannon ball by water when it strikes it in a very oblique direction.[16]

Pringle then presented a calculation of the fireball's speed: 30 miles a second (108,000 miles an hour). "It seems almost incredible," he wrote, but there was "sufficient *data*," and that's what the data showed. He was correct. Even, he said, if the bright meteor was traveling at only half that speed, it was "above 100 times swifter than the mean celerity of a cannon ball, and nearly equal to that of the earth in its orbit round the sun."

Pringle was especially interested in the loud explosion heard by a farmer five minutes after the passage of the fireball. Pringle thought that the delay between the observer seeing the light and hearing the sound indicated that the meteor was more than 41 miles high.

Some people had reported a hissing noise as the meteor passed, but Pringle dismissed that claim as "a deception . . . which frequently connects sound with motion." The altitude calculations clearly indicated that the fireball was so distant that several minutes would have to pass before a sound could reach the observer.

Pringle regretted that the "excellent naturalist" Edmond Halley had turned back to the concept of earthly exhalations to explain meteors. Where are the experiments to prove such a notion, he asks. He lists a number of logical objections and notes ironically that within the last century *comets* were still thought to be vapors from Earth rising high in the atmosphere where they caught fire.[17] (It was Halley who had proved that comets were astronomical bodies orbiting the Sun.)

Pringle believed that a few meteors actually do crash into Earth (although, he said, none have yet been found), but most meteors are deflected back to space by our atmosphere. Pringle expressed no concept that meteors could be entirely vaporized by friction due to their high speed through the air.

Pringle then added up the evidence and adopted the daring conclusion that Halley had reached and later abandoned. Meteors move with the

Meteors and alcoholism

* John G. Burke: *Cosmic Debris: Meteorites in History* (Berkeley: University of California Press, 1986), page 217.

A meteor myth from the Philippines warns of the dangers of drinking. The story says that shooting stars are the souls of drunkards. Each night they return to Earth and sing, "Do not drink, do not drink." Every day the souls try to climb to heaven, but fail. Each night, down they fall.*

speed of celestial bodies. They fly in all directions. Meteors therefore must have a motion of their own, not imparted to them by Earth. They must come from space and orbit some gravitational center beyond Earth.

Such objects, Pringle felt, must have a purpose. Newton after all had proposed that comets provide fuel for the Sun and stars and replenish or purify the atmospheres of the planets.[18] Pringle offered a purpose for meteors: that during their fiery passage through the Earth's atmosphere, "they may supply some subtle and salutary matter or remove from it such parts as begin to be superfluous or noxious to the inhabitants of the earth."

For Pringle, the concept that meteors must be beneficial was less science than faith. Two centuries later, scientists would begin to recognize the chemical contributions that meteors have brought to Earth.

An American voice

Twenty years after Pringle, an American scientist joined the quest to measure the altitudes of meteors and consider the implications. David Rittenhouse was a self-educated scientist who used his mechanical and artistic craftsmanship to build exquisite clocks and other instruments for surveyors, meteorologists, and astronomers. But he was not just an instrument maker. He improved Benjamin Franklin's famous stove. He used his own instruments to survey the borders of Pennsylvania and New York for the new United States after he had used his political skill to help write documents that united the country in its drive for independence and after he had used his engineering skill to design Delaware River defenses and produce saltpeter and guns to help win the war. He also experimented with telescopic sights for rifles and artillery.

David Rittenhouse: Courtesy American Philosophical Society

Rittenhouse's favorite instruments were telescopes and he was one of the first to use spider webs as cross hairs for sighting. He established an observatory in Philadelphia to make regular observations of comets, planets, satellites, eclipses, and – meteors.

Half an hour after sunset on October 31, 1779, a friend of his in Williamsburg, Virginia saw a fireball and reported its position to him. Rittenhouse, 365 miles away in Philadelphia, had seen the trail left by this meteor and had recorded its position. Using the angular difference in the meteor's apparent position and the distance between the viewing sites, Rittenhouse concluded that the meteor was 61 miles high.

The altitude of that meteor filled Rittenhouse's mind with a series of ideas. Meteors could not coalesce from the air at a height of 50 to 60 miles, he said, because up there the atmosphere was too thin. Instead, he asked rhetorically, "May not these stars be bodies altogether foreign to the earth

and its atmosphere accidentally meeting with it as they are swiftly travers-
ing the great void of space?"[19]

Rittenhouse, like Pringle and Halley, was well ahead of his time. He realized that meteors are partially controlled by gravity because no meteors
were ever observed to move upward. But gravity was not the only factor in
a meteor's motion. If gravity alone ruled meteor motion, then all meteors
would only fall vertically – directly down. But obviously they didn't.
Meteors moved horizontally as well as vertically, indicating that they had a
velocity of their own before they encountered the Earth.

In a 1780 letter to his friend Benjamin Franklin, who was serving during
the Revolutionary War as the American rebels' ambassador to France,
Rittenhouse wrote about meteors: "I would suppose them to be bodies
altogether foreign to this Earth, but meeting with it in its annual orbit, are
attracted by it, and on entering our atmosphere, take fire . . . "[20]

Struck by the sight of the fireball trail and still more by the implications
of its height, Rittenhouse concluded his report to his friend in
Williamsburg by wondering if meteors might have still greater consequences: "They may perhaps produce some important and necessary
effects in the atmosphere surrounding this globe, for the welfare of man
and its other innumerable . . . inhabitants.

For Rittenhouse, as for Pringle, two centuries would pass before his
insight would be appreciated.

The Great Spirit at war

A trading post owner in Skunk Grove, Wisconsin saw the meteors of 1833
because of the Indians that his store served. He recorded on the back of a
receipt how the Indians reacted to the sight.*

* Receipt from John Gardner
and Lewis Benton to McCarty
& Walker for merchandise,
Skunk Grove, Racine County,
Wisconsin; with notes on the
back regarding a meteor
shower and Indians' reaction
to it, November 12 and 13,
1833. State Historical Society
of Wisconsin, Madison: File
1833 November 1.

November 12 & 13 1833. Witnessed an unusual phenomenon in the heavens.
The Indians were much horrified and awoke me about 12 Oclock by looking
into my cabin and exclaiming the Great Spirit is at War. Get up and give us
powder and lead so that we may join the Great Spirit in the fight. – But
before I could dress myself they broke open my store and helped themselves to powder & lead and fired away for dear life for they supposed they
would all die soon. When I got out the Heavens presented a most sublime
spectacle. There was a perfect meteoric Shower which lasted until sunrise
and of all sizes from the size of a snowflake to a Hogs Head.

Aztec record of a meteorite that fell, leaving a long smoke trail. Meanwhile a battle takes place on a mountain. *Codex Telleriano-Remensis*, folio 42r. Courtesy Anthony F. Aveni

A realization about meteorites

The man who made the pivotal discovery that meteorites were rocks that had fallen to Earth from space – and hence opened the door to the idea that *meteors* also might be extraterrestrial – was a musician, an inventor of

American Indian meteor lore

Native Americans had a wide variety of beliefs about meteors. For the Blackfeet of Montana and the Kawaiisu of California, meteors were omens of sickness or death. The Shawnee Indians north of the Ohio River thought that meteors were beings who were fleeing from danger or the wrath of an enemy. The Eastern Pomo Indians of north-central California thought that meteors were fire dropping from heaven. The Wintu and Chumash Indians of California believed that meteors were people's souls on their way to the afterlife.

Other Indians had less polite explanations of meteors. The Nunamiut Eskimos, the Koasati Indians of Louisiana (formerly living in Tennessee), and most southern California tribes saw meteors as the falling feces of stars. The Kiliwa Indians of Mexico's Baja peninsula believed that meteors were the fiery urine of the constellation Xsmii. Although the identification of this constellation may not be known, the urine image suggests meteors coming from a single source. Might these Indians have noticed that meteor showers have radiants?

Five southern California Indian tribes said that bright meteors – fireballs – were an evil spirit named Takwich who lived in the mountains. After dark, Takwich would wander the heavens, looking for a person far from his tribe. When he found someone, he would dart across the sky, carry him off to his home, and eat him.*

The Pawnees, who lived in Kansas and Nebraska, had songs about the Sun, Moon, and stars, but their favorite songs were about meteorites, because they had flown down from the heavens to them. In 1909 four Skidi Pawnees explained a hide painted with stars – the famous Pawnee chart of the heavens – to University of Chicago astronomer Forest Ray Moulton at the request of George A. Dorsey, curator of anthropology at the Field Museum of Natural History in Chicago. Moulton copied down what the Indians said and sent it to Dorsey. Concerning meteors, Moulton wrote: "A meteor is a star visiting. It also indicates a message from friends in the direction from which it came."§

* Gary W. Kronk: "Meteors and the Native Americans," http://www.maa.mhn.de/Comet/metlegends.html; and Travis Hudson: "California's First Astronomers," pages 11–41 in E. C. Krupp, editor: *Archaeoastronomy and the Roots of Science*, AAAS Selected Symposium 71 (Boulder, Colorado: Westview Press for the American Association for the Advancement of Science, 1984). Takwich appears in the lore of the Serrano, Cahuilla, Cupeño, Luiseño, and Ipai tribes. Takwich may, however, indicate ball lightning rather than a bright meteor. See Hudson for sources.)

The suggestion that urine (meteors) from a constellation may indicate Kiliwa recognition that meteor showers have radiants is mine.

§ Von Del Chamberlain: *When Stars Came Down to Earth: Cosmology of the Skidi Pawnee Indians of North America* (Los Altos, California: Ballena Press, 1982), pages 44–45 and 227.

musical instruments, and a pioneer in acoustics. The ancestors of Ernst Florens Friedrich Chladni (Ka-LAHD-knee) moved from Hungary to Germany in the 17th century. He had a bit of the gypsy in him. Chladni, wild and dark in appearance, spent most of his life on the move, making his living by traveling about Europe performing on musical instruments he had created. He also entranced the public – and scientists – by demonstrating the world of acoustical waves he had discovered. He would sprinkle sand on thin copper plates of different shapes, sizes, and suspension points and then draw a violin bow across a corner of the plate so that the sound vibrations would cause the sand to rearrange itself into unusual patterns and striking figures. He was musician, magician, and mad scientist. On a visit to Paris in 1809, Napoleon attended one of his demonstrations.

Ernst Chladni. Deutschen Staatsbibliothek

In 1794, Chladni read about a 1600-pound iron mass found in Siberia. He compared its description and chemical composition with analyses of other peculiar iron masses found at different sites around the world and reached an unconventional conclusion. These iron oddities were so much alike that they must have had the same origin.

The Siberian witnesses swore that their relic dropped from heaven. Instead of dismissing such a claim as fantasy as scientists had almost always done, Chladni was inclined to trust these witnesses, even if they were uneducated common people. At least it was a starting point.

Chladni noticed that the temperature required to shape the pits in these iron masses was beyond current human technology and different from the rocky products created on Earth by lightning strikes or volcanic eruptions. It was hard to believe that dense iron masses weighing a ton or many tons could be formed by columns of air rising into the sparse upper atmosphere where they would mysteriously condense rather than disperse. And these iron masses were frequently found hundreds of miles from iron mines, yet they weighed many tons.

Instead, said Chladni, these iron masses are related to fireballs – bright meteors. The fireballs show they are dense by surviving the intense heating due to the friction of their passage through the atmosphere.[21] They show they are heavy by falling according to the law of gravity. No force on Earth can impart to them the speed they demonstrate.

Their unearthly speed and composition lead to only one conclusion, said Chladni. Fireballs are meteorites falling to Earth. They truly do fall from the sky because they are formed in outer space where they have been traveling until attracted to the Earth by gravity. Fireballs are sometimes heard and sometimes observed to hit the ground. Thus meteorites come from fireballs.

Chladni then linked shooting stars (common meteors) to fireballs and meteorites. Shooting stars, he said, are the same as fireballs, but farther away and traveling at greater speed so that gravity does not bend their courses earthward as much. Because they are farther from Earth, the atmosphere they encounter at such high altitudes is too rarefied for much fire, so common meteors have only short, brief, faint trails.

Chladni called attention to the pitted surfaces of iron meteorites. These pits suggest that the iron masses contained bubbles of gas that, when heated in the atmosphere, expanded and burst away fragments of the meteorite. That would account for the sparks scattered by fireballs.

If these iron masses were created when lightning hits the ground, Chaldni wondered, why are meteorites all so similar even though lightning strikes extraordinarily different terrain? The lightning would have to completely change the nature of the existing rock. But where lightning has been observed to strike, no such iron meteorites have been found.

Therefore, said Chladni, the origin of the meteorites is the same as the origin of the planets. Stars have been observed to form and die, he said, thinking of exploding stars. Should we not expect that planets also form and die?

The orbits of the other planets show they are very massive, so we should expect they contain large amounts of iron – like the Earth, like meteorites. Might the meteorites be fragments of planets that experienced an explosion or an external shock?

Chladni went a step further. These meteorites contain not only iron but also sulfur, silicon, manganese, and other chemical elements. If iron meteorites are indeed fragments of shattered worlds, Chladni wisely observed, then the elements they contain must be common on all planetary worlds.[22]

In a subsequent article, Chladni reviewed a report of a fireball seen on July 24, 1790 in France. The fiery display, brighter than the full moon, was followed 2½ to 3 minutes later by a "dreadful clap of thunder" that rattled windows and knocked utensils off kitchen walls. Yet there was no movement of the ground beneath the observers' feet. The effect must have been produced by a "violent concussion of the atmosphere." The cannonlike noise rolled and echoed around them for four minutes.

A few miles north of them, the fireball was not only seen and heard, but many stones fell from the sky, some weighing as much as 50 pounds.

Chladni repeated his interpretation of a meteorite as a large solid body from space passing through the atmosphere as a fireball, where it shatters and falls to earth in pieces. He regretted that critics had ridiculed his interpretation as romantic, without trying to refute his arguments, which fit the facts better than any other theory. About the only serious refutation offered, Chladni said, was that a large mass falling from a great height at a great speed would hit the ground and sink to the center of the Earth. Nonsense, said Chladni. Meteorites break apart.[23]

Chladni ended his article by suggesting how his theory could be proved or disproved. Shooting stars are the fainter versions of fireballs. A few can be seen every clear night. If two observers positioned themselves some miles apart and kept records of the times and paths of meteors they saw, they would be able to determine the height of the meteors.[24]

The inspiration of Chladni

That idea excited Georg Christoph Lichtenberg. Germany had no more distinguished university than Göttingen and no more distinguished professor than Lichtenberg, Göttingen's professor of everything.[25] Lichtenberg saw instantly that the height and speed of shooting stars would settle the question of their origin – vapors rising from the Earth or objects in space colliding with the Earth. He bestowed the project on two of his best students, both destined to be professors, Johann Friedrich Benzenberg and Heinrich Wilhelm Brandes. Between September 11 and November 4, 1798, Benzenberg and Brandes observed from sites 9¼ miles (15¼ kilometers) apart.[26] They recorded a total of 402 meteors. Comparing notes after each night's watch of the time and path of each meteor, they found that they could be confident that 22 of these meteors were identical, and hence the slightly different background against which they had seen each of those shooting stars could allow them to triangulate that meteor's altitude. The height at which the meteors disappeared ranged from 6½ up to 140 miles (10.5 up to 226 kilometers). Their range for meteor burnout was far too broad because the distance between their observing sites was still too short, so that an error of one degree in an instant, naked-eye meteor sighting resulted in an altitude error of many miles.[27] However, the mean height of meteor burnout was 55 miles (89 kilometers), a meaningful number. Benzenberg and Brandes did not conclude that meteors came from space, but the height implied at least that the atmosphere extended beyond previous measures.[28]

Benzenberg and Brandes were not the first to measure the height of meteors, but they were the first to undertake a *program* to observe many meteors and calculate their height.[29]

Stones from the sky

In 1794 Chladni had proposed that meteorites come from space, that fireballs are the passage of meteorites through the Earth's atmosphere, and that shooting stars are essentially faint fireballs. As time passed Chladni was frustrated that his work attracted offhand rejection rather

The legacy of Professor Lichtenberg (1742–1799)

Georg Christoph Lichtenberg

Only five of his seventeen brothers and sisters survived childhood. Georg Christoph Lichtenberg, the youngest, just barely survived, born with an obvious deformity of the spine. He was a bright little fellow and pursued astronomy and history, literature and meteorology, geometry and art at the University of Göttingen – and other subjects as well, especially experimental physics.

He went from student to professor. A special professorship in experimental physics was created for him, the first in Germany.

Students flocked to his lectures, where he urged them to be bold in their conceptions and skeptical about what they were taught or read. He encouraged them to play with ideas, as long as the outcome could be tested experimentally.

Lichtenberg sought to resolve scientific debates about the shape of the Earth, the amount of lava a volcano belches, how rain forms, and the mathematical concept of fairness in a gambling game. Perhaps prodded by his physical deformity, Lichtenberg was a pioneer in the psychological concepts of subconscious motivation, repression, and compensation. He noticed and carefully described how facial expressions and gestures indicate moods.

As entertaining and inspiring as Lichtenberg was to students, he was a pungent satirist of bigotry masquerading as science. He slashed the popular theory of Johann Kaspar Lavater that the shape of your body is the outward expression of your soul.

In 1777, he developed the first electrostatic recording process. He induced an electrical charge in specific patterns on a nonconducting plate, sprinkled oppositely charged dark powder on the plate, then pressed the plate against a piece of paper and transferred the pattern of particles to the paper as an image. Almost two centuries later, Chester F. Carlson created a practical application for Lichtenberg's discovery. Twenty companies rejected the idea before the Haloid Company took on its development and changed its name to Xerox.

than serious attention. But the seeds he had planted were germinating and within nine years his central tenet – meteorites from space – had grown so formidable that his opposition withered.

Three factors contributed to his victory. One was work by chemist Edward Hopper, who in 1802 analyzed a number of iron meteorites and found substantial quantities of nickel in all of them. They more closely resembled each other than the terrain in which they were found, suggesting that these meteorites had a common origin that was not of this Earth.[30]

Jean Baptiste Biot

A second factor was the evergrowing, everspreading population of Earth, which meant that ever more meteorite falls were observed and specimens recovered.

The final nail in the coffin bearing the concept of the earthly origin of meteorites was hammered in by Jean Baptiste Biot. At about 1 p.m. on April 26, 1803, approximately 3,000 meteorites rained down on L'Aigle, France. Within a week, news and fragments of this queer event were selling on the streets of Paris. A member of the Institut, an elite French scientific society, who lived in L'Aigle reported by letter about the fireball, explosion, and shower of stones. The Institut had specimens analyzed, and they were similar to other meteorites.

Biot, a professor of physics at the Collège de France, wanted more data.[31] He sought and received a government commission to investigate the oddity. Over a period of nine days, he interviewed observers and delineated the elliptical area in which the meteorites fell. On July 17, 1803, he read his report to his colleagues at the Institut. All the meteoritic pieces were alike chemically. None were similar to local minerals or to slag from foundries. Stones had indeed fallen from the sky . . . and had crushed the opposition.

These things Olmsted knew at least in outline form as he pressed forward into the new terrain of meteoric *astronomy*.

NOTES

1. Johann Friedrich Benzenberg: *Die Sternschnuppen* (Hamburg: Perthes, Besser and Maure, 1839), epigraph: "Der Mond ist ein unartiger Nachbar, dass er die Erde mit Steinen begrüsst."

2. Aristotle: *Meteorology*, translated by F. W. Webster, in *The Complete Works of Aristotle*, Jonathan Barnes, editor (Princeton, New Jersey: Princeton University Press, 1984), volume 1, book I, section 4, lines 7–10. Phrasing slightly modified by me. Almost all of

Aristotle's commentary on shooting stars is found in book I, section 4.

3. Aristotle: *Meteorology*, book I, section 4, lines 32–34.

4. Virgil: *Georgics*, book 1, lines 365–367. This translation is by John Dryden in *The Works of Virgil* (London: Oxford University Press, 1961), page 58.

5. Lucius Annaeus Seneca: *Naturales Quaestiones*, translated by Thomas H. Corcoran (Cambridge, Massachusetts: Harvard University Press, and London:

William Heinemann, 1971). Seneca published *Natural Questions* about A.D. 62. His comments on meteors appear primarily in book 1, section 1.1–12; book 1, section 14.1–15.6; and book 2, section 55.2–3.

6. Pliny the Elder: *Natural History*, translated by H. Rackman (Cambridge: Massachusetts: Harvard University Press, and London: William Heinemann, 1938), volume 1, book 2, section 112.

   Seneca agreed with Pliny that shooting stars can be extinguished in clouds. See *Naturales Quaestiones*, translated by Thomas H. Corcoran (Cambridge, Massachusetts: Harvard University Press, and London: William Heinemann, 1971), book 2, section 55.2–3.

7. London: William Griffith, 1563. Reprinted as *A Goodly Gallerye: William Fulkes' Book of Meteors (1563)*, edited by Theodore Hornberger (Philadelphia: American Philosophical Society, 1979).

8. Theodore Hornberger, editor, in his introduction to *A Goodly Gallerye: William Fulkes' Book of Meteors (1563)* (Philadelphia: American Philosophical Society, 1979), page 19.

9. William Fulkes: *A Goodly Gallerye: William Fulkes' Book of Meteors (1563)*, edited by Theodore Hornberger (Philadelphia: American Philosophical Society, 1979), page 33.

10. Isaac Newton: *Opticks or A Treatise of the Reflections, Refractions, Inflections & Colours of Light* (originally published in 1704; this quotation from the 4th edition, London: William Innys, 1730; reprinted New York: Dover, 1979), page 380.

11. Edmond Halley: "An Account of several extraordinary Meteors or Lights in the Sky," *Philosophical Transactions* of the Royal Society of London, volume 29, number 341, 1714, pages 159–164.

12. Edmond Halley: "A Discourse of the Rule of the decrease of the hight of the Mercury in the Barometer, according as Places are Elevated above the Surface of the Earth; with an attempt to discover the true reason of the Rising and Falling of the Mercury, upon change of Weather," *Philosophical Transactions* of the Royal Society of London, volume 16, number 181, 1686, pages 104–116.

13. Martin Beech chronicles Halley's pioneering work and retreat on meteors in "Halley's Meteoric Hypothesis," *Astronomical Quarterly*, volume 7, 1990, pages 3–18.

14. Edmond Halley: "An Account of the Extraordinary Meteor seen all over England, on the 19th of March 1718/9, With a Demonstration of the uncommon Height thereof," *Philosophical Transactions* of the Royal Society of London, volume 30, number 360, 1719, pages 978–990.

15. John Pringle: "Some Remarks upon the Several Accounts of the Fiery Meteor (which Appeared on Sunday the 26th of November, 1758) and Upon Other Such Bodies," *Philosophical Transactions* of the Royal Society of London, volume 51, 1759, pages 259–274.

16. Pringle thought that the differences reported in the meteor's direction of flight indicated that upper atmospheric winds were deflecting the fireball. He was ahead of his time in conceiving of winds so high in the atmosphere, but he was wrong that the pressure of these winds could visibly alter a bright meteor's direction of travel, although winds rather quickly change the trail that remains.

   Pringle also offered cautiously an estimate of the size of the fiery body, with the admonitions that "its dazzling brightness would occasion some deception" and "that imagination is so apt to enlarge such objects." But even after adjustments for exaggeration, Pringle concluded that the fireball was 1½ miles in diameter – perhaps a thousand times too large.

17. Pringle did not mention that almost two centuries earlier, in 1577, Tycho Brahe had established that comets were far more distant than the Moon.

18. Arthur Jack Meadows: *The High Firmament: A Survey of Astronomy in English Literature* (Leicester: Leicester University Press, 1969), page 125.

19. This and subsequent quotations and conclusions are from an exchange of letters between John Page and David Rittenhouse about a fireball. These letters were published in David Rittenhouse: "Account of a Meteor," *Transactions of the American Philosophical Society*, volume 2, 1786, pages 173–176. Rittenhouse's letter was dated December 4, 1779 and he read it to the American Philosophical Society meeting on May 2, 1783. The letter does not mention the earlier work of Halley or Pringle.

   For biographical information on Rittenhouse, see Brooke Hindle: *David Rittenhouse* (Princeton: Princeton University Press, 1964) and David Parry Rubincam and Milton Rubincam II: "America's Foremost Early Astronomer," *Sky & Telescope*, volume 89, number 5, May 1995, pages 38–41.

20. Rittenhouse's letter to Franklin is dated December 31, 1780. Rittenhouse got the incendiary mechanism wrong however. He thought meteors "take fire and are exploded something in the manner steel filings are on passing thro' the flame of a candle." Brooke Hindle, editor: *The Scientific Writings of David Rittenhouse*, New York: Arno Press, 1980.

21. Chladni envisioned fireballs ignited by the heat of friction and burning by conventional processes.

22. Ernst Florens Friedrich Chladni: *Ueber den Ursprung der von Pallas gefundenen und anderer ihr ähnlicher Eisenmassen, und über einige damit in Berbindung stehende Naturerscheinungen* (Riga: Johann Friedrich

Hartknoch, 1794). A reprint of the entire book, with commentary and biographical information, appears as E. F. F. Chladni: *Über den kosmischen Ursprung der Meteorite und Feuerkugeln (1794)*, with commentary by Günter Hoppe (Leipzig: Geest & Portig, 1979). A contemporary English-language summary of Chladni's book appears as [Ernst F. F.] Chladni: "Observations on a Mass of Iron found in Siberia by Professor Pallas, and on other Masses of the like Kind, with some Conjectures respecting their Connection with certain natural Phenomena," *Philosophical Magazine*, volume 2, October 1798, pages 1–8.

23. Chladni thought that when a meteorite was heated in its atmospheric passage, gases inside it inflated the body so much that it exploded.

24. [Ernst F. F.] Chladni: "Account of a remarkable fiery Meteor seen in Gascony on the 24th of July 1790; by M. Baudin, Professor of Philosophy at Pau. With some Observations on Fire-Balls and Shooting-Stars," *Philosophical Magazine*, volume 2, December 1798, pages 225–231; reprinted and translated from *Magazin für das Neueste aus der Physik*, volume 11.

25. Chladni had a series of discussions with Lichtenberg about meteors and acoustics. Chladni credited Lichtenberg for stimulating his research.

26. Initially Benzenberg and Brandes observed from 6 miles (10 kilometers) apart, but they quickly realized that the shooting stars were so high that a longer baseline was necessary to measure their altitude with accuracy.

27. Friedrich Bessel: "Vorläufige Nachricht über eine die Berechnung der Sternschnupper betreffende Arbeit," *Annalen der Physik*, volume 47, 1839, pages 525–527.

28. J[ohann] F[riedrich] Benzenberg and H[einrich] W[ilhelm] Brandes: "Versuch die Entfernung, die Geschwindigkeit und die Bahn der Sternschnuppen zu Bestimmen," *Annalen der Physik*, volume 6, 1800, pages 224–232.

29. Charles Olivier was not correct in crediting Benzenberg and Brandes with being the first to measure the altitude of meteors: "These figures gave the first approximate idea as to the stratum of the earth's atmosphere in which meteors appear and disappear." Charles P. Olivier: *Meteors* (Baltimore: Williams & Wilkins, 1925), page 6.

Benzenberg and Brandes thought that 17 of the 22 meteors were traveling horizontally, while 2 were descending and 2 were ascending. Lichtenberg was proud of his students and wrote to Benzenberg that "in such a short time you have accomplished more in this science [meteorology] than all other physicists since the creation of the world or certainly since the Flood and the time of Aristotle." John G. Burke: *Cosmic Debris: Meteorites in History* (Berkeley: University of California Press, 1986), page 68. Lichtenberg subscribed to the Aristotelian notion that meteors were part of meteorology – they rose as vapors from the Earth until kindled in the upper atmosphere. The two meteors that seemed to rise rather than fall disproved, Lichtenberg thought, the idea that meteors came from space, and he was especially pleased.

Brandes returned to the meteor altitude problem in 1823 and carried out even more extensive observations, but his work attracted little attention.

30. For a more complete account of early attempts to explain meteorites, see John G. Burke: *Cosmic Debris: Meteorites in History* (Berkeley: University of California Press, 1986).

31. Biot was trying to prove that meteorites come from the Moon.

CHAPTER 4

The November meteors in history

There was not a space in the firmament equal in extent to three diameters of the moon which was not filled with burning stars.

AIMÉ BONPLAND *quoted by Alexander von Humboldt (1799)*[1]

O NE OF the first missions of Olmsted and other scientists who observed or heard about the November 1833 meteors was to search records of all kinds to see if a display of that magnitude had been seen before. They found various reports of a great downpour of meteors in the predawn hours of November 12, 1799 in the western Atlantic Ocean, from South America through the Bahamas north to Greenland. But no one at the time had paid much attention to them.

The sight from South America

Three hundred years after Columbus, two young men embarked on what they intended to be, and what they succeeded in making, Europe's *scientific* discovery of America. They compiled the first survey of the abundance and novelty of animal and plant life and mineral resources in South and Central America.

On the day in 1799 when their ship sailed from Spain to the Americas, Alexander von Humboldt, a 30-year-old German naturalist wrote, "I must find out about the unity of nature." He and his partner, 26-year-old French botanist Aimé Bonpland, landed in the city of Cumaná, Venezuela, part of the vast Spanish colonial empire in the New World.

Over a period of five years, Humboldt and Bonpland gathered a wealth of botanical, zoological, geological, geographical, meteorological, oceanographic, astronomical, and anthropological information. But it was more than just a trove of data. Humboldt especially was intent on finding the interrelations between life and its environment – the beginnings of ecology.

As of sundown on November 11, 1799, Humboldt and Bonpland had been in South America for four months. They had already collected around Cumaná more than 1,600 plants and found 600 new species and recorded observations of wildlife and the skies while arranging for boats to explore along the Orinoco River and the Rio Negro. They had experienced

Alexander Humboldt and
Aimé Bonpland in Ecuador.
Artwork by F. G. Weitsch,
Berlin, 1806. Courtesy
American Philosophical
Society

a partial eclipse of the Sun (on October 28), an earthquake (on November 4), and even an attack (on October 27) by a club-wielding native who gave Bonpland a concussion. Their sojourn was already rich in adventure.

It was Bonpland who saw them first. He had risen, Humboldt said, to enjoy the freshness of the air. It was 2:30 in the morning, November 12, 1799.

Shooting stars. Bolides – "the most extraordinary luminous meteors." Everywhere. Thousands of them.

Bonpland, despite an aching head, had the presence of mind to notice that "from the first appearance of the phenomenon, there was not in the firmament a space equal in extent to three diameters of the moon which was not filled every instant with bolides and falling stars."

Many of these meteors were not pinpoints streaking through the sky, but balls of fire – some larger than the disk of the Moon. As they fell without a sound,[2] some spewed "darted sparks of vivid light," some "seemed to burst as by explosion," while others left "behind them phosphorescent bands."

"Almost all the inhabitants of Cumaná witnessed this phenomenon," Humboldt wrote, "because they had left their houses before four o'clock to attend the early morning mass." "They did not behold these bolides with indifference;" Humboldt understated, "the oldest among them remembered that the great earthquakes of 1766 were preceded by similar phenomena."[3]

Native fishermen in the outskirts of Cumaná reported that the shower of falling stars had started at one o'clock. They were still faintly visible fifteen minutes after sunrise.

Humboldt and Bonpland made no specific mention that the meteors were radiating from a particular constellation in the sky, but their account vaguely suggests a radiant: all the meteors were coming from the northeast (where Leo was rising) and "those meteors might be compared to the blazing sheaves which shoot out from fireworks."[4]

The sight from Florida

Humboldt and Bonpland were not the only ones in the right place at the right time. Andrew Ellicott, also on a voyage of discovery, was destined to be a witness.

Three years earlier, on September 16, 1796, Ellicott left his wife and children to lead a United States government mission to survey and appraise the Ohio and Mississippi River valleys and the coast of the Gulf of Mexico. The land bordering the Gulf of Mexico was the Spanish colony of Florida, which included the peninsula and the Gulf coastline all the way to New Orleans. The expedition was to last four years.

Ellicott started from his home in Philadelphia and made his way overland to Pittsburgh, chronicling his observations as he went. He decried the slavery he saw in the Baltimore area: poor cultivation was the result. Free men work better, he said. The dwellers along the Ohio River downstream from Pittsburgh annoyed him. The land was rich. Feeding themselves should have been no problem. Yet these people were a filthy lot, making their money by selling alcohol to river travellers. He admired the great Indian earth mounds just past Wheeling and then again near Marietta and Gallipolis, Ohio.[5]

Nearing the mouth of the Mississippi, he observed:

The natives of the southern part of Mississippi are generally a sprightly people, and appear to have a natural turn for mechanics, painting, music, and the polite accomplishments, but their system of education is so extremely defective that little real science is to be met with among them.[6]

Ellicott recognized the importance of a port city where the Mississippi met the ocean, but found that New Orleans was poorly situated. Yet he could find no better site. Increasing numbers of Americans were pushing into Louisiana, then owned by Spain. Border disputes put Ellicott constantly at odds with the Spanish governor. But Ellicott praised the man as a ruler concerned with the welfare of all those he ruled – and a good family man.

At New Orleans, Ellicott and his expedition abandoned their riverboats for sailing ships and worked their way along the Mississippi, Alabama, and Florida coasts, exploring each river to see how far it was navigable and to appraise the value of the land.

On November 5, 1799, he and his men reached Duck Key, Florida, where they were attacked by a swarm of mosquitoes so vicious that everyone leapt into the water to avoid being sucked dry.

Ellicott then headed up the east coast of Florida, where he was awestruck by the Gulf Stream and recognized its usefulness to ships.

In 1800 he returned to Philadelphia and his family, presented his report to the government, and in 1803 published an account of his journey, just after the United States purchased Louisiana and the vast territory west of the Mississippi River from France. Ellicott ended his book with reverence for the country he had explored and the hope that this realm would be "the future theatre of free governments . . . May the now peaceful shores of the Mississippi never be made vocal with the noise of the implements of war and may its waters never be dyed with human blood!"[7]

It was on that voyage that Andrew Ellicott saw a sight that had nothing to do with determining the border between the new United States and the Catholic Kingdom of France and Spain. But the sight was of monumental significance. Ellicott's ship was just off Key Largo on November 12, 1799:

> At about two o'clock in the morning, I was called up to see the shooting of the stars (as it is vulgarly termed). The phenomenon was grand and awful; the whole heavens appeared as if illuminated with skyrockets, flying in an infinity of directions, and I was in constant expectation of some of them falling on the vessel. They continued until put out by the light of the sun after daybreak. This phenomenon extended over a large portion of the West India islands and was observed as far north as St. Mary's [Georgia, on the Florida border], where it appeared as brilliant as with us.[8]

From Greenland to the Bahamas

Reports of other sightings of the November 1799 meteors turned up. A Moravian Church missionary in Labrador, well to the north on the east coast of Canada, reported that in the early morning of November 12, 1799 colonists and natives in two settlements saw a "strange appearance in the air that terrified the Eskimos": fireballs of great size and brightness flying in all directions. This same eerie sight was also seen from two settlements in Greenland, 100 miles from one another.[9]

In Charleston, South Carolina, David Ramsay, a medical doctor, reported

Another 1799 sighting at sea

Sailors on the American brig *Nymph* also saw the meteors from a position about 600 miles east of Cape Canaveral, Florida. They were headed home to Newburyport, Massachusetts from the island of Santo Domingo (now Hispaniola). Upon reaching port, Captain Joseph H. Woodman recounted the experience in the *Newburyport* (Massachusetts) *Herald and Country Gazette* on December 20, 1799:

> ... at half past 1 o'clock in the morning, ... [I] observed the stars to shoot in great numbers from every point of the compass, and at two o'clock, the whole atmosphere appeared to be full of stars, I may say thousands of thousands, shooting and blazing in every direction in a most alarming manner, and so continued till daylight ...

When he reached Martha's Vineyard, he was amazed to find that captains on several other vessels had seen the meteors too although they were considerably north of him.

The event was so strange that the newspaper publisher introduced the account of the meteors with a note that "Captain Woodman's established character for veracity would not admit a doubt of its truth in the minds of any acquainted with him."

> a singular but splendid phenomenon. Instead of a few solitary meteors sporting along the sky, which is not unfrequent, they appeared in countless numbers darting incessantly in all directions ... the fears of some timid persons were so excited by the ... light darting in all directions that they apprehended the day of judgment & conflagration of the world to be at hand."[10]

This account and others from the Bahamas reported the weather before and after the meteor fall, under the impression that one causes or at least influences the other.

From Nassau, in the Bahamas, came a report of

> a most splendid phænomenon, from about twelve o'clock till near sunrise; the atmosphere was one continued blaze of fire-balls & what are commonly called falling stars, fifty at least might have been counted in the space of one minute, darting in every direction & bursting with a glare which often exceeded the brightness of the moon which was then nearly in the zenith.
>
> Sometimes they appeared so near that people imagined there could be heard a hissing noise as they passed along, leaving a long train behind

Meteor shower of November 12, 1799 as seen by Andrew Ellicott off the southern coast of Florida. Edwin Dunkin: *The Midnight Sky*, 1891

Meteor shower of November 12, 1799 as seen by Moravian missionaries in Greenland Thomas Milner: *The Gallery of Nature*, 1849

them, which would remain fixed for several minutes, presenting a beautiful variety of colors & gradually approaching to the appearance of a slender white cloud curling at length by the motion of the atmosphere; sometimes where a kind of explosion would take place, a ring of smoke would form like that from the mouth of a cannon.[11]

The nearly forgotten display

The spectacle of November 13, 1833, when the stars fell, immediately reminded Olmsted and others of the November 1799 celestial fireworks that Alexander von Humboldt told of in his very popular *Personal Narrative of Travels to the Equinoctial Regions of America, During the Years 1799–1804*. But most scientists did not immediately recall the unusual number of meteors seen in Europe before dawn on November 13, 1832 – just one year earlier. As European and American scientists combed history books, observing journals, and newspapers, they began to realize that November meteors had been seen more recently than 1799, although only Alfrède Gautier in Geneva, Switzerland had gone out of his way at that time to collect reports of the event and sort out the reliable ones.[12]

In Geneva, the meteors flashing behind broken cloud cover gave the effect of lightning. When the clouds cleared, the effect was of silent fireworks, with lingering trails of all colors.

In Neuchâtel, Switzerland, as elsewhere, moonlight prevented an appreciation of the full extent of the display, but even so, there were 30 meteors a minute on the average, a spectacle bright enough "that one could easily have read by its light." For more than two hours, and especially between 4 and 5 a.m., witnesses saw "innumerable stars forming a sort of flaming rain."

The *Portsmouth* (England) *Herald* got a little carried away – "A thousand meteors glowed continuously in all directions . . ." – but the sight was strange enough that the driver of a night carriage from London had trouble controlling his horses.

"At Glossop [England], these luminous apparitions appeared so frightful that several persons going to their early morning work were struck by terror and panic and returned home until the cause of their fears disappeared."

In his summary and effort at interpretation, Gautier complained that so little was known about the *celestial* bodies which are close to Earth. By using the word celestial, Gautier seemed to sense that these meteors came from space, not Earth. He cited efforts by Brandes and Benzenberg in 1798 and Brandes in 1823 to measure the height and speed of meteors and used

Two firsts: the earliest sighting of the Leonid meteors and the earliest observation of a meteor shower radiant

In A.D. 902, King Ibrahim ben Ahmed ruled much of North Africa and reached out for Europe. His dynasty had driven the Christians out of Sicily. In one battle, he trapped a bishop and the remnants of his people in a church and burned it down. He then crossed to the Italian mainland and besieged Cosenza. It was then that the king died suddenly: October 14 on the Julian calendar (October 30 on the modern Gregorian calendar). According to one chronicler, "On the night when King Ibrahim ben Ahmed died, an infinite number of stars scattered themselves like rain to the right and left."* Another chronicler reported that the shooting stars "moved as if they had been darted through the atmosphere from a culminating point . . . On account of this phenomenon, this year was called the year of the stars."§ This mention of a culminating point may be the earliest recorded observation of a radiant in a meteor shower, although the position of the radiant among the constellations is not given.

These accounts of the year 902 mark the earliest definite sighting of the great meteor barrage that Olmsted and others in 1833 saw radiating from Leo.

* Cited by Hubert A. Newton: "Relation of Meteorites to Comets," *Nature*, volume 19, February 6, 1879; pages 315–317; 340–342. A reprint exists, paginated 1–12.

§ Cited by Edward C. Herrick: "Contributions towards a History of the Star-Showers of Former Times," *American Journal of Science and Arts*, volume 40, number 2, April 1841, pages 349–365. See also J. A. Condé: *History of the Dominion of the Arabs in Spain* (translated by Mrs. Jonathan Foster), volume 1 (of 3) (London: Henry G. Bohn, 1854), page 403.

that altitude (5 to 40 miles [8 to 65 kilometers], he incompletely recalled) to doubt that shooting stars are Earth exhalations that ignite high in the atmosphere.

Gautier wisely judged that the meteor events of November 12, 1799 and November 13, 1832 were the same. But then he backed away from a celestial cause. He explained the meteors as electrical discharges in the rarefied and dry upper atmosphere of Earth, perhaps related to the aurora.

Later, following the 1833 display, Edward Herrick at Yale College collected other accounts of the 1832 meteor downpour. One story told that on the Vienne River in central France near Limoges, workers who were laying the foundation of a bridge

> perceived in the sky shooting stars which amused them very much at first, but some hours later the number of shooting stars multiplied so greatly that, in the end, the spectators were seized with horror, and the terror was so great that they abandoned their work to go say their last goodbyes to their families, saying that the end of the world had come.

Their supervisor had much difficulty stopping them.[13]

So the great November shooting star show of 1833 was not a one-night stand, astronomers were finding. It had been seen in earlier years, awakening, frightening, and awing people everywhere on Earth north of the equator. Would it be seen again?

1202 and 1366: Celestial grasshoppers and the air in flames

* Edward C. Herrick: "Contributions towards a History of the Star-Showers of Former Times," *American Journal of Science and Arts*, volume 40, number 2, April 1841, pages 349–365.

§ W. S. Rada and F. R. Stephenson: "A Catalogue of Meteor Showers in Medieval Arab Chronicles," *Quarterly Journal of the Royal Astronomical Society*, volume 33, 1992, pages 5–16. Translations slightly modified.

¶ J[ohn] R[ussell] Hind: "The Meteor-Shower of November," *The Times* (London), November 12, 1866, quoting from Alexander von Humboldt's *Cosmos*.

The meteors that Denison Olmsted saw in 1833 had also fallen on October 19, 1202 (October 26 on the modern Gregorian calendar). An Arab account said:

> The stars rushed across the heavens . . . like grasshoppers in a field. This continued until dawn. The inhabitants cried out with terror and fervently implored the mercy of the Most High.*

In Yemen, the meteors fell in such abundance "that a person could notice an uncountable number in one glance." The observer remained calm enough to notice that "the well-known stars did not move from their positions."§

This stellar exhibition returned in force on October 23, 1366 (October 30 on the Gregorian calendar). According to a Portuguese chronicler, "there was in the heavens a movement of the stars such as men never before saw or heard of . . . they fell from the sky in such numbers and so thickly together that . . . the air seemed to be in flames, and even the earth appeared as if ready to take fire."¶

NOTES

1. Alexander von Humboldt and Aimé Bonpland: *Personal Narrative of Travels to the Equinoctial Regions of America, During the Years 1799–1804*, translated and edited by Thomasina Ross, volume 1 (of 3) (London: Henry G. Bohn, 1852–1853 [volume 3]), page 352.

2. Humboldt did not mention hearing any noise from the meteors.

3. If the 1766 meteor date had been around November 11, it would have strongly encouraged the interpretation that such unusual meteor outbursts recur at 33- or 34-year intervals. But Humboldt furnished no month and day for the 1766 meteors, and no one in 1799 or 1833 postulated a third-of-a-century periodic recurrence of a meteor storm.

4. Humboldt credits this observation comparing the November 1799 meteor shower to the appearance of fireworks – which might suggest a radiant – to the Count of Marbois, who was living in (unjust) exile in Cayenne, French Guiana. Humboldt and Bonpland themselves did not report observing a radiant nor mention any other observer who did.

   Olmsted might have read of Humboldt's meteor account in *Personal Narrative of Travels to the Equinoctial Regions of America, During the Years 1799–1804* in a 5-volume English translation published in 1825 or possibly in *Voyages aux régions équinoctiales du nouveau continent, fait en 1799–1804*, the original French edition of 3 volumes, published between 1814 and 1819. The account ends in March 1801. Humboldt planned to complete the project, but for reasons unknown destroyed the almost completed manuscript, according to Douglas Botting: *Humboldt and the Cosmos* (New York: Harper & Row, 1973).

   Humboldt had expected Bonpland to collaborate on these volumes, but Bonpland found that he disliked organizing and writing. He managed to complete only one short section on botany, full of careless errors. Bonpland returned to South America and went native. Publication of the expedition's findings, full of meticulous color drawings, exhausted Humboldt's inheritance and left him in financial ruin. But it also propelled him to the height of scientific fame.

5. Andrew Ellicott: *The Journal of Andrew Ellicott* (Chicago: Quadrangle Books, 1962; originally published in Philadelphia in 1803), pages 2, 7–8, 9, 10, 13.

6. Andrew Ellicott: *The Journal of Andrew Ellicott* (Chicago: Quadrangle Books, 1962; originally published in Philadelphia in 1803), page 135.

7. Andrew Ellicott: *The Journal of Andrew Ellicott* (Chicago: Quadrangle Books, 1962; originally published in Philadelphia in 1803), page 300.

8. Andrew Ellicott: *The Journal of Andrew Ellicott* (Chicago: Quadrangle Books, 1962; originally published in Philadelphia in 1803), page 249.

9. The observer noted that such a distance could allow the approximate altitude of these meteors to be calculated, but no calculation was made. "Aus den Tagebüchern der Missionarien der evangelischen Brüdergemeine," *Annalen der Physik*, volume 12, 1803, pages 217–218 of 206–223. The article is signed at the end only by the initial "A." Alexander von Humboldt had mentioned this observation prior to the November 1833 meteor display in his *Personal Narrative of Travels to the Equinoctial Regions of America, During the Years 1799–1804*. After the November 1833 meteor show, Edward C. Herrick recorded this display in his notebook of historic sightings (Yale University Library, Manuscripts and Archives Department), as did Hubert A. Newton: "On November Star-Showers," *American Journal of Science and Arts*, volume 37, May 1864, pages 377–389.

10. Edward C. Herrick's notebook listing accounts of historic meteor showers, in the Yale University Library, Manuscripts and Archives Department. Herrick found this account in David Ramsay: *The History of South Carolina* (2 volumes) (Charleston, 1809), volume 2, page 307.

11. Edward C. Herrick's notebook listing accounts of historic meteor showers, in the Yale University Library, Manuscripts and Archives Department. He recorded other observations from the Bahamas as well. Herrick copied a portion of a letter from Exuma in the Bahamas that wisely noted: "The awfulness of this extraordinary scene was probably lessened by the presence of the moon, which was just past the full & shone very bright."

12. Alfrède Gautier: "Notice sur les météores lumineux observés dans la nuit du 12 au 13 novembre 1832," *Bibliothèque universalle des sciences, belles-lettres, et arts*, volume 51, October 2-December 2, 1832 [read at the December 6, 1832 meeting of the Society of Physics and Natural History of Geneva], pages 189–207.

13. Herrick found this account in a letter from a retired Captain Tharaud to François Arago published in *Comptes Rendus*, volume 6, number 16, October 16, 1839, pages 502–503. Edward Herrick's meteor record notebook in the Yale University Library, Manuscripts and Archives Department.

   Herrick recorded other notable sightings of

November meteors in 1832. An observer on the island of Mauritius, east of Madagascar off the east coast of southern Africa, reported shooting stars too numerous to count crossing the sky in all directions, some so dazzling they cast a shadow. Herrick found this account in *L'Institut*, Supplement to number 218, August 1837, page 276.

In the Malvern Hills in western England, despite bright moonlight, meteors rained down at a rate of about ten a minute, some as bright as Roman candles. It was best described as a "shower of fire." Herrick found portions of this account in *The Literary Gazette* (London), number 830, December 15, 1832, page 794; and *London and Edinburgh Philosophical Magazine and Journal of Science*, volume 3, number 13, July 1833.

Denison Olmsted logged a report of the 1832 shower seen by a Captain Hammond aboard a ship near Mocha, a Red Sea port: "It appeared like meteors bursting in every direction. . . . On landing in the morning I inquired of the Arabs if . . . the like had ever appeared before: the oldest of them replied that it had not." Denison Olmsted: "Observations of the Meteors of November 13th, 1833," *American Journal of Science and Arts*, volume 26, number 1, April 1834, page 136 of pages 132–174. Olmsted found this report in the *Salem Register*.

CHAPTER 5

The critics attack

A "meteoric rise" in a man's career may well merit congratulations, but the approval should be tempered by thought of some other connotations of the simile. Even a moderately bright meteor can generate energy at a rate of thousands of horse-power, but its life is inevitably brief and fleeting and its trajectory is always down to earth, never upward.   DONALD W. R. McKINLEY (1961)[1]

THE EXCITEMENT created by the grand starfall of November 13, 1833 and Olmsted's hypothesis that the Earth crossed the orbit of a swarm of meteor particles every year led to considerable scientific expectation as November 13, 1834 neared.

Alexander Bache, a great-grandson of Benjamin Franklin, observed from Philadelphia and recorded, near dawn, after moonset, shooting stars appearing at the rate of about 30 an hour, which he termed "no remarkable display of meteors of the kind witnessed in 1833."[2] *(Today, a display of 30 meteors an hour would be listed as moderate among the major annual meteor showers.)*

Bache and his scientific colleague and friend James P. Espy watched the meteors to see if they appeared to radiate from the constellation Leo, as Olmsted had claimed for the meteors in 1833. Bache and Espy plotted the trails of five meteors and traced them backward. Only three of the five diverged from Leo, Bache reported. *(Today, astronomers understand that there are always "sporadic" meteors to be seen, ones that do not belong to an identified meteor stream and that therefore do not share the radiant of the shower that they seem to be part of. On a clear, moonless night away from city lights, about eight sporadic meteors can been seen every hour.)*

The number and motion of the November 1834 meteors did not impress Bache. As a founder and president of the American Philosophical Society, based in Philadelphia, Bache had the Society minutes record that a recurrence of the 1833 meteor shower in November 1834 had not been seen. He submitted his article on the failure of the November meteors to return in 1834 to Benjamin Silliman for the *American Journal of Science and Arts* eight days after November 13, 1834.

Bache was pleased with his results and had been expecting such an outcome, which he took to undermine Olmsted's hypothesis on the extraterrestrial origin of the 1833 meteor shower. He mentioned in his article that he and Espy had seen the meteors of 1833 and that they had been watching for meteors throughout the summer of 1834 to establish expected numbers of meteors on ordinary nights, so that he was "compe-

Alexander D. Bache. Courtesy American Philosophical Society

tent, as far as experience could render me so, to the task which I had under-taken."

"It will be interesting," he smirked, "to have information on this subject from different quarters of our country as having a direct bearing upon the explanation of the meteoric phenomenon of last year."

He got more than he bargained for in reply. In the very next article in that issue of Silliman's Journal, Alexander Twining reported from West Point what sounded like a recurrence of the meteors on a reduced scale.

Despite the brightness of a nearly full moon, Twining saw about 30 meteors in 25 minutes, a rate of more than 60 per hour *(which is slightly better than the usual maximum of any annual meteor shower today)*. All but four of the meteors Twining saw emanated from the same fixed point and that radiant was in the constellation Leo, as in 1833. The exact position of the radiant was harder to determine in 1834 because of the relative scarcity of meteors compared to 1833.

"I have not formed a decided opinion whether this whole display is to be considered a slight recurrence of the meteoric phenomenon of November 13, 1833 or not," Twining wrote 2½ weeks after the event. However, "It certainly possessed, on a greatly diminished scale, the same general character."[3]

No doubt Silliman prodded his colleague Olmsted to respond to Bache's attack and held off publication as long as he could to include a reply. Olmsted supplied Silliman with a response in January 1835 and it appeared in the Miscellanies section at the end of the January 1835 issue, accompanied by a note from Silliman that Olmsted had promised a more extended article for a future issue.[4]

"On the morning of the 13th of November," Olmsted reported, without referring to Bache, "there was a slight repetition of the meteoric shower which presented so remarkable a spectacle on the corresponding morning of 1833." Anticipating some recurrence of the phenomenon because of his hypothesis of the periodic nature of the meteor cloud's visit, Olmsted had alerted friends in New Haven and other cities to watch for meteors.

The presence of a bright Moon until nearly 4 a.m. "permitted only the larger and more splendid meteors to be seen," Olmsted noted, but he counted 155 in the three hours from 2:15 to 5:15, for a single observer rate of about 50 an hour *(about the rate expected of the August [Perseid] and December [Geminid] meteors, the best annual meteor showers at present)*. "The *number* of meteors," he wrote, "though small compared with last year, was evidently much above the common average."

Furthermore, "The *directions* of the meteors were more remarkable than their number, and afforded more unequivocal evidence of the iden-tity of the phenomenon with that of last year. They appeared, as before, *to*

Plotting of meteors on November 13, 1866, demonstrating the shower's radiant. Four sporadic (non-shower) meteors are also shown. H. N. Russell, R. S. Dugan, and J. Q. Stewart: *Astronomy*, volume 1: The Solar System (revised), 1945

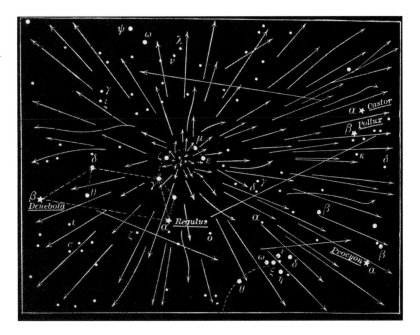

radiate from a common center, and that center was again in the constellation *Leo*."

Whereas Bache had trouble discerning the radiant of the meteors he and Espy observed in Philadelphia, Olmsted reported that he had Elias Loomis, a tutor at Yale, keep track of meteor trails and later trace them on a celestial globe to ascertain the radiant. As in 1833, the radiant held its position among the stars as Leo rose higher in the predawn hours, reconfirming Olmsted's proposal that the meteors came from beyond the Earth's atmosphere.

At the same time that Olmsted and Twining found themselves in conflict with Bache in the *American Journal of Science and Arts,* Bache's close friend James P. Espy, chairman of the Meteorological Section of the Franklin Institute in Philadelphia, published his refutation of Olmsted in the *Journal of the Franklin Institute*. The haughty and denigrating tone that pervades the article is evident in its title: "Remarks on Professor Olmsted's Theory of the Meteoric Phenomenon of November 12th [*sic*], 1833, denominated Shooting Stars, with some Queries towards forming a just Theory."[5]

Espy was angry. He was a meteorologist. Meteors were part of meteorology. Always had been. The idea that the meteors of the 1833 deluge were part of astronomy, travelled in space and collided with the Earth, violated his territory. He didn't say it outright, but his article oozed resentment that Olmsted had crossed the boundary into territory he knew nothing about and in which he, Espy, was an expert.

James P. Espy. William Jay Youmans, editor: *Pioneers of American Science*, 1896

The meteor debate

Espy and Olmsted never met in a debate, but if their articles are placed side by side, one can easily imagine the sound and content of the confrontation.

ESPY

You say that meteors travel in a nebulous swarm along an elliptical orbit that carries them close to the Sun and then outward to the orbit of the Earth. If so, the meteors would approach the Earth on the sunward side – the daytime side – and none would be seen in the night. Yet you, I, and every witness saw with his own eyes that the meteors on November 13, 1833 fell at night.

OLMSTED

My point is that these meteors orbit the Sun and are not in orbit around the Earth. I proposed an approximate orbit with a period of about 182 days so that the meteor swarm would go around twice and then, on its second orbit, a small region of the swarm would collide with the Earth to provide a meteor shower. If the orbit of the meteors stretched out just beyond the orbit of Earth, then, as they began to return to the Sun, the meteors could strike the Earth on the nighttime side of the planet.

ESPY

While you adjust your orbit and diddle with your numbers, I will point out other flaws in your ideas. How, Professor Olmsted, can you explain the curved appearance assumed by some meteor trails?

OLMSTED

Long meteor trails that started off straight became sinuous and curved and moved from west to east. I can only conclude that there must be a wind at high altitude.

ESPY

And at what altitude do you imagine this wind?

OLMSTED

Based on several reports of the same bright meteor seen by two or more observers many miles apart, Alexander Twining and I independently calculated that the meteor must have burned out at an altitude of about 30 miles (50 kilometers).

ESPY

And how high are atmospheric winds? No, you are an astronomer, not a meteorologist. I will tell you. Up to two miles high or a little higher.[†] There is no atmosphere at 30 miles.[‡]

OLMSTED

I must say that winds at such great height puzzle me too. I may have to place the burnout of meteors much lower.

ESPY

And well you should. A lot lower. At 30 miles, to explain the motion you and others observed, the winds would have to be blowing at 300 miles per hour (500 kilometers per hour). An absurdity.

And another: you, my good professor, claim that meteors burn up in the Earth's atmosphere but cannot explain how this happens.

OLMSTED

I have proposed that a meteor falls into the Earth's atmosphere at high speed, which compresses the atmosphere, creating high temperatures, which burn up the meteor. These meteors must be composed of rather light material – of low density – that is easily consumed by heat, for as best I can determine no meteors of the 1833 shower struck the ground. They are not the same composition as aerolites – the celestial rocks – that fell on France in 1803.

ESPY

And did you not calculate that the temperature of the burning meteors would be 46,080 degrees centigrade? I thought at first you had made a typographical error, but, no, you persist in this error and in your misunderstanding of the atmosphere throughout your article.

[†] By 1900, high-altitude winds of over 200 miles per hour (320 kilometers per hour) had been measured from the ground by the movement of cirrus clouds. G. D. Swezey: "The Present Status of Meteoric Astronomy," *Proceedings of the Nebraska Academy of Science*, volume 7, November 1901, pages 79–89. However, these jet stream winds 5 to 9 miles (8 to 14 kilometers) high were not generally recognized until World War II when American B-29 bombers encountered them en route to Japan. Of course, meteors burn up far above the jet stream.

[‡] Espy was wrong. The atmosphere is very thin at 30 miles (50 kilometers), but even at 100 miles (160 kilometers), it is dense enough to cause orbiting spacecraft to fall out of orbit within weeks. It is this thin upper atmosphere that creates the friction that causes fast-traveling meteoroids to heat up and vaporize. There are winds within this tenuous upper atmosphere.

If a meteor had plunged deep into our atmosphere, as you claim, compressing the air before it, the compressed heated air would have expanded, and the recoil would have bounced the meteor back out into space. Yet no one saw the meteors reverse direction.

These reasons allow us to safely infer that the meteors of the 13th of November 1833 were not bodies falling toward the Earth by gravity, and in fact were not falling at all, either from inside or outside our atmosphere. Instead, we must conclude that the meteors were certainly less than 10 miles (16 kilometers) high and hence a part of our atmosphere and weather.

And I can prove that the meteors were in fact much closer to the ground than 10 miles. If the meteors were 10 miles high or more, observers in different locations would have seen the same meteor at the same time, but against a different background of stars.

OLMSTED

Indeed, this is just what we observed and that is how we calculated the height that meteors burn out to be about 30 miles.

Trail left by meteor on November 17, 1966, demonstrating high-velocity winds at altitudes of 50 to 70 miles (80 to 110 kilometers) (1-minute exposure). Jet Propulsion Laboratory/ NASA, Table Mountain Observatory; photograph by James W. Young

ESPY

But did you really? One observer says he saw a fireball at 4:15 a.m. Another says he saw a fireball at 4:30 –

OLMSTED

Give or take a few minutes. You know how clocks vary.

ESPY

You give or take a few minutes. I will not. The reports that you, my friend Professor Olmsted, have collected and published clearly reveal your problem: they agree on virtually nothing. Some say the meteor shower began two hours or more before midnight. Others at midnight or 1:00. Others at 3:00 or 4:00. Some say the meteors were all or mostly in the northwest, others say the southeast, others say they were evenly spread about the sky. Some say the meteors had colors, others say they were white. Some people say they saw the meteors radiate from a fixed point, but they cannot agree where – was it the zenith, 10 of 20 degrees from the zenith, half way up the southeastern sky? Others say they saw no point of radiance at all.

The key is that if this shower of meteors had been 10 miles or higher, observers over a wide region would have seen the same sights. Yet they agree on almost nothing. That's because nothing was seen in common. That's because the meteors were close to Earth. And that means that the meteors could not have penetrated the atmosphere, for, traveling at the speeds that you, Professor Olmsted, claim, they would have burned up before getting so close to the ground.

And now the final proof. I must thank you, Professor Olmsted, for providing much of it in the reports you collected. I have provided the rest, with the help of my friend Professor Bache. Remember how many observers reported the noise the meteors made – the popping, the hissing? Those sounds, so nearly simultaneous with the light of the meteors, proves that the meteors could not have been many hundreds of yards – I say *yards* – from the observer – certainly not over half a mile.

Even if there had been no other witness, who could doubt so patient and minute an observer as Mr. James N. Palmer – cited by you, yourself, Professor Olmsted. Mr. Palmer heard the meteors popping, which absolutely and decisively proves to me and any unbiased friend of science that the meteors were in very close proximity to the observer, at least those which were heard by Mr. Palmer.

Why was the noise not heard everywhere? Because the meteors had to be very close to be heard.

The evidence that the meteors made a popping or hissing noise is too widespread and too respectable to be doubted; and this settles the question of their very close proximity to the surface of the Earth at the time of their explosion; and proves positively their atmospheric origin.

With our minds now freed from wild speculations and visionary theories, the true cause of meteors is easy to discern. Meteors do not come from a celestial source. Instead, meteors arise in our own atmosphere.

Recall from the reports from all over the country that you, Professor Olmsted, have published. Notice how often the aurora was seen in conjunction with the meteors.

OLMSTED

In perhaps one case other than Mr. Palmer.[†]

† See chapter 2. Darius Lapham of Cincinnati thought he saw the aurora during the meteor shower.

The November 13, 1833 issue of the Boston *Mercantile Journal* published a one-paragraph notice of "an extraordinary colored illumination" seen before sunrise that day, but it was not called an aurora, although the northern lights were a relatively familiar sight in Boston. Stories about the display in subsequent issues of the *Mercantile Journal* all identified what was seen as meteors.

A fraudulent report

James N. Palmer, a surveyor living in New Haven, read Olmsted's report of the meteors in the paper. Several weeks later he dropped by Yale College to see Olmsted and discuss what he had seen. Olmsted was very impressed – too impressed – with Palmer, whom he called "ingenious," and accorded Palmer's account considerable weight in his analysis of the event.

Palmer gained Olmsted's trust by parroting some of the basic observations that Olmsted had made and reported in his newspaper article and by claiming to have conducted scientific experiments during the shower. He made most of it up, gauging by his inconsistencies and absurdities, but that's easier to see in hindsight.

Olmsted's notes* showed that Palmer testified that

From 7 o'clock in the evening, I had noticed a reddish vapor, which first appeared low in the south but gradually rose up the southern sky to the zenith. It was very thin, but still obscured the smaller stars. This vapor continued to prevail during the earlier parts of the meteoric display. I retired to rest about 12 o'clock. At 2 o'clock, a man in my employment discovered the meteors through the window of his chamber and immediately called me.

Considering the phenomenon as electrical, I immediately made some experiments to ascertain the electrical state of the atmosphere. I held my silk pocket-handkerchief at one end in my right hand and drew it swiftly

*Transcribed from third to first person and condensed. Denison Olmsted: "Observations on the Meteors of November 13th, 1833," *American Journal of Science and Arts*, volume 25, number 2, January 1834, pages 383–385 of 363–411; continued under the same title in volume 26, number 1, April 1834 (although bound as "July 1834"), page 168 of pages 132–174.

ESPY

Auroras or luminous clouds or strange lights in the sky – not everyone is skilled at recognizing the aurora.

There is a significant reason why meteors and auroras accompany one another. They are two aspects of the same phenomenon. Meteors are the aurora seen close up and hence distinct. The aurora is meteors seen far off so that the distant meteors cannot be seen singly.

OLMSTED

But even allowing for clouds and bright skies to be called the aurora, very, very few people saw such an aurora, but many people saw the meteors.

ESPY

That's because in those cases the aurora was there but it was too faint to be seen.

No one yet knows for certain what the aurora is, but I strongly suspect that the aurora is electrical. It depends on the overlapping of two currents of

through my left hand. It emitted a very unusual number of electric sparks. On turning on a small machine, I found the sparks, which are usually short and feeble, much longer and more intense than I had ever seen them before. When I presented silk threads to an iron bar that stood on the ground leaning against the house, the threads were strongly attracted toward the iron.

I next examined my compass and found the needle more unsteady than ordinary, but as nearly as I could, I judged the declination of the needle to be the same as usual.

From 3 to 4 o'clock the air was still, but at 4 o'clock a strong gust of wind blew for a short time from the northwest and, immediately afterwards, the meteors increased astonishingly. These gusts returned at moderate intervals, each time occasioning a perceptible increase in meteors. I heard at different times a number of slight explosions, which usually resembled the noise of a child's pop-gun, and was not unlike that of a fire-rocket. They were followed by a peculiar odor observed by all of us – four men – which one compared to the smell of sulfur and another to that of onions. The meteors which afforded these sounds all passed along in a northwest direction. Two of them had a well-defined nucleus the size of a tea cup. They severally afforded so much light that I could distinguish the color of a man's beard. They passed below the tops of the trees at the distance of 25 rods from the place where I stood, giving a "pop" just before they reached the trees.

One appeared to strike the barn, and gave a louder pop than any of the others. An auroral light resembling daybreak appeared constantly in the east from the time when my observations commenced.

air of different electrical tensions. Remember those luminous clouds that people saw? Meteors are particles of condensed vapor separate from one another which we see when electricity passes from one particle to another.

The objections I have made appear fatal to the theory advanced by you, my friend Professor Olmsted.

Another eruption

In one of his articles on the 1833 meteor storm, Olmsted had cited an observation of meteors seen in England on November 13, 1832 by a Reverend W. B. Clarke. Olmsted had used that account and others to advance his theory that meteors come from outer space and revolve in a cometlike orbit around the Sun.[6]

Clarke was angered by this interpretation of his observations. In an article published in September 1834 in the British *Magazine of Natural History*, he scolded Olmsted for attributing an extraterrestrial origin to the

A report from Canada

Samuel Strickland was 19 years old when he left his parents in England and set sail across the ocean to settle in Ontario – Canada West. There he married, lost his wife during the birth of his first son, remarried, and began a new family. In his book *Twenty-Seven Years in Canada West; or, The Experience of an Early Settler*, he recorded a memorable night nine years after he arrived.*

I think it was on the 14th of November 1833 [it was actually early morning on the 13th], that I witnessed one of the most splendid spectacles in the world. My wife awoke me between two and three o'clock in the morning to tell me that it lightened [lightninged] incessantly. I immediately arose and looked out of the window, when I was perfectly dazzled by a brilliant display of falling stars. As this extraordinary phenomenon did not disappear, we dressed ourselves and went to the door, where we continued to watch the beautiful shower of fire till after daylight.

These luminous bodies became visible in the zenith, taking the northeast in their descent. Few of them appeared to be less in size than a star of the first magnitude; very many of them seemed larger than Venus. Two of them, in particular, appeared half as large as the moon. I should think, without exaggeration, that several hundreds of these beautiful stars were visible at the same time, all falling in the same direction, and leaving in their wake a long stream of fire. This appearance continued without inter-

* London: Richard Bentley, 1853. The title page shows "Major Strickland" as author and Agnes Strickland as editor.

showers of meteors in 1833, 1832, and 1799. "Plainly," he said, they are of "volcanic origin." If they had come from comets, Clarke sniffed, "Why may we not suppose that the meteors . . . seen on August 10, 1833 . . . were cometic fragments?"[7] Thus, while attempting to ridicule Olmsted, Clarke had identified (and rejected) the correct source of the August (Perseid) meteors and all meteor showers. Not until 33 years later would the data be available to confirm what Clarke laughed off as a preposterous joke.

As for his hypothesis that meteors come from volcanoes, it did not faze Clarke that he could find no reports of volcanic eruptions anywhere in the world just prior to November 13, 1833 or 1832 or November 12, 1799. After an eruption, he declared, "volcanic vapours may float for a considerable time in space before they become sufficiently condensed to assume a solid form . . ." Alternatively, Clarke argued, "such vapours may issue a long time before the actual outbreak of the volcano." To prove his case, Clarke noted that "Bocket Kaba, a volcano of Palambang," had a "most dreadful" eruption only 11 days *after* the 1833 meteor storm in America. Therefore meteors must come from volcanoes, said Clarke. Clarke's hypothesis could not be

mission from the time I got up until after sunrise. No description of mine can give an adequate idea of the magnificence of the scene, which I would not willingly have missed.

This remarkable phenomenon occurred on a clear and frosty night, when the ground was covered with about an inch of snow. Various accounts appeared in the newspapers at the time as to the origin of this starry shower. It was, however, generally considered that it was not meteoric, since its elevation must have been far above our atmosphere; for these stars were visible on the same night all over the continents of North and South America [actually not south of the equator]. Besides, it is a well known fact that more or less of these luminous bodies have been seen on or about the 14th of November, likewise on the 14th of August, provided the weather be clear, than at any other time.

Strickland stayed in Canada West for 27 years. Although he sustained many hardships, he considered that he had enjoyed comparatively good fortune. But then in 1851 he lost his youngest son before his second birthday; then his new son-in-law died, leaving his oldest daughter a widow just before she was to deliver their first child. The baby died shortly after birth. Then his wife died, just after giving birth to his fourteenth child, a daughter. He gave the infant to his widowed daughter who had just lost her baby and her husband, and that helped to revive her spirits, but then this baby died too, at the age of five months. Strickland and his widowed daughter left Canada and returned to England.

wrong: any volcano erupting at any time – past, present, or future – could cause any meteor shower.

Clarke was very pleased with himself. "If I be wrong and Professor Olmsted be right," he said, "the earth will in all probability enjoy the celestial carnival again in November, when meteors will fly about as sweet-meats and *bons-bons* do at Naples and Rome."

On November 13, 1834, Olmsted, his students, and his contacts in other cities observed a shower of meteors, again radiating from Leo. Although the shooting stars were less numerous than in 1833, the recurrence fit his theory that meteors were celestial particles orbiting the Sun like a comet and colliding with the Earth. Clarke read Olmsted's January 1835 journal report that provided the proof Clarke had specifically demanded in his 1834 article: the meteors had returned.[8] So was he ready to adopt Olmsted's theory? Of course not. Clarke churned out another rambling article for the March 1835 issue of Loudon's *Magazine of Natural History*.

There was no need to invoke the celestial realm, the Reverend Clarke

Celestial fungus?

In the same issue of the *American Journal of Science and Arts* in January 1834 with Olmsted's first article on the great meteor shower – and printed ahead of his – was an article on the November 1833 falling stars by prominent geologist Edward Hitchcock.

Hitchcock noticed that the shooting stars were radiating from a point in the sky, but he did not identify its position among the constellations or that it moved westward with the stars. He agreed with Olmsted that the meteors were very high because there were clouds at Amherst College in Massachusetts where he taught and no meteor passed in front of a cloud. Yet the fact that the meteor trails were gradually distorted by winds told him that the display was occurring within the Earth's atmosphere.

Hitchcock himself heard no sound from the meteors and strongly doubted claims of snapping or crackling or any other noises because, in spectacles of this sort, "there is so much room for the play of an excited imagination."

Hitchcock's hunch was that the meteors were guided by the Earth's magnetic field along magnetic lines of force, racing toward the Earth's magnetic poles. There was, he felt, "a very strong and remarkable resemblance between the [meteor shower] and the *aurora borealis*." The aurora, he noted, is also above the clouds and may be carried by the wind.

Edward Hitchcock. William Jay Youmans, editor: *Pioneers of American Science*, 1896

insisted. The Earth and its atmosphere are periodically "deranged." God created volcanic eruptions, earthquakes, and storms on our planet for wise purposes unknown to man. It cannot be "that so wonderful and so complicated a world should be dependent for the necessary *corrections* in its phenomena upon a foreign and independent body." If this invocation of God and natural law on Earth offends anyone, Clarke concluded, how much less likely it is that we are encountering "a comet at full speed" or that we are "being shot at by an aerolitic [meteoritic] popgun from the moon or the dogstar." Considering such speculations, said Clarke, "it is at least evident who stands the best chance of being deemed *lunatic.*"9

As Olmsted continued to elaborate his theory in the *American Journal of Science and Arts*, Clarke matched him ten words for one in Loudon's *Magazine of Natural History*. Clarke felt he could show that somewhere on Earth a volcanic eruption or earthquake coincided with or preceded or followed a meteor shower. Therefore, said Clarke, meteors must be of volcanic origin. "[M]any a man has been found guilty, condemned, and

Thus meteors and the aurora, Hitchcock hypothesized, have the same origin – are "modifications of the same appearance." He worried about this conclusion, though, because the aurora tends to arise in the north and proceed toward the zenith, whereas the November meteors went in the opposite direction: they fell downward from high in the sky. Maybe, he concluded, the meteors were the opposite of the aurora borealis – the aurora australis, the southern lights.

It was not one of Hitchcock's better theories or scientific publications.

Hitchcock's article, mistaken in most of its astronomy, was, however, helpful in another way. Knowing that Olmsted had received several reports of meteors falling to earth and leaving a gelatinous residue, Hitchcock added a postscript to his article in which he recalled that in 1819 Rufus Graves of Amherst had reported a meteor that had deposited a gelatinous body in the woods. When Hitchcock moved to Amherst in 1825, Graves took him on a "damp, sultry morning in August" to the site near the first fall where a new meteor was supposed to have descended the previous night and left a similar deposit. "But," said Hitchcock,

> I recognized it in a moment as a species of gelatinous fungus which I had sometimes met with on rotten wood in damp places during dog days. . . . The day being a warm and damp one, I predicted that similar funguses might spring up within twenty-four hours; and, in fact, two others appeared before . . . evening . . . whose vegetable character was still more unequivocal; thus settling the question in my own mind . . . *

* Edward Hitchcock: "On the Meteors of Nov. 13, 1833," *American Journal of Science and Arts*, volume 25, number 2, January 1834, pages 354–363. In 1833, Hitchcock was professor of chemistry and natural history at Amherst College. In 1844, he was elected president of Amherst and served for 10 years. He was one of the founders of the American Association of Geologists, predecessor of the American Association for the Advancement of Science. He was also a charter member of the National Academy of Sciences. One of his most interesting projects was the study of the fossil tracks of large animals left in the sandstone of the Connecticut River Valley. He thought the tracks were produced by early giant birds. Now they are known to be the footprints of dinosaurs, the ancestors of the birds.

executed, according to law and *justice*, upon very much less proof than I have, in these papers, cited to establish" that the Earth produces meteors, not "any harlequinading comet in the universe."

A lack of respect

Clarke was absolutely incensed at the implications of an aggregation of debris in a cometlike orbit encountering his home planet. He took it as a personal insult against the Earth.

> It is, methinks, treating one of the most important (to us, at any rate) of the solar planets with little respect to suppose it only a receptacle for the waste contributions of a body that keeps such a mysterious secrecy, and favours us but once a year with a notification of his existence . . .

How odd, Clarke huffed, that in 2½ millennia of astronomical observations only recently have comets hurled meteors at us when "from the days of Pythagoras to the present hour, . . . some of them have advanced near enough to the earth to have whisked their tails in our faces had they been so unpolitely inclined."

What a chemist saw

W. E. Aiken, M.D., professor of chemistry and natural philosophy at Mount St. Mary's College, observed the meteors in Emmitsburg, Maryland, about 50 miles (80 kilometers) northwest of Baltimore, and wrote about them in a local paper.*

> The scene was altogether brilliant beyond conception. Instead of the usual random course of such meteors, these described paths diverging like the radii of a circle from a central space. This point was in the neck of Leo, near the star Gamma. From this center as a radiating point, the meteors proceeded in numbers exceeding the visible stars and in intensity of light often rivaling the full moon. No noise of any kind accompanied them.
>
> It would be difficult for one who had not witnessed the grand exhibition to conceive the effect of this uninterrupted succession of innumerable meteors. It could only be compared to one grand and continuous discharge of fireworks occupying the whole visible heavens.
>
> A good refracting telescope directed to the center whence the radiating meteors diverged discovered nothing peculiar. While directing the telescope to other points, many meteors darted across the field of vision, but their relative motion over so small a space was too rapid to admit of satis-

* Condensed from Aiken's account reported by Denison Olmsted: "Observations on the Meteors of November 13th, 1833," *American Journal of Science and Arts*, volume 25, number 2, January 1834, pages 373–375 of pages 363–411.

Later in his ramblings, Clarke quoted John Herschel, the greatest living authority in British astronomy, who correctly suggested that zodiacal light was the reflection of sunlight off of particles stripped from millions of comets passing close to the Sun. The implication, clearly, was that the Earth could occasionally pass through some of this debris. Clarke scoffed,

> Could we believe that a comet had any thing to do with these meteors, we might almost venture to suppose that Biela's [Comet] would satisfy us, for [quoting Herschel] "its orbit, by a remarkable coincidence, very nearly intersects that of the earth; and had the latter, at the time of its passage in 1832, been a month in advance of its actual place, it would have passed through the comet; a singular rencontre, perhaps not unattended with danger." (*Herschel's Astronomy, Lardner's Cyclopedia,* xliii. p. 309. art. 484.)[10]

But Clarke dismissed Biela's Comet as a possible source of any meteors because "Biela's comet has no tail, and no appearance of a solid nucleus (therefore, no pyrotechnic establishment with it) . . ."

Today, astronomers know (as Herschel suspected) that comets gradually lose their tails after repeated visits to the Sun have vaporized their ices and driven their gases and embedded dust into space. Biela's Comet was not always tailless. It became tailless over the centuries because of all the

factory examination. It was the prevailing opinion of those who witnessed them that the meteors were solid masses. But there seemed to be a difference between solid meteorites which at different times have fallen to the ground from the heavens and those appearances known as shooting stars, visible every night in the year, which do not strike the ground. It appears to me most probable that there may sometimes occur in the upper atmosphere the production and ignition of gaseous matter similar to the flames we see at the surface of the earth and call will-o'-the-wisp. If it is objected that we know of no gas high in the atmosphere that would produce such results, I would answer that this is no objection as long as we are ignorant of the composition of the will-o'-the-wisp and similar meteoric lights so often seen over low grounds. Those who prefer it, however, will consider all meteors as solid masses, and will then have the liberty of regarding them as the exuviae of lunar volcanoes, or perhaps as juvenile earthly comets, or supposing them to arise, in the words of a learned author, "from the fermentation of the effluvia of acid and alkaline bodies which float in the atmosphere." Yet a profound thinker has said, "He that knew not what he himself meant by learned terms cannot make us know anything by his use of them, even if we beat our heads about them ever so long." So I advise you not to beat your head long about the latter supposition.

material the Sun extracted from it and set adrift, each particle in its own cometlike orbit in the solar system – litter through which the planet Earth might plow so that collisions with these particles would illuminate the night sky as meteors.

Poor Clarke. In the realm of meteors, he constantly confronted and repudiated the correct answers and clung fiercely to his own ideas which were always wrong.[11] Astronomers in 1867 discovered that Comet Biela was responsible for a meteor shower of its own, with its radiant in Andromeda. The meteors created by Comet Biela came each November about two weeks after the famed Leo-radiating ones. Comet Biela was also responsible for the notable meteor storms on November 27, 1872 and 1885, although the Andromeda-radiating meteors are essentially extinct today.

Mindsets and other obstacles

Why didn't scientists recognize earlier that meteor showers return annually? Closed minds was probably the leading culprit, but two *natural* phenomena conspire to disguise the yearly cycle.

One problem that nature poses is clouds. Even partly cloudy conditions are likely to conceal the presence of a meteor shower. If skies are partly, mostly, or completely cloudy half or more of the time a shower returns, it is hard to recognize it as annual.

A second natural problem is moonlight. Most meteors are faint. A few meteors can be seen every hour on any clear night. Sky darkness is crucial to gauge the extent of a meteor shower. If a meteor shower occurs at night while the Moon is visible and bright, any phase from broad crescent to full, only the brightest meteors will be seen, and they may be so few in number that they will give no strong hint that dimmer meteors are falling by the dozens or hundreds.

The Moon's cycle of phases is 29½ days, which does not fit evenly into the cycle of the Earth's year, 365¼ days. If the sky is moonless on, say, November 17, 1998, the next year, on November 17, 1999, the Moon will be in a waxing gibbous phase (past first quarter, on its way to full) and will not set until well after midnight. The next year, November 17, 2000, the Moon will be at last quarter, still bright and up from midnight to noon, interfering with the best time to view most meteor showers: midnight to dawn. The next November 17, 2001, however, will be almost moonless.

So the clouds and the Moon's cycle of phases make it difficult to detect the existence of annual meteor showers. Still, scientists would have noticed the annual return of meteors if their minds had not been clouded by the preconception that meteors were part of meteorology. Until 1833, almost all scientists took it for granted that meteors, which were known to take place high in the atmosphere, were part of the ever-changeable weather, rather than part of astronomy, with its steady orbital rhythms. The early popular and scientific observations of meteors almost always included the temperature, wind, and humidity before and after the shower, trying to link shooting stars to the weather as if one caused the other. An observer who associated meteors with odd weather conditions was unlikely to think in terms of yearly cycles or any predictable return at all. If scientists had conceived of meteors as an astronomical phenomenon, they would have been thinking of orbits and dependably regular returns.

The view from Ohio

Olmsted received an article from the *Ohio State Journal and Columbus Gazette* in which an observer identified only as J. L. R. reported what he saw in Worthington, Ohio. Olmsted later identified J. L. R. as John R. Riddell of the Reformed Medical College of Ohio.

A numberless multitude of shooting stars were constantly marking the cloudless sky with long trails of light. As seen from this place, they seemed to proceed from a point in the heavens a little west of Gamma in the constellation Leo.* This observation was made at five o'clock. They were not generally visible in their course through a greater arc than 20 or 20 degrees. Each meteor in its course left a pale phosphorescent train of light, which usually remained visible for some minutes. Occasionally one would seem to burst into flames and burn with increased energy, illuminating the face of terrestrial nature with a degree of brightness and splendor inferior only to sunshine. But this effect would be of merely momentary duration, for the substance of the meteor would be rapidly consumed, leaving a broad luminous way which would perhaps remain distinctly visible for twenty minutes, while the wind or some other cause would appear to waft it gently eastward, so modifying its form as to give it the irregular outline of a cloud.

If observations have been made at different and distant places, I think it will be determined that these subtle and mysterious bodies (if bodies they be) first became visible in the aerial regions high above the grosser strata of our atmosphere. As witnessed from this place, they seemed to diverge from a common center. But I have no doubt that this apparent divergence was an illusion and that their true courses were nearly parallel. A luminous spot or ring would frequently appear for a moment near the point whence they seemed to emanate, which was unquestionably occasioned by a coincidence of the course of the meteor with the line of observation.

Respecting the origin of these meteors, let him speculate who pleases.[§]

* Riddell originally said Delta, but corrected it to Gamma in his letter to Olmsted on December 18, 1833.
§ Condensed from Denison Olmsted: "Observations on the Meteors of November 13th, 1833," *American Journal of Science and Arts*, volume 25, number 2, January 1834, pages 377–378 and 406 of pages 363–411.

In the years after 1833, meteors were the rage in scientific journals as they made the contentious transition from the domain of meteorology to astronomy. Most of the great astronomers of the day contributed their thoughts. And names like Herrick and Quetelet began to appear ever more often.

NOTES

1. D[onald] W. R. McKinley: *Meteor Science and Engineering* (New York: McGraw-Hill, 1961), page 3.
2. A[lexander] D[allas] Bache: "Meteoric observations made on and about the 13th of November, 1834," *American Journal of Science and Arts*, volume 27, number 2, January 1835, pages 335–338.
3. Alexander C. Twining: "Meteors on the morning of November 13th, 1834," *American Journal of Science and Arts*, volume 27, number 2, January 1835, pages 339–340.
4. Denison Olmsted: "Zodiacal Light," *American Journal of Science and Arts*, volume 27, number 2, January 1835, pages 416–419.
5. James P. Espy: "Remarks on Professor Olmsted's Theory of the Meteoric Phenomenon of November 12th [*sic*], 1833, denominated Shooting Stars, with some Queries towards forming a just Theory," *Journal of the Franklin Institute*, volume 15, numbers 1, 2, 3, and 4; January, February, March, and April 1835; pages 9–19, 85–92, 158–165, 234–238.
6. Denison Olmsted: "Observations on the Meteors of November 13th, 1833," *American Journal of Science and Arts*, volume 26, number 1, April 1934, pages 132–174.
   The Reverend William Branwhite Clarke (1798–1878) went to Australia in 1839 and laid the foundations of Australian geology.
7. W[illiam] B[ranwhite] Clarke: "On the Meteors seen in America on the Night of Nov. 13, 1833," *Magazine of Natural History, and Journal of Zoology, Botany, Mineralogy, Geology, and Meteorology*, edited by J. C. Loudon, volume 7, September 1834, pages 385–390. This quotation as printed read: "Why we may not

suppose . . . ?" This phrasing appears to be a typographical error. I interchanged "we" and "may."
8. Denison Olmsted: "Zodiacal Light," *American Journal of Science and Arts*, volume 27, January 1835, pages 416–419.
9. W[illiam] B[ranwhite] Clarke: "On certain recent Meteoric Phenomena, Vicissitudes in the Seasons, prevalent Disorders, &c., contemporaneous, and in supposed connection, with Volcanic Emanations. No. 6," *Magazine of Natural History*, edited by J. C. Loudon, volume 8, March 1835, pages 129–161.
10. W[illiam] B[ranwhite] Clarke: "On certain recent Meteoric Phenomena, Vicissitudes in the Seasons, prevalent Disorders, &c., contemporaneous, and in supposed connection, with Volcanic Emanations. No. 7," *Magazine of Natural History, and Journal of Zoology, Botany, Mineralogy, Geology, and Meteorology*, edited by J. C. Loudon, volume 8, August 1835, page 431 of pages 417–453.
11. Edward C. Herrick of Yale College, co-discoverer of the annual return of Perseid meteors each August, was contemptuous of Clarke. In a January 17, 1838 letter to his friend Professor Elias Loomis at Western Reserve College in Hudson, Ohio (before it moved to Cleveland), Herrick commented on articles in Loudon's *Magazine of Natural History* for December 1837 by Clarke and H. H. White: "They both come to the conclusion that there was no meteoric shower in November 1837. Their observations are however poor enough. One of them appears to have gone to bed at midnight; & the other probably looked out a window." (Yale University Library, Manuscripts and Archives Department.)

CHAPTER 6

The discovery of the August meteors

A new planetary world is beginning to reveal itself to us.

FRANÇOIS ARAGO (1837)[1]

EDWARD CLAUDIUS HERRICK was a bookworm. His father was a Yale graduate and founder of a girls' school. His mother was a descendant of one of Yale's founders. The Herricks lived in New Haven, Connecticut, the home of Yale. But young Edward did not go to Yale. He did not go to college. His parents felt that his chronic eyelid inflammation would not allow him to succeed in higher education.[2] So in 1827, at the age of 16, Edward became a clerk in a bookstore that served Yale students and faculty and was also the college's publishing house.

Everyone in New Haven with intellectual interests stopped by the bookstore and Herrick reveled in conversations with professors such as astronomer Denison Olmsted and chemist Benjamin Silliman. The young clerk worked hard and in 1835, at the age of 24, he became one of the owners of the bookstore. But during the next three years, the business failed and left Herrick broke.

On the evening of August 9, 1837, just as his business was teetering toward collapse, Herrick observed an unusual number of meteors in the clear night sky. From people who stayed up later that night, he heard that the meteors were even more numerous and brilliant after midnight.

American astronomers in 1837 were still gripped by the excitement of the meteor deluge four years earlier, November 13, 1833, when more than a thousand shooting stars radiated from the constellation Leo each *minute*. They were startled by Olmsted's discovery that these meteors were traveling together in an orbit in space before they collided with the Earth and that more than ordinary numbers of shooting stars could be seen every mid-November. But abundant meteors on a specific date in *August*? That was odd.

Herrick, alert and opportunistic, wondered if these August meteors might be an annual affair. He started looking for reports that these meteors had been seen in 1837 from places other than New Haven. He searched historical archives for evidence that the meteors had been seen in previous years around August 9. Herrick found seven references to significant meteor displays in August, from 1029 in Egypt to 1833 in England.

Edward C. Herrick. Courtesy Yale Picture Collection, Yale University Library

Following Olmsted's example, Herrick wrote an article for the October 1837 issue of Silliman's *American Journal of Science and Arts* in which he proposed a second annual meteor shower, listed his evidence, and requested information from anyone who had seen the display.[3]

While his first article was being published, Herrick found four more accounts of meteors in other years plummeting through the skies on August 9 or 10. He was convinced. "There generally occurs on or about the 9th of August in every year," he wrote, "a remarkably large number of shooting stars."[4]

In this second article, published in January 1838, Herrick came to other correct conclusions as well:

- The August meteors remain near their peak for about three days, and off-peak meteors span perhaps two weeks.
- Like those of November, the August meteors have a "starting point," a spot in the sky from which they seem to radiate. Herrick could not yet fix its position among the stars. The meteors were not abundant enough – several visible at the same time traveling in different directions across the sky – to make a radiant obvious.
- The August meteors are more numerous than the November meteors almost every year, except on rare, overwhelming occasions when the November meteors pour down in torrents the August meteors can never match.
- In addition to November and August, there was probably a third annual meteor shower around April 30. Herrick found only three cases of it, in 1095, 1122, and 1803, but those late Aprils experienced mighty storms of meteors.[5]

Herrick discarded previous notions that meteors were meteorological – part of the weather, like clouds, lightning, and thunder – or terrestrial, debris falling back to earth after the eruption of a volcano. "Shooting stars are without doubt cosmical or celestial bodies," he wrote, "and not of atmospheric or terrestrial origin."[6]

Heinrich Wilhelm Brandes of Germany had measured that meteor velocities reached 36 miles per second (58 kilometers per second) – twice the orbital velocity of Earth. That, said Herrick, was so fast that it had important implications for the orbit of a meteor particle. Objects under the influence of the Sun's gravity can reach only a certain maximum velocity at the distance of the Earth – 26 miles per second (42 kilometers per second) – the same speed whether they are inbound for or outbound from a passage close by the Sun. If the object is orbiting the Sun the same direction as Earth, that means it can barely catch up with the Earth as our planet coasts around the Sun at 18½ miles per second (30 kilometers per second).

But if meteors dive into our atmosphere at 36 miles per second, that must mean that they are crashing nearly head on into the moving Earth. The meteor particles must be traveling around the Sun in the opposite direction of Earth – retrograde motion.[7]

Could these high-speed head-on collisions gradually slow the Earth, Herrick wondered, so that it would spiral in ever closer to the Sun? In each fall to a lower orbit, the Earth would gain speed and, in the shorter orbit, it would complete its journey around the Sun sooner, so a year would be shorter. But Herrick found no observations indicating that the length of the year was gradually shortening. The implication was that meteors, despite their numbers and their speed, have negligible mass, and thus impart little momentum to the Earth.

However, Herrick realized, over eons of time, the amount of matter deposited in the atmosphere of Earth by meteors "must be very considerable."

Herrick was at full gallop now. He calculated the number of meteors that would be visible to the human eye each day if the sky were always dark and if the entire sky could be watched. Herrick and his friends watched the whole sky in shifts from dusk till dawn at different times of the year, counting the background sprinkle of meteors – or, as Herrick called them, "straggling shooting stars."[8] A small group of observers could see at least 20 meteors each hour on the average. That meant 480 meteors per day from a single spot. Herrick estimated that from a single spot he and his companions could see meteors 100 miles away in all directions, so observers could cover an area (πr^2) of almost 32,000 square miles. Approximately 6,150 such circles would cover the Earth, so 6,150 times 480 would be the number of bright meteors that – seen or not – fall each day over the entire Earth. This conservative calculation came out to about 3 million meteors. (A recent estimate is 90 million.[9]) And that did *not* include meteor showers. "Shooting stars are wonderfully numerous," Herrick decided.

That surprising abundance of shooting stars led to another conclusion. "The source of these meteors," said Herrick, thinking of bank withdrawals for his failing business, "must be of vast extent to be able to sustain for thousands of years such incessant and enormous drafts."

What could that source be? It would have to account for both annual meteor showers (of different intensities) and the everyday slow drizzle of shooting stars that did not seem to be part of a shower.

"It is not impossible," said Herrick, "that these meteoric showers are derived from nebulous or cometary bodies which, at stated times, the earth falls in."[10] It was another correct insight, one that Olmsted had broached. If comets were flying dust and rubble heaps without a solid nucleus, the Earth could pass through the head of a comet and experience only a splendid meteor display. The missing nuance (provided by Fred Whipple in 1950) was that comets do have a small solid nucleus – a dirty snowball a few miles across. The meteor particles are debris dislodged from this dirty snowball by vaporization of the ice. The particles are then pushed away from the nucleus by the minute but steady pressure of light from the Sun. The hypothesis that meteors have a cometary origin would be confirmed 28 years after Herrick's article when the connection between meteor and comet orbits would be demonstrated.

Pleased with his achievement, Herrick wrote up his new evidence for the annual meteor shower in August he had discovered, including his theory of meteors, and gave the paper to Benjamin Silliman for publication in his *American Journal of Science and Arts.*

Benjamin Silliman
William Jay Youmans, editor:
Pioneers of American Science,
1896

Less than two weeks later, Herrick received crushing news. He was not the first to realize that meteors fall in large numbers every August 9 or 10.

Adolphe Quetelet, the Belgian statistician and director of the Brussels Observatory, had noticed mid-August meteors very tentatively six months earlier. His attention had been called to meteors by François Arago of France, who dominated European science at this time with his skill in discerning important scientific problems and suggesting experiments to solve them. What, asked Arago in the wake of the November 1833 downpour of meteors, constituted a *shower* of meteors and what was an ordinary, everynight drizzle? On a typical night with good conditions, how many meteors an hour would a careful observer see?

The problem was ideal for Quetelet, who founded the Brussels Observatory but whose passion was statistics. He examined previous work and made observations of his own. In a speech to the Royal Academy of Sciences and Arts of Brussels on December 3, 1836, Quetelet gave his answer: averaging over the entire night and averaging over the entire year, a single observer should expect to see eight sporadic (random, non-shower-associated) meteors per hour. That figure is still good today. There was no mention in his December 3 talk of an *annual* meteor shower in August, but Quetelet said in response to the one and only comment on his talk that he had planned to say more about August meteors but omitted it to shorten his speech.[11]

François Arago

In his 1836 annual report of the Brussels Observatory, as he detailed his research on the rate of sporadic meteors, Quetelet had more to say about August meteors, but just barely. He presented the idea almost in passing and very timidly: "I thought I also noticed a greater frequency of these meteors in the month of August (from the 8th to the 15th)."[12] One sentence; that was all.

But by the March 4, 1837 session of the Royal Academy of Brussels, Quetelet had accidentally found records in the Brussels Observatory of

Adolphe Quetelet: father of statistics

JACQUES SAUVAL, PH.D., Department of Astrophysics, Royal Observatory of Belgium

Adolphe Quetelet. Courtesy Observatoire Royal de Belgique; special thanks to Peter A. Grognard, Embassy of Belgium

When Adolphe Quetelet died in 1874 at the age of 78, the most illustrious European scientists came to Brussels to pay tribute. Quetelet had contributed to many different fields: mathematics, astronomy, meteorology, physics, sociology. He had even founded one: statistics.

Lambert Adolphe Jacques Quetelet earned the first doctoral degree awarded by the University of Ghent in 1819 and was immediately appointed professor of mathematics at the Royal Athenaeum of Brussels. Beginning in 1823, he proposed and worked diligently for the creation of a national observatory. In 1834, the Royal Observatory of Brussels was completed with Quetelet as the first director. At first his staff was only two or three underpaid assistants.

An 1832 paper by Quetelet launched a new field: the sociology of crime. Quetelet provided statistics on the influence of sex, education, profession, climate, and age on the likelihood that a person would become a criminal. How environment contributes to crime still exerts a strong influence on criminology. In his paper, Quetelet advocated improving the environment as a way to prevent crime: "The costs of prisons, guards, and scaffolds are paid with frightful regularity. Efforts should be made to drastically reduce this budget."

Quetelet's most famous work is *On Man and the Development of His Faculties: A Treatise on Social Physics* (1835). His scientific approach to moral phenomena was quite new, making him a founder of sociology. His conception of the *average man* generated great controversy. His data on crime and other "moral statistics" prompted fierce debates on whether human beings had free will or were products of their environments.

Quetelet's name is recognized in mathematics for his collaboration with Germinal Dandelin on theorems for conic sections and also for the Quetelet Index (QI), which is a measure of obesity. It is defined as QI = weight (in kilograms) divided by height (in meters) squared. If the QI is larger than 30, the person is obese. (Do not confuse QI with IQ!)

Quetelet had artistic as well as scientific talents. He and Dandelin composed the words for a one-act opera.* A newspaper reviewer in Ghent, Belgium, where the opera premiered in 1816, liked the work, but audiences were small and the opera, now lost, received only two performances.

Quetelet also wrote poetry that was widely published. One poem was a tribute to André Grétry, an 18th century French comic opera composer. In this "Eulogy for Grétry," Quetelet calls the composer the "vanquisher of

*The opera's name was *Jean Second ou Charles Quint dans les murs de Gand* (John Second or Charles V Within the Walls of Ghent), with music by Charles Ots. Charles V was the emperor of the Holy Roman Empire who abdicated in 1555. John Everaerts, also called Jean Second (1511–1536), was a Dutch poet and coin scholar.

time" and envisions for Grétry nearly divine acclaim. I will leave judgment on Quetelet's literary merit to others.

Another of Quetelet's hobbies was art, especially painting portraits. When he taught geometry, he used his artistic skills to make exceptionally clear drawings. That and the clarity of his explanations made Quetelet a very popular teacher.

With scientific talent, artistic interests, hard work, good looks, and leadership graces, Quetelet was often called upon to be an administrator, but his heart was never in this domain. His annual reports were almost always late. Yet he retained his observatory directorship and high posts in scientific societies because people respected him as a scientist and as a man. His worldwide correspondence with 2,500 scientists and men of letters constitutes a gold mine for the history of science. He liked to laugh and to talk, but he also listened to others, which is a rare quality. He put everyone at ease.

Quetelet held no hard feelings for French scientist François Arago. In a December 1836 letter, he had asked Arago to announce to the Academy of Sciences in Paris that scientists should watch for a meteor shower around August 10, 1837. Arago forgot. After the meteors returned, Quetelet reminded Arago, and Arago publicly apologized for his oversight. In September 1837, Arago paid Quetelet a visit. They toured the observatory and Brussels, but not the nearby field where Napoleon lost his final battle a generation earlier, a subject painful to a Frenchman like Arago. As he departed, Arago said, "Thank you for not having spoken about your shooting stars and the Battle of Waterloo." As always, Quetelet had behaved with tact and lack of bitterness.

In 1855, at the age of 59, Quetelet suffered a stroke that greatly affected his short-term memory. He soon returned to work with the same eagerness as before, but his writings now required enormous editing. The end of his sentences often had no relation to the beginning. Yet his collaborators and assistants revised and rewrote as best they could out of fondness and loyalty to this man. Following Quetelet's stroke, his son Ernest, an army officer, began work as an assistant at the observatory, concentrating on double stars. After his father's death in 1874, Ernest served two years as the observatory's interim director.

exceptional meteor displays on August 10 in 1834 and 1835 to accompany the meteor shower he had seen on August 10, 1836. He called for scientists to watch the sky on August 10, 1837 to see whether these meteors would appear again.[13]

By the October 7, 1837 meeting of the Academy, Quetelet had tracked down 18 records of exceptional meteor displays in early and mid-August in other years, 16 of them between 1806 and 1837. "The night of August 10 ranks alongside that of November 13," Quetelet said, at last confident.[14]

By the end of 1837, Herrick had already published his discovery of the August meteors and his second article providing additional confirmation and offering his theory of meteors was in press. It was at that awkward moment that Herrick received news of Quetelet's report to the March 4 (*not* October 7), 1837 meeting of the Royal Academy of Brussels in which he announced three dates that meteors had been seen around August 10.

Herrick was a little dismayed that Quetelet had leapt to his conclusion on the basis of only three occurrences of the August meteors: 1834, 1835, and 1836. Herrick had found 11 previous showers dating back more than 800 years.[15]

Herrick was disappointed that he was not the first to recognize the annual return of the August meteors. But, even more, he was now worried about his scientific reputation. He had announced in a scientific journal a discovery that had been made and published earlier by someone else.

Silliman, founder and editor of the *American Journal of Science and Arts*, helped bail Herrick out by immediately publishing his acknowledgement of Quetelet. Silliman wedged the note into the Miscellanies section at the end of the second issue of the January 1838 volume, the volume that contained Herrick's first two articles in which he had proposed the annual nature of the August meteors. "At the time when the last number of this Journal was published," Herrick pleaded, "I was not aware that any person in Europe or elsewhere had ever advanced the idea of a meteoric shower in August."[16]

It had been a case of independent discovery. But Herrick, if not the first in print, was eager to stake his claim with a prediction: "We have now," he wrote, "an August meteoric shower in five successive years (1833 to 1837 inclusive), and there seems to be little risk in predicting its recurrence on or about the 9th of August."

Herrick pushed his meteor studies forward as hard as he could, constantly complaining about the lack of resources available to him. Historical research, he said, "must be left to those who have access to libraries more extensive than this city contains."[17]

Herrick's work and the earlier meteor observations and theories of Denison Olmsted, Alexander Twining, and Elias Loomis – all with Yale degrees – did not go uncontested. Joseph Lovering, a young professor of mathematics and natural philosophy at Harvard, launched an attack on the Yale astronomers because of the "crude state" of meteorology. Lovering mocked the idea of "the appalling spectacle of falling stars presented on the morning of 13th November 1833," as if "the earth in its revolution had encroached upon a nest of meteors."[18] He clung to the old belief that meteors were a weather phenomenon.

Lovering refused to believe that meteors return as a shower every year. Instead, he said, "meteoric appearances are much more common every night than has been imagined" and "no season of the year is especially provided: that about the same average number can be seen every fair night, . . . an equal and uniform distribution of meteors throughout the year."

The attack only spurred Herrick on. He pored over books looking for further confirmation of the August, November, and April showers. He couldn't be happier. He was organized, resolute, tireless. He moved across the Yale campus with speed and utmost efficiency, always taking the most direct route, regardless of whether there was a path, a fence, or a hedge.[19] He stumbled onto another annual meteor shower about December 7.[20]

This was fun: discovering annual meteor showers.[21] Herrick called for worldwide observations all night year round to determine when meteors were most abundant and most rare. He also offered some advice. "Shooting stars *must always* be watched in the open air: observations through a window can not be trusted."

Herrick then summarized in an article what he thought had been proved about the nature of shooting stars as of the end of 1838:

(1) *Meteors are small bodies of various sizes, compositions, and densities in orbit around the Sun that become luminous because of the heat generated as they pass through the Earth's atmosphere. Herrick liked Charles Augustin Coquerel's description of meteors as microscopic planets.*

(2) *Sporadic meteors and meteors in showers belong to the same family as brighter meteoric objects like fireballs and bolides, and to meteorites.*

(3) *The Earth is pelted by meteors every hour of every day, but is bombarded more intensely at certain parts of its orbit (certain times of the year) than at others.*

(4) *Most meteors move in broad rings with the particles frequently clustered in one part of the ring. Occasionally the Earth crosses this ring where the particles are densely clustered, causing great meteor showers.*[22]

Now August was approaching and Herrick looked forward to his favorite and most dependable meteors. What Herrick and three friends concentrated on this year, 1839, was the location of the radiant, the spot in the sky from which the meteors appear to come. They decided the August meteors radiated from the constellation Perseus – and they were right.

But again Herrick was not the first. This time he was five years too late.

On August 11, 1834, the *Cincinnati Daily Gazette* inadvertently became a scientific journal when it published a letter to the editor from John Locke, a physician by training who was now a girls' school founder and headmaster. (He was about to receive an appointment as professor of chemistry at

Sears Cook Walker
William Jay Youmans, editor:
Pioneers of American Science,
1896

the Medical School of Ohio and begin, at the age of 43, a highly productive career as a physicist, geologist, and scientific instrument maker.) Locke had seen a meteor shower on the evening of August 9 and, impressed by how Denison Olmsted had discovered that the November meteors less than a year earlier all seemed to radiate from a point in the constellation Leo, Locke observed these August meteors carefully and detected that they too had a radiant. It was in Perseus (true), near the star Algol (about 17 degrees too far south).[23]

The readers of the *Cincinnati Daily Gazette* took scant interest in Locke's scientific assessment and his request for additional observations.

And because Locke had chosen to publish scientific observations only in a small newspaper on the Western frontier rather than in a scientific journal, his discovery remained unknown to subsequent admirers of the August meteors such as Herrick and Quetelet. But Locke continued to read the scientific journals and was angry when Herrick and Quetelet gained acclaim for the discovery of the annual return of the August meteors and eventually for determining their radiant. He wrote to astronomer Sears Walker in Philadelphia, sending copies of his newspaper articles. Walker wrote back, congratulating him on his discovery. Locke then sent a letter to the *Cincinnati Daily Gazette*, published on August 14, 1841, in which he said:

> Sears C. Walker, Esq., the astronomer of Philadelphia, has awarded to me the credit of having discovered in 1834 the radiant point of the meteors which appear annually on or near the 10th of August. I had also discovered in 1835 their periodical return. They have since been noticed by the philosophers of Europe.

Locke omitted any mention of Herrick.

Locke's next move was to send this newspaper article to Silliman, for whom he had served as a lab assistant during the 1815–1816 school term prior to earning his medical degree in one year at Yale in 1819. Silliman passed Locke's claim and snub on to Herrick.

Herrick's immediate response was to write a notice for the *American Journal of Science and Arts* that quoted the brief article in the Cincinnati paper in which Locke proclaimed himself the true discoverer of both the annual return and the radiant of the August meteors.[24] Herrick might have pointed out Locke's unorthodox way of announcing scientific findings. Herrick could have pointed out that Locke had only two observations – 1834 and 1835 – on which to base his assertion that the meteors he had seen returned *every* August. Herrick could have – but he didn't. He quoted Locke and was silent about the matter of credit.

So, the existence of a shower of meteors every August had three independent discoverers.

Well, not three discoverers after all, it turned out. Thousands.

The earliest discoverers were anonymous and their feat lay buried in an English farmer's almanac. Both Quetelet and Herrick chanced upon it. Bravely, Herrick acknowledged, "The annual occurrence of a meteoric display about the 10th of August appears to have been recognized for a very

A scientist on the frontier

John Locke (1792–1856), the American scientist, was not a direct descendant of John Locke (1632–1704), the English philosopher.

The American Locke was born in New Hampshire. John's father Samuel was a millwright whose skill in designing and building water-powered mills was in great demand and kept the family moving from state to state in New England until the father built mills of his own in Maine, where the boy grew up. John shared his father's interest in machinery and new technology. The young man was fascinated by break-throughs in electricity, especially efforts to capture and store the power of lightning in batteries for future use. He built 20 zinc and silver batteries and linked them together with brine-soaked cloth in place of wire, but the system was too weak to give him a shock even when he wet his fingers and touched the poles. So Locke stuck pins through his skin and touched them to the poles. His hands and whole body snapped backward from the powerful jolt. He was very pleased.

Locke left home after a quarrel with his father about John's desire for college training. Locke worked his way through a year at Yale as Benjamin Silliman's chemistry lab assistant. He left school for medical and botanical apprenticeships and then returned to Yale for one year to receive his degree as a doctor of medicine in 1819. But Locke failed as a physician in private practice. "He found patients," said a friend, "but they could give thanks more easily than money." Discouraged, he gave up medicine and began teaching in a girls' school in Vermont. He consoled himself by writing books on botany and English grammar, but he felt suffocated by religious intolerance in New England.

Like many a misfit, he sought a new life on the American frontier. After teaching one year at a girls' school in Lexington, Kentucky (1821), he opened one of his own in Cincinnati (1822). He succeeded as a teacher because of his unorthodox classroom style. He favored class discussion over formal lectures and appeals to reason over harsh discipline.

Locke was headmaster of the Cincinnati Female Academy he founded when in 1834 he observed and wrote about meteors in August streaming out of Perseus.

John Locke. Courtesy Archives & Rare Books Department, University of Cincinnati

great length of time." Thomas Furley Forster of London had recorded it in 1827 in his *Pocket Encyclopædia of Natural Phenomena*.[25] "According to Mr. T. Forster," Herrick reported in October 1839, citing Quetelet, "a superstition has 'for ages' existed among the Catholics of some parts of England

Science – it was Locke's passion, particularly inventing new instruments for precise measurement. His inventions and encyclopedic knowledge earned him a professorship in chemistry at the Medical College of Ohio in Cincinnati in 1835. But his proficiency extended far beyond chemistry and astronomy. He made a geological survey of Ohio in 1838, where he discovered the largest trilobite found up to that time – 19½ inches long. In 1847 he surveyed the upper peninsula of Michigan and conceived how a canal could be built to bypass the Sault Sainte Marie (the Saint Mary's River Falls), thereby allowing boats to sail directly between Lake Superior and Lake Huron. In 1855, the year before Locke died, the first of the Soo Canals was built as he had recommended.

But Locke's greatest satisfaction lay in his inventions. They included a gravity escapement for regulator clocks, a pocket compass, and a device that allowed observatory clocks to send an accurate time signal across telegraph wires. This invention made possible precise calculation of longitude differences between cities and the synchronizing of astronomical observations. The U.S. Naval Observatory installed Locke's system in 1848 and Congress awarded him $10,000.

Most people found Locke pleasant, even friendly, but he had ferocious powers of concentration. He was usually thinking too hard to care what food he ate. He consumed whatever lay on his plate or was directly in front of him. He never asked for the meat or potatoes at the other end of the table. He didn't carry an umbrella; it was inconvenient. He got wet instead.

His religion was "natural theology." Nature reveals truth and beauty, as desired by God. When we study the work of creation, we are studying the Creator.

Locke died at the age of 64 from coronary artery disease and a series of strokes that paralyzed his right side. He left behind a wife and ten children. His medical school colleague Marmaduke Burr Wright wrote a 69-page tribute to Locke for the Cincinnati Medical Society, which concluded with the results of the autopsy on Locke. Locke's brain showed "softening" of the left hemisphere and one of the examiners exclaimed that "His aorta is as large as that of an ox."*

* M[armaduke] B[urr] Wright: *An Address on the Life and Character of the Late Professor John Locke, Delivered at the Request of the Cincinnati Medical Society* (Cincinnati: Moore, Wilstach, Keys, 1857). Wright was Locke's assistant, holding the title of associate professor of chemistry. Another detailed account of Locke's life appears in George Mortimer Roe, editor: *Cincinnati: The Queen City of the West* (Cincinnati: C. J. Brehbiel, 1895.) Locke's biography is the longest in this book of Cincinnati's "principal men and institutions." Neither of these biographies mentions Locke's discoveries about the August meteors.

The discoverers of the first annual meteor showers 1834–1840

by Jacques Sauval, Ph.D,[a] Department of Astrophysics, Royal Observatory of Belgium

Shower nights in 1834–1840	Name of meteor shower	Year discovered	Discoverers (those in bold awarded priority by Jacques Sauval)
November 11–12	Leonids	1834	**Denison Olmsted**
		1834	**Alexander Twining**
		1835	**François Arago**
		1836	Wilhelm Olbers
August 9–10	Perseids	1834	*John Locke[c]*
		1836 (37?)	**Adolphe Quetelet**
		1837	Thomas Forster
		1837	François Arago
		1837	Wilhelm Olbers
		1837	Johann Benzenberg
		1837	*Edward Herrick[d]*
April 20–26	Lyrids	1835	**François Arago**
		1838	**Edward Herrick**
		1838	Johann Benzenberg
		1839	Adolphe Quetelet
January 2	Quadrantids (Boötids)	1839	**Edward Herrick**
		1839	**Adolphe Quetelet**
		1841	Louis Wartmann
October 8–15	Orionids	1839	**Johann Benzenberg**
		1839	**Edward Herrick**
		1839	**Adolphe Quetelet**
December 7	Andromedids (Bielids)	1838	**Johann Benzenberg**
		1839	**Edward Herrick**
		1839	Adolphe Quetelet

Notes: [a]Jacques Sauval: "Quetelet and the Discovery of the First Meteor Showers," *WGN: the Journal of the International Meteor Organization*, volume 25, number 1, 1997, page 30 of pages 21–33. See also Jacques Sauval: "Adolphe Quetelet et les étoiles filantes," *Bulletin Astronomique de l'Observatoire Royal de Belgigue*, volume 11, number 1, 1996, pages 67–82 (in French and German).
Two modifications by Mark Littmann have been put in bold italic typeface and footnoted.

[b]ABL = *Annuaire du Bureau des Longitudes*
AJS = *American Journal of Science and Arts*
AROB = *Annuaire de l'Observatoire de Bruxelles*
BARB = *Bulletins de l'Académie Royale des Sciences et Belles-Lettres de Bruxelles*
CMP = *Correspondance Mathématique et Physique*

Catalogue = Adolphe Quetelet: "Calalogue des principales apparitions d'étoiles filantes," *Mémoires de l'Académie Royale des Sciences et Belles-Lettres de Bruxelles*, volume 12 (1st edition), 1839, pages 1–56.

CR = *Comptes Rendus des Séances de l'Académie des Sciences* (Paris)

[c]I have added bold type with italics. Sauval disagrees because Locke published only in his local newspaper although he was a scientist and aware of scientific journals.

[d]I have added bold type with italics. Sauval disagrees. Both Herrick and Quetelet agreed that their discoveries were independent.

Reference[b]	Comet creating those meteors and its period
AJS 26, 132–174 AJS 26, 320–352 CR 1, 395; ABL 1836, 293 Benzenberg: *Die Sternschnuppen*, 1839, vii	Tempel-Tuttle ~33 years

Cincinnati Daily Gazette, Aug. 11, 1834
BARB III, 412; AROB 1837, 272; CMP IX 184, 432–441, 1837
CMP IX, 448–453; 467–468
CR Oct. 16 (V, 553)
CMP IX, 392–419
CMP IX 388–391
AJS 33, 176–180
	Swift-Tuttle
	~120 years

ABL 1836, 297 (footnote) AJS 34, 398; 35, 366; 36, 358 *Die Sternschnuppen*, 1839, 253 *Catalogue* 1839, 23	Thatcher (1861I) ~415 years
AJS 35, 366 *Catalogue* 1839, 26 BARB VIII, 226	Machholz 1 5.2 years
Die Sternschnuppen, 1839, 244 AJS, 35, 366 Catalogue 1839, 25	Halley ~76 years
Die Sternschnuppen, 1839, 331 AJS 35, 366 *Catalogue* 1839, 25	Biela 6.6 years

and Germany that the *burning tears* of St. Lawrence are seen in the sky on the night of the 10th of August; this day being the anniversary of his martyrdom."[26]

St. Lawrence was tortured and killed in Rome on August 10 in the year 258, during the reign of the anti-Christian emperor Valerian. While being roasted on an iron grating, Lawrence said to his tormenters, "It is well done; turn it over and eat it." He was then dispatched by a sword.

"The peasants of Franconia and Saxony have believed for ages past that St. Lawrence weeps tears of fire which fall from the sky every year on his fête (the 10th of August)," Herrick wrote, quoting a Brussels newspaper.[27] "This ancient popular German tradition or superstition has been found within these [past] few years to be a fact which engages the attention of astronomers. The inhabitants of Brussels can bear witness that on the night of the 10th this year, St. Lawrence shed [an] abundance of tears."

Many people had reported in scientific journals that they had seen numerous meteors during the first half of August. But no scientist before Locke, Quetelet, and Herrick had recognized that these meteors were an annual event. No scientists, just thousands of European peasants.[28]

Postscript

Herrick continued to tend his August meteors with great faithfulness and to report their activity in Silliman's Journal all the remaining years of his life.

In 1838, soon after his first scientific articles appeared in print, Herrick lost his bookstore.[29] But Yale was so impressed with his scientific scholarship that it awarded him an honorary master of arts degree. Five years later, Yale built a new library building and made Herrick college librarian. It was a pleasant irony for a man whose eye trouble kept him from attending college and who had complained in his scientific articles about the poor libraries in New Haven. Herrick spent the next 15 years vigorously developing the Yale library collections. He never married. He never took a vacation.

Later he took over the preparation of the Yale publication that described the careers of Yale graduates and faculty who had died. Herrick was so organized and efficient that he wrote his own death notice a few days before he died in 1862 at the age of 51. He gave it and others he had written to the printer with a note that read: "Onward the ceaseless current sweeps—."[30]

NOTES

1. Quoted by Edward C. Herrick in "On the Shooting Stars of August 9th and 10th, 1837; and on the Probability of the Annual Occurrence of a Meteoric Shower in August," *American Journal of Science and Arts*, volume 33, number 1, October 1837, pages 176–180. Please see chapter 7, endnote 1 for a citation by Quetelet.

2. T. A. Thacher: "Edward C. Herrick," *New Englander*, volume 21, October 1862, pages 820–859. This information appears on page 823. Other biographical information about Herrick appears in an unsigned "Obituary," *American Journal of Science and Arts*, volume 34, July 1862, pages 159–160, and in [anonymous]: *Obituary Record of Graduates of Yale College Deceased from July, 1859, to July, 1870* (New Haven: Yale University, 1871).

3. Edward C. Herrick: "On the Shooting Stars of August 9th and 10th, 1837; and the Probability of the Annual Occurrence of a Meteoric Shower in August," *American Journal of Science and Arts*, volume 33, number 1, October 1837, pages 176–180.
   During this period, the *American Journal of Science and Arts* was published in two "volumes" each year – January and July – composed of four "issues" – April, July, October, and January. The April and July issues formed the July volume; the October and January issues formed the January volume. Herrick's article came out in the October 1837 ("number 1") issue of the *American Journal of Science and Arts* although it was marked as the January 1838 volume.

4. Edward C. Herrick: "Further proof of an annual Meteoric Shower in August, with remarks on Shooting Stars in general," *American Journal of Science and Arts*, volume 33, number 2, January 1838, page 354–464. Herrick noticed T. Forster's encyclopedia and its reference to an annual meteor shower in August before news of Quetelet's discovery and mention of Forster arrived.

5. See also Edward C. Herrick: "On the Meteoric Shower of April 20, 1803, with an account of observations made on and about the 20th April, 1839," *American Journal of Science and Arts*, volume 36, number 2, July 1839, pages 358–363. This article appeared in the journal immediately after his follow-up article on the December meteors.

6. "I prefer *shooting* to *falling* stars," Herrick wrote to Elias Loomis in a letter dated October 30, 1838, "because the first implies that the body has a projective force of its own, apart from that caused by the earth's attraction." Letter in the Yale University Library, Archives and Manuscripts Department.

7. The fastest meteors, the Leonids in November, actually travel 44 miles (71 kilometers) per second. The Earth is orbiting the Sun at 18.5 miles per second (29.8 kilometers per second). A meteor stream in gravitational freefall at the Earth's distance from the Sun will be traveling 26.1 miles per second (42.1 kilometers per second). So if they collide head on, the velocity of encounter is 44.6 miles per second (71.9 kilometers per second).

8. Herrick to Elias Loomis in a letter dated October 30, 1837. Yale University Library, Archives and Manuscripts Department.

9. Gerald S. Hawkins: *Meteors, Comets, and Meteorites* (New York: McGraw-Hill, 1964), page 64. D. W. R. McKinley: *Meteor Science and Engineering* (New York: McGraw-Hill, 1961), page 140, gives the figure 100 million. Both estimates are for meteors of visual magnitude +5 or brighter.

10. Herrick was consistent in linking meteors to comets. In a letter to Elias Loomis dated January 17, 1838, he offers this explanation for why the November meteors in 1837 were so sparse: "[T]he moon pulled the comet away from us, or at least drew it so high that the earth could not induce many meteors to leave their celestial seats!" Yale University Library, Archives and Manuscripts Department.

11. Adolphe Quetelet: "Étoiles filantes," *Bulletin de l'Académie Royale des Sciences et Belles-Lettres de Bruxelles*, December 3 session, volume 3, number 11, 1836, pages 403–413.
   A. Jacques Sauval (personal communication, August 27, 1996) has discovered that by December 1836, Quetelet was aware of six previous occurrences of a meteor shower in early to mid-August: 1353, 1717, 1815, 1819, 1823, and 1836.

12. Adolphe Quetelet: "Étoiles filantes," *Annuaire de l'Observatoire de Bruxelles pour l'An 1837*, 1836, pages 268–273.
   Quetelet's article was summarized in *L'Institut* (Paris), number 218, August 1837, page 256. Notice of Quetelet's tentative hypothesis also appeared in the *London and Edinburgh Philosophical Magazine and Journal of Science*, volume 11, 3rd series, September 1837, pages 270–273 as "On the Height, Motion, and Nature of Shooting Stars," in translation "from the *Annuaire de l'Observatoire de Bruxelles* for 1837" [meaning a report on research in the preceding year, 1836]. Here, in that translation, is the relevant sentence in context:

It would seem that a cause exists which produces from about the 8th to the 15th of November more frequent appearances of shooting stars. I have also thought that I remarked a greater frequency of these meteors in the month of August.

13. [Adolphe] Quetelet: "Étoiles filantes," *Bulletin de l'Académie Royale des Sciences et Belles-Lettres de Bruxelles*, 1837, séance du 4 mars, volume 4, number 3, pages 79–81. The key passage, written as session minutes, reads: "Il [Quetelet] persiste à croire que ce n'est pas le milieu de novembre seul qui soit remarquable par le grand nombre d'apparitions de ces météores, mais que le milieu du mois d'août, et particulièrment le 10, mérite aussi de fixer l'attention." Quetelet never quite says that the shower is annual.

14. Adolphe Quetelet: "Étoiles filantes," séance du 7 octobre, *Bulletin de l'Académie Royale des Sciences et Belles-Lettres de Bruxelles*, volume 4, number 9, 1837, pages 376–380. In his communication, reported in the third person, Quetelet offers the first brief listing of meteor showers of previous years, promising a longer catalog soon. He published valuable catalogs of meteor showers of previous times in 1839, 1842, and 1861. These catalogs stimulated others.

For more information on Quetelet's contributions to meteor science, see A. Jacques Sauval: "Adolphe Quetelet et les étoiles filantes," *Bulletin Astronomique de l'Observatoire Royal de Belgique*, volume 11, number 1, September 1996, pages 67–82 (in French and German). I am grateful to Dr. Sauval for sharing with me a copy of his article before its publication and for his helpful comments on a condensed version of this chapter that appeared as "The Discovery of the Perseid Meteors" in *Sky & Telescope*, volume 92, August 1996, pages 68–71.

15. Herrick in a letter to Elias Loomis dated October 30, 1837: "I am perfectly aware that a great many more observations than we yet have are necessary to establish the doctrine of the annual occurrence of a meteoric shower in August, & I do not mean to fall into the error of generalizing from one or two instances only. I cannot but consider the August meteors quite as certain as the November ones." Letter in the Yale University Library, Archives and Manuscripts Department.

16. Edward C. Herrick: "On Meteoric Showers in August; supplementary to Art. XX," *American Journal of Science and Arts*, volume 33, number 2, January 1838, pages 401–402.

17. Edward C. Herrick: "Additional Observations on the Shooting Stars of August 9th and 10th, 1837," *American Journal of Science and Arts*, volume 34, number 1, April 1838, pages 180–182. In his obituary for Herrick, T. A.

Thacher comments on the Yale library in the 1830s: "The college library was kept in the attic of the chapel and was opened only once or twice a week for an hour at a time by a professor who kept the key." T. A. Thacher: "Edward C. Herrick," *New Englander*, volume 21, October 1862, page 824 of pages 820–859.

18. J[oseph] Lovering: "Meteoric Observations made at Cambridge, Mass.," *American Journal of Science and Arts*, volume 35, number 2, January 1839, pages 323–328.

19. T. A. Thacher: "Edward C. Herrick," *New Englander*, volume 21, October 1862, pages 825–826 of pages 820–859.

20. Edward C. Herrick: "Report on the Shooting Stars of December 7, 1838, with remarks on Shooting Stars in general," *American Journal of Science and Arts*, volume 35, number 2, January 1839, pages 361–368. (Denison Olmsted and two of his sons, observing separately, also contributed their sightings to Herrick.) Also, Edward C. Herrick: "Additional Account of the Shooting Stars of December 6 and 7, 1838," *American Journal of Science and Arts*, volume 36, number 2, July 1839, pages 355–358.

This December meteor shower was the Andromedids (also known as the Bielids), an intense shower in Herrick's time, but now extinct because the orbit of the meteor particles has been shifted by planetary perturbations so that it no longer crosses the orbit of the Earth. A "new" meteor shower, the Geminids, now occurs in early December.

21. See, for example, Herrick in a playful mood in a letter to Elias Loomis on January 17, 1838, responding to Loomis' account of the November meteors in 1837: "You must recollect that I am not in the least degree responsible for the November shower: this I cheerfully relinquish to other hands if I can be allowed to take charge of all the others which the year affords." Letter in the Archives and Manuscripts Department of the Yale University Library.

22. Herrick provided a list of dates of great meteor showers in history: 686, 29, and 25 B.C.; 532, 558, 750, 765, 901 [902?], 935, 1095, 1096, 1122, 1799, 1803, 1832, and 1833 A.D. Edward C. Herrick: "Report on the Shooting Stars of December 7, 1838, with remarks on Shooting Stars in general," *American Journal of Science and Arts*, volume 35, number 2, January 1839, pages 361–368.

Based on the dates he had collected of great meteor displays, Herrick observed in passing that "the cycle of the November shower seems to be, without much doubt, 33 or 34 years." Of the five orbits Olmsted suggested for the November meteors, Olmsted still favored a half-year period.

Herrick also differed with Olmsted on another matter. The zodiacal light, Herrick said, is not the cause of meteors. Edward C. Herrick: "Report on the Shooting Stars of December 7, 1838, with remarks on Shooting Stars in general," *American Journal of Science and Arts*, volume 35, number 2, January 1839, pages 361–368.

23. The radiant of the August meteors was independently discovered by George C. Schaeffer of New York on the night of August 9–10, 1837. He reported his observation in "Notice of the Meteors of the 9th and 10th of August, 1837, and also of Nov. 12th and 13th, 1832," *American Journal of Science and Arts*, volume 33, number 2, January 1838, pages 133–135.

    Schaeffer placed the radiant just across the Perseus boundary into Camelopardalis, only about 7 degrees northwest of the actual position near the northern limit of Perseus. Schaeffer observed an average of 35 meteors per hour for almost 7 hours, from 8 p.m. to nearly 3 a.m., when "not having taken any precaution to ensure wakefulness," he fell asleep.

24. Edward C. Herrick: "Observations on the Shooting Stars of August 9 and 10, 1841," *American Journal of Science and Arts*, volume 41, number 2, September 1841, pages 399–400.

25. Thomas Furley Forster: *The Pocket Encyclopædia of Natural Phenomena* (London: J. Nichols and Son, 1827), pages 39–41 and 298. The subtitle of Forster's book is: "For the use of mariners, shepherds, gardeners, husbandmen, and others; being a compendium of prognostications of the weather, signs of the seasons, periods of plants, and other phenomena in natural history and philosophy."

    The key passages are: "In the month of August meteors of all kinds are more common than at any other time of the year" (pages 40–41) and in the "Rustic Calendar" section that concludes *The Pocket Encyclopædia of Natural Phenomena*, under August 10: "Falling Stars and Meteors most abound about this time of year" (page 298). Herrick called attention to these sections of Forster's work in his "Further proof of an annual Meteoric Shower in August, with remarks on Shooting Stars in general," *American Journal of Science and Arts*, volume 33, number 2, January 1838, pages 354–364.

26. Edward C. Herrick: "Report on the Shooting Stars of August 9th and 10th, 1839, with other facts relating to the frequent occurrence of a meteoric display in August," *American Journal of Science and Arts*, volume 37, number 2, October 1839, pages 325–338.

    Ruth Freitag, Library of Congress, points out that Italian countryfolk also had a long tradition of connecting the martyrdom of St. Lawrence (San Lorenzo) to the annual meteors in August, as recorded by Paolo Toschi: "Lorenzo, santo, martire," IV. "Nel folklore," *Enciclopedia Cattolica* (Città del Vaticano: Ente per l'Enciclopedia cattolica e per il Libro cattolica, [1951]), volume 7, page 1543.

    Robert Hopkins, a technical and science writing student at the University of Tennessee, points out that English folk knowledge of the annual return of meteors in August is recorded in *Summers Last Will and Testament*, a play by Thomas Nashe published in London in 1600. At lines 644–662, Autumn is criticizing Orion:

> . . . heavens circumference
> Is not enough for him to hunt and range,
> But with those venom-breathed curs [*Canis Major and Canis Minor*] he leads,
> He comes to chase health from our earthly bounds:
> Each one of those foul-mouthed mangy dogs
> Governs a day (no dog but hath his day),
> And all the days by them so governed [*August especially*],
> The Dog Days hight [*are called*]; infectious fosterers
> Of meteors from carrion that arise,
> And putrefied bodies of dead men,
> Being naught else but preserved corruption.
> 'Tis these that, in the entrance of their reign,
> The plague and dangerous agues have brought in.
> They arre [*growl?*] and bark at night against the Moon,
> For fetching in fresh tides to cleanse the streets.
> They vomit flames and blast the ripened fruits:
> They are death's messengers unto all those
> That sicken while their malice beareth sway.

Thomas Nashe: *The Works of Thomas Nashe*, volume 3, edited by Ronald B. McKerrow (Oxford: Basil Blackwell, 1958), pages 253–254. I have modernized the spelling and punctuation and supplied notes in brackets.

27. Actually, this linkage of the August meteors to St. Lawrence cannot have gone back "ages," but instead probably less than two centuries, as an anonymous reviewer kindly pointed out:

> It would be interesting to establish just when the European peasants started calling the Perseids 'St. Lawrence's Tears' since precession means the August 10 date for the shower's maximum would only have been correct from the early 1700s. If we say that meteor rates would be particularly noticeable for two to four days across this peak (which is certainly true modernly, albeit the difference between peak and near peak rates is clearly obvious even to casual observers now), we could perhaps go back to the 1630s and 1640s, when the peak date would have fallen on or about August 7, but not much earlier. By the 11th century, the correct date would have been in late July, for instance.

28. Quetelet considered himself to be the discoverer of the

annual return of the Perseid (August) meteors, with Herrick as a later independent co-discoverer. He resented all other claims:

Once the periodicity of the 10th of August was announced and accepted, some wanted to appropriate it for their benefit. Others asserted that it was not new; that they found traces of it in all peoples and in all times. It was known by the Irish, the Greeks, and even the Chinese. So be it, but why did they not say so earlier?

Quotation given by A. Jacques Sauval: "Adolphe Quetelet et les étoiles filantes," *Bulletin Astronomique de l'Observatoire Royal de Belgique,* volume 11, number 1, September 1996, from Adolphe Quetelet: *Sciences Mathématiques et Physiques chez les Belges au commencement du XIXème siècle.* Bruxelles: Thiry-Van Buggenhoudt, 1866, page 575.

29. Herrick refused to shield himself through bankruptcy. With hard work and the painful sacrifice of his scientific research, he repaid his debts over the next five years. T. A. Thacher: "Edward C. Herrick," *New Englander,* volume 21, October 1862, page 838 of pages 820–859.

30. *Obituary Record of Graduates of Yale College Deceased from July, 1859, to July, 1870* (New Haven: Yale University, 1871). Herrick did not graduate from Yale, so technically his obituary should not have appeared in this publication. It appeared, however, as the first notice in the volume with a note from the Yale administration that Herrick's "important service . . . for the College renders it obviously fitting that, although only an honorary graduate of the Institution, his name should be here enrolled."

CHAPTER 7

1866: The first predicted return

......a spectacle which cannot be imagined or forgotten.

TIMES (London) November 15, 1866

M ETEORS WERE the rage after 1833. They were democratic, visible to anyone lucky enough to look outside in the wee hours of the morning on the right day under clear skies, in the right part of the world. A meteor storm did not require a telescope, which only astronomers and a few wealthy people had. Meteors were democratic to astronomers as well; they were available without the foremost lens or mirror that allowed only the few best-equipped observatories to make all the newsworthy and career-advancing discoveries. Above all, meteors were the rage because they were . . . spectacular. Almost everything else in the heavens had majesty, and beauty, to be sure, but meteors offered *action*. Meteors were showmen, and meteor storms were, well, superstars.

Meteor storms and annual meteor showers had been seen before. Once astronomers began to look, the historical records were full of them. Yet, before, who could believe such stories? They must be exaggerations. Or they were meteorological freaks, gone in a flash, unlikely to be seen again in a lifetime.

Now meteors were astronomical. Some returned annually. Some returned periodically in droves. There was order, perhaps meaning. They were susceptible to study, maybe understanding. Meteors were getting real attention. François Arago of France was enchanted by the way a swarm of millions of tiny bodies followed a path around the Sun like the planets and asteroids and comets. He captured the mystery of meteors precisely when he said, "A new planetary world is beginning to reveal itself to us."[1]

The November meteors of 1833 had returned in subsequent years, though feebly. Would they come again in great torrents as they had in 1833 in America, 1832 in Europe, 1799 in America? Wilhelm Olbers in Germany noticed the interval between the recent November meteor storms and suggested that the inhabitants of Earth might have to wait until 1867 to once again see the remarkable phenomenon of those earlier years.[2] Edward Herrick searched for records of great meteor showers in history and found that many had occurred at the right time of year to be the same

Wilhelm Olbers. Courtesy Archive of the Archenhold Observatory

as those that radiated from Leo. Looking at the dates, Herrick observed, "The cycle of the November shower seems to be, without much doubt, 33 or 34 years."[3] Herrick had once again made an important discovery, independently, second.

Julius Robert Mayer

Meteors as solar fuel

While astronomers were discovering annual meteor showers and predicting their returns, other scientists were discovering the bones of huge animals that were significantly different from any living animals on Earth. In 1841, some of the bones of those extinct animals were grouped together and given the name dinosaurs. Those bones and the layers of terrain visible in mountainsides persuaded many scientists that the Earth was much more than 6,000 years old, the age that Christian theologians had derived from the Bible.

If the Earth was older than 6,000 years, astronomers had a problem. How could the Sun shine that long? Could the Sun be burning like a lump of coal? Gauging by how much solar energy the Earth received and calculating how much missed the Earth, the Sun could not survive by chemical burning for more than a few thousand years. A different mechanism was needed to explain how the Sun could shine steadily for millions of years.

Astronomy was still vibrating from the dazzling downpour of Leonids in 1833 and from the discovery of other annual showers that followed. Could meteors solve the fuel problems of the Sun?

In 1848, Julius Robert Mayer, a German physician and co-discoverer of the law of the conversation of energy, proposed that meteoroids falling continually on the Sun imparted the kinetic energy that the Sun then radiated away as light and heat. Meteoroids could not be adding any significant mass to the Sun or the Sun's increasing gravity would pull the planets inward, increasing their speeds. A year on Earth would be shorter by a measurable fraction of a second. But no records indicated that the length of a year on Earth was decreasing. The mass of meteoroids, Mayer said, was being offset by the Sun's radiation, sending corpuscles of light (as Isaac Newton had proposed) off into space.[4]

In 1854, Hermann von Helmholtz offered a different theory: that the Sun was shining by gravitational contraction. The Sun's huge mass was creating enormous gravitational pressure on its interior. The higher the pressure, the higher the temperature. Helmholtz calculated that the Sun could provide the energy it does if it shrank by 200 feet (60 meters) a year – less than the length of a football field. Such shrinkage was too slight to be observable.

Hermann Helmholtz

Simon Newcomb

The Helmholtz contraction theory gathered more support than the meteoroid infall theory, but Mayer's concept continued to be viable until late in the 19th century. Attacks by Mattieu Williams in 1870 and Simon Newcomb in 1883 weakened support for meteoroids as solar fuel. Williams sneered that if the Sun's energy came from the impact of meteoroids, the Earth would have to be getting hit so often that we would be tripping over meteoroids wherever we stepped. Instead, meteorites were so rare that they were hoarded by museums. Thus, he wrote, "the meteoric bombardment hypothesis appears in its true colours as a monstrous physical absurdity."[5]

Newcomb also considered how many impacts the Sun would require to shine as it does. Were that the case, Newcomb said, the Earth would be experiencing a pelting so constant and intense "that its whole surface would be made hot by the force of the impacts, and all life would be completely destroyed."[6]

Such criticism dampened support for meteoroid infall as the source of the Sun's energy. The Helmholtz contraction theory won the day, only to fail when the Earth was recognized to be not several million years old but more than a billion. Gravitational contraction could light the Sun for no more than 25 million years.

The mystery of how the Sun and all the stars were able to emit such vast quantities of energy for eons of time was not solved until Albert Einstein's special theory of relativity in 1905 gave scientists the concept that mass could be converted into energy: $E = mc^2$.

As for Mayer's meteoroid infall theory, it died for the wrong reason. Williams and Newcomb jeered at the implications of the Earth being deluged by, Newcomb figured, 910 tons of meteorites a year. They were wrong. The Earth is hit by 240 times that much material annually.[7]

A man named Newton

Hubert A. Newton. Courtesy Yale Picture Collection, Yale University Library

Hubert Anson Newton had a surname that suggested promise and distinction. He entered Yale College[8] at 16 and emerged four years later with a Phi Beta Kappa pin and first prize in mathematical problem solving. He continued studying math on his own for 2½ years, then returned to Yale as a tutor in 1853. That year the chairman of the mathematics department fell terminally ill and Newton was named to chair the department. He was 23. When he was 25, Yale conferred on him the rank of full professor, one of the youngest ever, and sent him to Paris to study geometry at the Sorbonne for a year with Michel Chasles. Newton responded by publishing three papers.

But upon his return to Yale, a different calling haunted him. From all around him came the siren song of elusive meteors, almost as delightfully intangible as airy mathematics. Denison Olmsted was there, collector and analyst of observations of the November 1833 meteor blizzard. He had been Newton's astronomy teacher and was still teaching.

Alexander Twining was there, trained at Yale, independent discoverer with Olmsted that the November 1833 meteors radiated from Leo and a wise interpreter of its significance. After nine years as a professor at Middlebury College in Vermont, Twining had returned to New Haven in 1848 and resumed his engineering practice. He continued to observe and publish articles on meteors while he invented the refrigerator as a means to make ice, although he never earned any money from it.

Elias Loomis. Courtesy Yale Picture Collection, Yale University Library

Elias Loomis, trained at Yale, would soon return after professorships in mathematics and natural philosophy at Western Reserve College in Ohio and the University of the City of New York. He returned to Yale in 1861 to replace the dying Denison Olmsted. He and Olmsted had been the first Americans to see Halley's Comet on its return in 1835 and the first to recalculate its orbit. He and Twining had worked together to measure the altitude of meteors.

Edward Herrick was there, Yale librarian and treasurer, and co-discoverer of the annual return of the meteors in August and discoverer of four other annual meteor showers.

These scientists at Yale, more than others, had transferred the study of meteors from the realm of meteorology to astronomy, despite bitter opposition from scientists at Harvard and in Philadelphia. Yale *was* meteor science and the first generation meteoriticists were nearing the end of their careers. Someone had to preserve and collect and analyze the meteor data. Someone had to carry on the Yale tradition.

Hubert Anson Newton became a meteor scientist, one of the first of the American second generation.

In a two-part article in the *American Journal of Science and Arts* in May and July 1864, Newton reviewed reports of 13 eye-popping meteor displays over the past thousand years that occurred in mid-autumn from October 13, 902 to November 13, 1833. Clearly the date of this returning shower was slowly migrating. The meteors did not return in exactly one year.[9]

Every year, said Newton, we on the moving Earth pierce the ring of meteor debris, each time in a new place. Only occasionally – about every 33.25 years – does the Earth punch through the denser swarm and see great quantities of shooting stars. Newton visualized an elliptical ring of tiny bodies where most of these particles are concentrated along 7 to 10 percent of the circuit, with the rest of the ring dotted irregularly with stragglers.

On the basis of his list of previous outbreaks of the meteors from Leo,

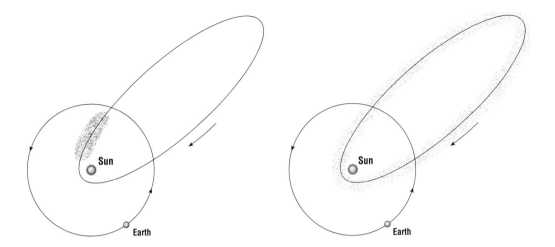

Meteoroid particles on an orbit around the Sun that the Earth crosses each year. *Left:* If the Earth passes through a swarm of tightly bunched particles, a meteor storm will be seen. But in subsequent years the Earth will miss the swarm and the meteors will be few until the swarm returns. *Right:* Over time, the meteoroids spread more evenly but more thinly along their orbit. Each year the Earth passes through the stream, providing a meteor shower. But a meteor storm is unlikely. Diagram by Will Fontanez and Tom Wallin, University of Tennessee Cartographic Services Laboratory.

Newton noted that the storm of streaks would most likely return November 14, 1866, to be best seen in the western Atlantic.[10] He cautioned that this suggestion was not a prediction because gravitational perturbations from other planets could alter the swarm's position and because of irregularities in the density of the swarm.[11]

Two years to go. Would Newton – and Olbers and Herrick – be right? The world began to prepare.

The return

In Boston, Washington, Chicago, and New York, the fire commissioners planned to ring the fire bells long and loud to alert citizens when the great meteor storm began. The bells would be rung, the New York commissioners explained, because "to all it will be a grand curiosity to be remembered and told of to our children's children."[12]

In New Haven, Hubert Newton, the most prominent of Yale College astronomers, and Elias Loomis, former student of Olmsted at Yale[13] and now his successor as professor of natural philosophy and astronomy, wrote lucid articles for newspapers about the expected return of the remarkable meteors seen in 1833, their significance to astronomy, and how to observe them. Both these letters to the editor, especially the shorter one by Loomis, were reprinted in newspapers everywhere across the country. The two authorities must have checked with one another before posting their letters. The articles differed in wording but there were no significant disagreements in background information or observing advice. Though they warned readers that viewing a giant meteor storm was not guaranteed, the message that came through was that this event was not to be missed.

Newton's original letter to the *New York Times*, published November 11, 1866, advised that not every 33rd year since 902 (the earliest recorded display) had experienced storms of November meteors. Further, "If the time of the shower falls this year between the mornings of the 13th and the 14th, it may appear only to those on the other side of the earth." Newton also pointed out that the display of shooting stars on November 13, 1832 had been visible only in Europe, whereas the great storm on November 13, 1833 was seen exclusively in North America.

In his letter to the editor of the *New York Evening Post*, Loomis also provided a caveat that the reappearance of a great display of shooting stars was by no means certain: "it is probable that there will be a repetition of the shower either this year or next in some part of the world, but we cannot be sure that the principal display will take place in the United States."[14] However, Loomis' and Newton's description of the 1833 shower and the probability of its recurrence often generated comment by newspapers that further stoked public excitement:

> Our imperfect knowledge of the meteoric ring and the distribution of the meteors therein . . . forbids the prediction of this phenomenon with anything of the certainty of an eclipse or the reappearance of a periodical comet. Nevertheless, the assurance of such a shower is very great, and no one who cares to witness so sublime and rare a phenomenon will fail to be watchful of the heavens after midnight on Tuesday morning next.[15]

Newton wrote: "A prodigious flight of meteors, the most imposing of its kind, will make its appearance, probably for the last time in this century, on Tuesday or Wednesday evening." In St. Louis, the *Daily Missouri Democrat* quoted this portion of Newton's article and went on to add: "Such a phenomenon awakens a deep interest among all classes of persons, and everywhere extensive preparations are being made to observe this sublime spectacle."[16]

Witnesses to the 1833 display shared their recollections. In the *Galveston News*, C. G. Forshey remembered that he saw the meteors because of mathematics:

> The writer is indebted to the difficulties of solving the binomial theorem in algebra for a most complete view of this spectacle. As a cadet at West Point, I was up and at my duties long before dawn when the flashes through my window attracted me. Upon looking out, the grandest spectacle of a lifetime broke upon me. . . .
>
> I soon had the whole corps of cadets and professors out on the plain, uttering exclamations of awe and wonder.

The profound Professor Courtney, who taught us philosophy and astronomy, was up and gazing when I reached his quarters. My appeal to him for solution was answered by the reply, "I know nothing more than you or a child, Mr. F., of this wonderful phenomenon."

At reveille, when the stars had disappeared from the sky, the meteors were still flashing.[17]

Newspapers reached into their files to recall the splendor of 1833 and to prepare the public for the celestial show. Expectations ran high. In its November 13, 1866 edition, the *New York Herald* announced its intention

to reproduce in our columns today a copious chapter of information relating to the meteoric exhibition of 1833. We think it will prove . . . an agreeable diversion to our readers from the clashing opinions and purposes of our grovelling earthly politicians. They, like the poor of the Twelve Apostles, are always with us; but the sight of one of the most awful of the revelations of St. John is the event of a lifetime, a century, or a thousand years.

The effort to educate and interest the public worked to perfection. Despite cold November weather, an almost full moon, and intermittent cloud cover, the American public turned out.

Trouble was, the meteors didn't.

The *New York Herald* reported on November 14:

The people of the city were wide awake last night waiting, in impatient squads and knots, for the promised exhibition of meteoric pyrotechnics; . . . Groups were to be seen on the house tops and in the streets . . . [they] were sadly disappointed as the hours passed without the exhibition.

The fire alarm bells never rang. "There was no great shower anywhere," said the *Daily Missouri Democrat*.[18] "Hour after hour passed, but the heavens made no sign," said the *New York Times*. "The weary watchers sat, sat and froze their very vitals in their conscientious adherence to the faith that was in them."[19]

There were a few meteors, but not many. In New Haven, Newton and his students counted 32 meteors an hour as the average maximum for individual observers.[20] Elsewhere the count was more like 15 to 20 an hour for single observers. Yes, it *was* a meteor shower, but a weak one, more like a drizzle. It offered shooting stars at only two or three times the rate that sporadic meteors could be seen on any clear night.

The *New York Times* noted that popular interest in the display of 1866 was widespread and intense. People evidently lost lots of sleep watching for the meteors, else "how can we account for the yawns, the heigh-hos, the dull eyes and plodding footsteps of today?"[21]

The *Boston Daily Advertiser* summed up the experience of American astronomers: "The savants at Cambridge and New Haven were, to use a vulgar but in this place rather apt expression, left out in the cold."[22] The failure of the November meteors to appear in great numbers, the *Advertiser* noted, was especially discouraging. "As the next flight of [these] meteors will not take place until the expiration of a period of thirty-three years, those of our citizens who were disappointed on the nights of the 13th and 14th will have occasion to exercise a good deal of patience."[23]

The popular press did an accurate and attractive job of informing their readers about the expected meteor shower in 1866, as they did in reporting the meteors seen in 1833. Astrological and other pseudoscientific inter-pretations were excluded. Misinformation, except for a hoax perpetrated by the *New York Herald*, was at a minimum. The pre-event coverage con-centrated on eyewitness sightings from 1833 and explanations of meteor science from experts, but written so that a nonscientist could understand.

Meteor showers were now known to be of astronomical rather than meteorological origin, caused by tiny particles orbiting the Sun in comet-like orbits through which the Earth plunged annually. The stream of parti-cles that created the November meteors was, however, of irregular density. About every 33 or 34 years, the Earth collided with a dense part of the stream, and the particles, traveling at immense speeds, burned up high in the Earth's atmosphere, without endangering people on the ground.[24]

When the *New York Herald* published a superficially plausible but bogus telegram from the Royal Observatory at Greenwich, England, reporting that a dazzling meteor shower had been seen, the American scientific community and many newspapers quickly condemned the tele-gram as a hoax.[25] The *Chicago Daily Tribune* did not hesitate to accuse the editors of the *New York Herald* of creating the stunt, remembering the 1835 Moon hoax the New York *Sun* had perpetrated, when it reported that John Herschel, from a telescope in South Africa, had seen life on the Moon.[26]

The American public bounced back quickly from the disappointment. Professor Loomis at Yale held out hope that the meteors might be seen in 1867 instead.[27] David Plum, heartened by that thought, wrote a poem for the *New York Evening Post* that concluded:

> Then bear us, O Earth, with our eyes upward gazing,
> To the place where the Star-God his fireworks displays;
> When countless as snowflakes are meteors blazing
> With their red, green and orange and amber-like rays.[28]

News from the Old World

As Americans tried to catch up on their sleep, reports began arriving by telegraph and mail from Europe. The meteors from Leo had returned in abundance as the astronomers predicted, just not in the New World.

"Special telegram to the New York Herald" – a hoax

"Brilliant Display Observed from Greenwich Observatory, England," trumpeted the headline in the *New York Herald* on November 15, 1866, as it reported in accompanying articles that the meteors were disappointingly sparse in the United States.

The *Herald* said it had received a telegram from astronomers in England. The telegram was plausible. In England, Scotland, and throughout Europe, the meteors fell in great numbers. The maximum did occur between 1 and 2 a.m. So perhaps the *Herald* had received a brief message. But most of the telegram just didn't sound like it was written by astronomers.

The expected meteoric showers were observed last night. At nine o'clock a few meteors fell; at eleven o'clock they had increased in number and size, and between one and two o'clock this morning the maximum was reached. The night was clear and the stars were out in great numbers. The whole heavens were brilliantly illuminated.

Why plural?

Size? Does writer mean brightness?

Not how astronomers usually describe observing conditions.

The showers of meteors were of great beauty and brilliancy, and radiated from the constellation Leo, near the star Gamma Leonis. Their direction was mostly from the east to the west. The paths of the meteors were from three to four degrees in the north. Near Ursa Major twenty or thirty were observed at one time, and crossing the zenith fifty or sixty more of unusual size and duration, the majority being larger than stars of the first magnitude. Several exploded from the vicinity of Jupiter; one, of immense dimensions, was colored red, blue, green, orange, and amber. Nearly all had trails of fire. Of two flaming from Leo at the same time, one crossed Beta Geminorum and the other Mars. Two more, one red and the other of an oriental sapphire color, crossed Alpha Orionis.

Astronomers usually omit æsthetic judgments.

Only while Leo is low in the east.

What is this supposed to mean?

So what? Astronomers want to know numbers per minute or hour and the path and time of spectacular fireballs.

Leo not the only radiant?

What does "immense dimensions" mean?

Not useful for altitude calculation unless exact time is given.

Such fine naked-eye color distinctions are meaningless.

Some of the meteors burst forth in splendor; one, breaking behind the rising clouds, flashed like sheet lightning, and another of emerald hue burst near Eta Leonis at fifteen minutes after two o'clock a.m., its trail of flame being visible for a minute and a half, and then faded away in brilliant nebulæ.	*Æsthetics, not astronomy.* *Telegram first said sky was clear.* *Too close to radiant to leave a trail.* *Not useful astronomically unless the path is described.*
At three a.m. they commenced to diminish gradually, until, at the present moment, they are all, meteors and stars, fading away in the morning light.	*Telegram previously said 2 a.m.*
We counted five thousand in one hour, nearly twelve thousand in all, with the naked eye.	*Meaningless unless we know how many people were counting.*

The Edinburgh *Scotsman*'s writer was overwhelmed:

Standing on Calton Hill and looking westward, with the Observatory shutting out the lights of Prince's Street, it was easy for the eye to delude the imagination into fancying some distant enemy bombarding Edinburgh Castle from long range; and the occasional cessation of the shower for a few seconds, only to break out again with more numerous and more brilliant drops of fire, served to countenance this fancy. . . . [F]rom no point in or out of the city was it possible to watch the strange rain of stars, pervading as it did all points of the heavens, without pleased interest, and a kindling of the imagination, and often a touch of deeper feeling that bordered on awe. The spectacle, . . . which truly should have been seen to be imagined, will not soon pass from the memories of those [who saw it].

American astronomer Benjamin Apthorp Gould, on surveying work in Valentia, Ireland, saw the meteors in great numbers, clearly radiating from Leo. He awakened his staff so that they could enjoy the sight.[29]

It was, as seen from Europe, a fine show. Those who watched, said the London *Times*,

were rewarded with a spectacle which cannot be imagined or forgotten . . . first one meteor then another shot across the sky. . . . The spectator had soon counted half a dozen; then he felt sure he had seen thirty; then six or seven in a minute; then they appeared faster than he could count them. Then there came two or three together; then not less than a dozen of all kinds. Some shot across the heaven, leaving long, bright, and lingering trains, the star itself seeming to explode and instantly disappear. Some darted as quickly and as bright, but without trains. Some struck the sight like sparks from a forge, everywhere at once. . . . There were times when it

Tracks of some meteors on November 14, 1866 plotted at the Royal Observatory, Greenwich. The tracks have been extended backward with lighter lines to delineate the radiant. The radiant is not a point but a small area Copyright Royal Astronomical Society; special thanks to Peter D. Hingley7.8. Earl of Rosse's 72-inch (1.8-meter) telescope at Birr Castle, Ireland. Robert S. Ball: *The Story of the Heavens*, 1886

seemed as if a mighty wind had caught the old stars, loosed them from their holding, and swept them across the firmament....

People have been a good deal taken by surprise. The apparition has been far out of the common range of ideas.[30]

The deluge of meteors was greatest for an hour between 1 and 2 a.m. on November 14, with a peak about 1:10.[31]

The London *Times* was filled with individual reports on the falling stars. An unnamed newspaper reporter was intrigued by the way the meteors radiated from one spot in the sky: "As the constellation Leo rose over the houses north of Paddington Green and cleared itself of the haze, the divergence of the meteor paths from a point within it became obvious..."[32]

J. M. Heath sent his observations to the *Times* – and his assessment:

The magnificent spectacle seen this morning in the skies has verified the predictions of astronomers... and has added one more to the known bodies, or groups of bodies, belonging to our planetary system,... [perhaps] the most magnificent of them all.[33]

The sight exceeded the expectations of astronomer Alexander S. Herschel, grandson of astronomer William Herschel and son of astronomer John Herschel: "The meteors... shot to all parts of the sky with a swift and stately motion most beautiful to behold, if not almost too wonderful and too surprising to describe."[34]

In fact, the meteors radiating from Leo stopped astronomers in their tracks. On the evening of November 13–14, 1866, Robert Ball, later royal astronomer of Ireland, was assisting William Parsons, later Earl of Rosse, with his 72-inch (1.8-meter) telescope, largest in the world at that time, at Birr Castle in Ireland. Ball recalled years later:

Earl of Rosse's 72-inch (1.8-meter) telescope at Birr Castle, Ireland Robert S. Ball: *The Story of the Heavens*, 1886

> I shall never forget that night. On the memorable evening I was engaged in my usual duty at that time of observing nebulæ with Lord Rosse's great reflecting telescope. I was of course aware that a shower of meteors had been predicted, but nothing that I had heard prepared me for the splendid spectacle so soon to be unfolded. It was about ten o'clock at night when an exclamation from an attendant by my side made me look up from the tele-scope, just in time to see a fine meteor dash across the sky. It was presently followed by another, and then again by others in twos and in threes, which showed that the prediction of a great shower was likely to be verified. At this time the Earl of Rosse . . . joined me at the telescope, and, after a brief interval, we decided to cease our observations of the nebulæ and ascend to the top of the wall of the great telescope from which a clear view of the whole hemisphere of the heavens could be obtained. There, for the next two or three hours, we witnessed a spectacle which can never fade from my memory. The shooting stars gradually increased in number until

Robert S. Ball

sometimes several were seen at once. . . . As the night wore on the constellation Leo ascended above the horizon, and then the remarkable character of the shower was disclosed. All the tracks of the meteors radiated from Leo. Sometimes a meteor appeared to come almost directly towards us, and then its path was so foreshortened that it had hardly any appreciable length, and looked like an ordinary fixed star swelling into brilliancy and then as rapidly vanishing. . . . It would be impossible to say how many thousands of meteors were seen, each one of which was bright enough to have elicited a note of admiration on any ordinary night.[35]

A few people had witnessed both the 1833 and 1866 displays and could compare them. Joseph Baxendell, a Fellow of the Royal Astronomical Society, saw the 1833 meteors off the west coast of Central America and the 1866 shower in Manchester, England. He judged "that the present display was far inferior to the former both in the number of meteors seen and in the brilliancy of the larger ones."[36]

The return of the November meteors in 1866 provided Europe with a display of shooting stars that was more than ten times as abundant as best of annual meteor displays. But the meteors of 1833 had mounted a sky show with a cast of falling stars that was more than ten times as numerous as in 1866.

Encore

A year later, on the morning of November 14, 1867, the Leonids finally made their appearance in the United States, somewhat unexpectedly. The U.S. Naval Observatory used five observers to count meteors. The single observer rate was perhaps 1,500 meteors an hour. It was not nearly what had been seen in 1833. It was not as abundant a display as Europe had seen in 1866. But it was "the most brilliant seen in this country since the great shower of 1833."[37]

The United States and the Far East saw yet another Leonid storm in 1868, but it was weaker than 1867. It would be the last major Leonid outburst seen in the United States for almost a century.

The prediction of a date when the November meteors would storm again and their return as predicted focused the attention of astronomers on shooting stars and set the stage for the discovery of the source of meteors.

NOTES

1. Arago is quoted by Edward C. Herrick in "On the Shooting Stars of August 9th and 10th, 1837; and on the Probability of the Annual Occurrence of a Meteoric Shower in August," *American Journal of Science and Arts*, volume 33, number 1, October 1837, pages 176–180. Arago is also quoted by Adolphe Quetelet in "Catalogue des principales apparitions d'étoiles filantes," page 5 of pages 1–63 in *Mémoires de l'Académie Royale des Sciences et Belles-Lettres de Bruxelles*, volume 12, 1939.

2. W[ilhelm] Olbers: "Noch etwas über Sternschnuppen, als Nachtrag," page 280 of pages 278–282 in H. C. Schumacher, editor: *Jahrbuch für 1837* (Stuttgart: J. G. Cotta, 1837). Sears C. Walker provides a translation in "Researches concerning the Periodical Meteors of August and November," *Transactions* of the American Philosophical Society, volume 8, 1843, pages 120–121 of pages 87–140.

3. Edward C. Herrick: "Report on the Shooting Stars of December 7, 1838, with remarks on Shooting Stars in general," *American Journal of Science and Arts*, volume 35, number 2, January 1839, second footnote on page 367 of pages 361–368.

4. David W. Hughes provides a good synopsis of this controversy in "The History of Meteors and Meteor Showers," *Vistas in Astronomy*, volume 26, 1982, pages 325–345; and "Sir John F. W. Herschel, Meteoroid Streams and the Solar Cycle," *Vistas in Astronomy*, volume 39, 1995, pages 335–346. Hughes points out in the Herschel article that Isaac Newton had suggested in *Principia* (1687) that the kinetic energy from infalling *comets* was the source of the Sun's light. For Mayer's biography and papers, see Robert Bruce Lindsay: *Men of Physics: Julius Robert Mayer: Prophet of Energy* (Oxford: Pergamon Press, 1973).

5. W. Mattieu Williams: "The Fuel of the Sun," *Nature*, volume 3, November 10, 1870, pages 26–27. Williams ended his article by arguing that there are insufficient comets and meteoroids to fuel the Sun – and other stars – for more than a few thousand years. After that, "all the lights of heaven must go out, eternal darkness must rest upon the face of the deep, and everlasting death must pervade the universe."

6. Simon Newcomb: *Popular Astronomy* (London: Macmillan, 1883), page 527.

7. David W. Hughes: "Earth – A Cosmic Dustbin," *Physics Review*, volume 1, 1992, page 22–26.

8. Yale officially became a university in 1887 when the Connecticut legislature approved its name change. However, Yale had been awarding doctoral degrees since 1861 (the first in the United States), so it might be said to be a university informally as of that date.

9. Newton noted that the forward migration of the date for the November meteors must mean that the meteor swarm moves in a retrograde orbit because the node of the meteor swarm is precessing in the direction opposite to the precession of the equinox on Earth, although both are caused by the gravitational influences of nearby bodies which have direct motion.

10. Newton also offered as less likely that the Leonids would return in abundance on November 14, 1864 in the western Pacific and Australia or on November 14, 1865 in western Asia and eastern Europe.

11. H[ubert] A[nson] Newton: "The original accounts of the displays in former times of the November Star-Shower; together with a determination of the length of its cycle, its annual period, and the probable orbit of the group of bodies around the sun" (in two parts), *American Journal of Science and Arts*, 2nd series; part 1: volume 37, number 111, May 1864, pages 377–389; part 2: volume 38, number 112, July 1864, pages 53–61.

12. The idea of ringing bells to alert the populace to the meteor shower may have been recalled from 1833, when the *Alexandria* (Virginia) *Gazette* reported that "They were very considerate in Charleston, Kenawha, on the occasion of the 'stars' falling.' The church bell was rung to awaken the citizens and give them an opportunity of 'seeing the sight.' We wish the same plan had been adopted here, and so bespeak it for the next occasion." (As repeated in the *National Intelligencer* [Washington, D.C.], November 27, 1833.)

13. Loomis entered Yale at age of 16 and graduated at 20 in 1830.

14. *New York Evening Post*, November 10, 1866.

15. *New York Evening Post*, November 10, 1866.

16. *Daily Missouri Democrat* (St. Louis), November 13, 1866.

17. As republished in the *New York Times*, November 11, 1866.

    As in previous chapters, punctuation, capitalization, and occasional spelling in quotations have been modernized to make understanding easier.

18. *Daily Missouri Democrat*, November 15, 1866.

19. *New York Times*, November 14, 1866.

20. Alexander Twining, observing from New Haven where he was now a professional engineer, saw 43 meteors an hour at the peak of the shower.

    Based on a bright meteor observed simultaneously in New Haven (by Newton) and in Williamstown, Massachusetts, Newton calculated its altitude and its speed – 46 miles per second (74 kilometers per second). Newton was astonished by the result. He included the speed in his report but added: "a result probably too great." He was wrong. He was the first person to calculate with reasonable accuracy the speed of the November meteors. Because they hit our planet head-on, the November meteors enter the Earth's atmosphere at 44 miles (71 kilometers) per second. Newton also calculated that the dense portion of the meteor stream was about 33,000 miles thick. H. A. Newton: "Shooting Stars in November, 1866," *American Journal of Science and Arts*, volume 43, January 1867, pages 78–88.

21. *New York Times*, November 14, 1866.

22. *Boston Daily Advertiser*, November 16, 1866.

23. *Boston Daily Advertiser*, November 16, 1866.

24. Only rarely did a newspaper report as plausible an old, discredited theory. One such paper was the *New York Tribune* on November 12, 1866, which cited a theory by Giambatista Beccaria (1716–1781) that said that meteors are, in an unspecified way, an electric phenomenon. But this mention was brief and followed a more detailed exposition of Olmsted's contribution. The *Tribune* also published a second discredited hypothesis: that the same orbiting particles that cause the November meteors to stream from Leo also cause the August meteors that radiate from Perseus.

25. Elias Loomis sent a new letter to the *New York Evening Post*, published November 16, 1866, in which he called the telegram "evidently spurious."

26. *Chicago Daily Tribune*, November 20, 1866. The 1835 Moon hoax was the work of Richard Adam Locke: "Great Astronomical Discoveries, Lately Made by Sir John Herschel, L.L.D., F.R.S., &c. At the Cape of Good Hope," *The Sun* (New York), series beginning August 25, 1835.

27. *New York Evening Post*, November 16, 1866.

28. David Plum: "Meteors," *New York Evening Post*, November 20, 1866.

29. H. A. Newton: "Shooting Stars in November, 1866," *American Journal of Science and Arts*, volume 43, January 1867, pages 78–88.

30. *Times* (London), November 15, 1866.

31. John Russell Hind, Bishop's Observatory, in a letter to the *Times* published November 15, 1966. Hind was the first English astronomer to find and recognize the planet Neptune after Johann Galle and Heinrich d'Arrest identified it in Berlin in 1846.

   H. A. Newton reported that Joseph Baxendell of Manchester, England said shower maximum occurred at 1:12 a.m.

32. *Times* (London), November 15, 1866.

33. *Times* (London), November 15, 1866.

34. *Times* (London), November 17, 1866.

35. Robert Stawell Ball: *The Story of the Heavens* (London: Cassell, 1886), pages 337–338.

36. H. A. Newton: "Shooting Stars in November, 1866," *American Journal of Science and Arts*, volume 43, January 1867, pages 78–88.

37. U. S. Naval Observatory [no author given]: *Observations and Discussions on the November Meteors of 1867* (Washington, D.C.: Government Printing Office, 1867).

CHAPTER 8

The meteor–comet connection

The demonstration of the identity of meteor swarms and comets is one of the
greatest physical triumphs of the century.   J. NORMAN LOCKYER (1890)[1]

STARRING

Giovanni Schiaparelli (1835–1910), Italian, director of the Brera
Observatory in Milan. His observations, beginning in 1877, of linear features on Mars that he called *canali* (channels) was mistranslated as canals
in English and led to almost a century of speculation, especially in the
United States, about intelligent life on Mars.

Daniel Kirkwood (1814–1895), American, professor of mathematics at
Indiana University. In 1857 he discovered and explained the gaps between
the orbits of the asteroids between Mars and Jupiter as due to resonance in
the gravitational pull of Jupiter.

Urbain J. J. Le Verrier (1811–1877), French, director of the Paris Observatory,
co-discoverer of Neptune in 1846. Through his extraordinary mathematical analysis of irregularities in the motion of Uranus, he showed that a
planet beyond Uranus must exist and where astronomers should point
their telescopes to find it.

Theodor Ritter von Oppolzer (1841–1886), Austrian, a young mathemati-
cian and lecturer on astronomy at the University of Vienna. He is best
remembered for his careful calculations of the paths and times of all 8,000
solar and 5,200 lunar eclipses from 1207 B.C. to A.D. 2161 in an era when a
computer was the contents of a human skull linked to a printer formed of
five digits.

Hubert A. Newton (1830–1896), American, professor of mathematics at
Yale College, who inherited the Yale tradition of meteor research.

AND FEATURING A HOST OF SUPPORTING PLAYERS, INCLUDING:

Angelo Secchi (1818–1878), Italian, Jesuit priest and astronomer, director of
the observatory of the Collegio Romano in Rome.

Christian A. F. Peters (1806–1880), German, director of the Altona Observatory in the outskirts of Hamburg and editor of the astronomy journal *Astronomische Nachrichten* (Astronomical News).

Carl F. W. Peters (1844–1894), German, Christian's son, also an astronomer. After his Ph.D., he served as his father's assistant and edited *Astronomische Nachrichten* briefly after his father's death.

John Couch Adams (1819–1892), English, professor of astronomy and geometry at Cambridge and director of the Cambridge Observatory. As a graduate student at Cambridge, he used irregularities in the motion of Uranus to accurately predict the existence and position of a planet beyond Uranus, but he could not interest astronomers to search for the object.

Edmund Weiss (1837–1917), Austrian, associate astronomer at the University of Vienna observatory. He received his early education in England. His twin brother Gustav became a distinguished botanist. Edmund participated in an expedition to observe the November meteors in 1899.

Heinrich d'Arrest (1822–1875), German, living in Denmark as observatory director and professor of astronomy at the University of Copenhagen. As a graduate student at the Berlin Observatory, he and Johann Galle found Neptune in 1846 based on calculations furnished by Urbain Le Verrier.

Johann Galle (1812–1910), German, living in what now is Poland as observatory director and professor of astronomy at the University of Breslau. He and Heinrich d'Arrest, while observers at the Berlin Observatory in 1846, discovered Neptune by using calculations sent to him by Urbain Le Verrier.

AND A CAST OF DOZENS, EVERY ONE A STAR-WATCHER.

Publishing obscurely

Daniel Kirkwood was a quiet man with keen scientific insight and a courageous disregard for established convictions. He also had a tendency to withhold his careful findings until even more data was available or to publish his discoveries in journals unlikely to be read by the scientific community – traits that deprived him of the acclaim that he might have received.

He made his greatest discovery in 1857 when he noticed that there were

Daniel Kirkwood. Courtesy
Mary Lea Shane Archives of
the Lick Observatory

gaps between the orbits of the asteroids and explained them as the result
of gravitational perturbations produced by the planets near the asteroid
belt – especially Jupiter. At that time there were only 50 asteroids known.
Unusual gaps between the orbits of minor planets in the 300 million miles
(500 million kilometers) between Mars and Jupiter were not obvious. He
wrote down his discovery in 1857 but didn't publish it until 1866 when 87
asteroids were known and it was possible for him to check his work against
lots of new data. At the same time, he took the opportunity to explain the
Cassini Division – the big gap in Saturn's rings seen through telescopes – as
also due to gravitational disturbances – this time produced by Saturn's
largest moons.[2]

While Kirkwood, chairman of the mathematics department at Indiana
University, was sitting on the discovery that eventually would be called the
Kirkwood Gaps, he published a separate paper on comets in the
December 1861 issue of the first volume of the *Danville Quarterly Review*,
signed at the end of the article only as "D.K."[3]

The article was clear, thorough, accessible to laymen, and contained for
the first time persuasive evidence of a connection between meteors and
comets.[4] It was a major scientific report of discovery – published in a
northern Kentucky Presbyterian journal of theology and some current
affairs (the just beginning Civil War), edited by "an association of minis-
ters."

On the basis of the modest number of comets seen from Earth through
recorded history that come as close to the Sun as Mercury, Kirkwood
gauged that there had to be more than 37 million comets in the solar
system, the great majority remaining invisible because they come
sunward no closer than the outermost planets.[5] Cometary cloudlike
masses must be very numerous and widely diffused in the solar system,
Kirkwood said.

The number of comets visible to the naked eye is few, only about 15 a
century. But since the invention of the telescope, Kirkwood noted, four or
five comets are discovered each year. Obviously there are many more faint
comets than bright ones. On this basis, there must be even more minute
comets falling sunward undetected all the time – a constant rain of these
tiny bodies falling on the Earth. These "drops" of "primitive nebular
matter" – Kirkwood's words – produce sporadic meteors, and he proposed
to call them "cometoids."

But how then could one explain the phenomenon of annual meteor
showers?

When Biela's Comet returned in 1845, Edward Herrick was the first of
many astronomers to observe that it had split into two pieces.[6] Kirkwood
reported five previous instances from astronomy history where twin

comets (even one triplet) were seen, probably the result of fragmentation. He asked, "May not the force that has produced *one* separation again divide into parts? And may not this action continue until the fragments become invisible?"[7]

Now, said Kirkwood, consider the meteors that fall in showers. They occur when the Earth in its orbit intersects the orbits of hosts of nebulous bodies. These rings of meteor particles are "very elliptical," and thus unlike planet orbits. They move instead in "cometary orbits." "May not our periodic meteors," suggested Kirkwood, "be the *debris* of ancient but now disintegrated comets, whose matter has become distributed around their orbits?"

Kirkwood was the first to make this suggestion. But no one in the mainstream of astronomy knew about it for the next five years.[8]

Newton again

In 1864, Hubert Newton at Yale traced November "star showers" back almost a thousand years and established the interval between them. That allowed him to predict that the meteors were likely to return at storm intensity in 1866. If the November meteors poured down in great numbers every 33.25 years, what then was their orbit? Obviously, the particles could be revolving with a period of 33.25 years. But four other orbits with shorter periods could account for outbursts of November meteors every 33.25 years equally well: orbits of 180.0 days, 185.4 days, 354.6 days, and 376.6 days. Newton favored the 354.6-day period but, most important, he pointed out that the correct orbit could be determined by calculating the planetary disturbances on each of the five orbits to see which could yield the observed migration in the date of the November meteors.[9] But Newton never got around to making this very difficult calculation.

In an August 6, 1864 speech to the National Academy of Sciences,[10] Newton reexamined the frequency of meteors that would be visible to the naked eye "if the sun, moon, and clouds would permit" – if there were 24 hours of darkness under clear skies. He doubled Herrick's calculation to 7½ million meteors a day.

In that speech, Newton talked extensively about the small solid bodies that travel in orbits around the Sun before they encounter the Earth and burn up high in the atmosphere as meteors. Those bodies, Newton thought, needed a name of their own, and he invented the name "meteoroids" for those objects prior to their fiery extinction as meteors.

Newton then proceeded to compare the space velocity of meteoroids with that of Earth and found that the meteoroids seemed to be traveling

1.414 (the square root of 2) times the Earth's orbital speed, indicating parabolic or near parabolic orbits – not almost circular like the planets but long ellipses or open (never-returning) orbits like *comets*.[11]

Newton was on the verge of a major discovery, but he didn't take the next step, didn't make the next leap soon enough.

Breakthrough

Meanwhile, a young Italian astronomer, unaware of Kirkwood's or Newton's work, was analyzing comet tails and how this debris is repelled from the head of the comet by the Sun. The particles are so scattered that stars are not dimmed when a comet tail passes in front of them. The particles continue to travel in an orbit like the comet from which they came. Giovanni Schiaparelli was incubating an idea.

The idea hatched when he observed the annual meteors of August in 1866. Two weeks later he wrote the first of four letters to Angelo Secchi, the Jesuit priest and astronomer who directed the observatory of the Collegio Romano. They were letters in that they began "Dear Mr. Director" and ended with something like "Please accept my best regards." But they were not chatty musings mixed with family news and observatory gossip. Between the greeting and the signature of these letters, three of them very long, were formal, dense technical journal articles with diagrams, lists, and mathematical formulae – clearly intended for publication. Secchi understood their intent and published them immediately in issues 8, 10, 11, and 12 of volume 5 of the *Meteorological Bulletin of the Observatory of the College of Rome* before 1866 ended.

Schiaparelli probably wrote his letters to Secchi because he knew of Secchi's own work on comets and meteors, similar to his. Secchi had observed how comets in 1861 and 1862 emitted jets of gases to produce their tails and how the vigor of the jets increased the closer a comet was to the Sun. Secchi had also measured the altitude and speed of meteors (not very successfully: too high, too fast) and used spectroscopy to reach the conclusion that falling stars, asteroids, and meteorites are much alike chemically.

Secchi would obviously be interested in Schiaparelli's investigations. But Schiaparelli probably had two other reasons in mind: the opportunity for his long articles to reach print *quickly* – and in Italian, to assert the quality of Italian science.[12]

In his first letter, dated August 25, 1866, Schiaparelli analyzed the average velocities of meteors, calculated from ten reports by different astronomers. The data, he said, show that meteor particles are moving in space at 1.447

Giovanni Schiaparelli

times the orbital velocity of the Earth. That number, Schiaparelli says, is highly significant. Allowing for slight observational errors, it is effectively the same as 1.414 times the orbital velocity of the Earth, a number that sends a pleasurable chill through the nervous system of astronomers. If an object is falling toward or away from the Sun solely under the influence of the Sun's gravity, the speed of that object when it crosses the orbit of a planet will be 1.414 (the square root of 2) times the orbital speed of that planet. It is the speed gravity imposes on a free-falling body, a speed that results in a parabolic orbit for that body, an open orbit on which a body will never return to the Sun. Faster, and the object flies along a hyperbolic orbit, never to return. But if the speed of the body is just a fraction of one mile an hour slower than a parabolic orbital speed, the body will travel a very long elliptical orbit. It *will* return to the Sun. Parabolic – or *near* parabolic – speeds are the space velocities observed in *comets*. Meteor orbits, Schiaparelli muses, are also parabolas, or very elongated ellipses.

But my letter is already too long, said Schiaparelli. I'll expand on my idea in another letter. "For now, with utmost respect, I am most devotedly yours, G. V. Schiaparelli."

The second letter

Three weeks later (September 16, 1866), Schiaparelli finished and mailed his second, even longer, letter about the *piogge luminose*, the luminous rain.

There are two types of objects in the solar system, Schiaparelli said, planets and comets. Planets come in all sizes, but may be distinguished by their nearly circular orbits, their revolution around the Sun almost in the same plane as the Earth, their revolution in the same direction as the Earth (true of the planets, asteroids, and almost all of the moons), and the concentration of all their material in large spheres.

Comets, on the other hand, travel highly elongated orbits that seem to come from beyond the solar system. These orbits are inclined at all different angles to the plane of the Earth's path. Some comets orbit the Sun in the same direction as Earth (direct motion); others in the opposite way (retrograde motion). Further, comets are more nebulous – tenuous.

To which class, Schiaparelli asked pointedly, do meteors belong?

Clearly, the paths of meteors resemble the paths of comets.

Denison Olmsted was right to class the November meteors as objects moving in space on a closed orbit, Schiaparelli said, but, based on comet–meteor similarities, Olmsted must be wrong in thinking that the November meteor swarm had an orbital period of half a year.

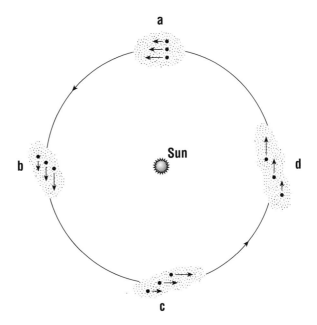

Bunched-up particles in orbit around the Sun will spread out along the orbit because those closer to the Sun revolve faster. Diagram by Will Fontanez and Tom Wallin, University of Tennessee Cartographic Services Laboratory.

Schiaparelli visualized meteors coming from cosmic clouds traveling in stellar space that intrude into the solar system. If they pass close by the Sun, they may be captured from parabolic paths to elliptical orbits, which would prevent them from returning to deep space. Each particle in these cosmic clouds is in a slightly different orbit at a slightly different distance from the Sun and hence is going a slightly different speed, with the ones closest to the Sun traveling fastest. In this way, the once-concentrated material in these cosmic clouds gradually spreads out along the orbit. When the Earth wallows through one of these spreading cosmic clouds, we collide with the particles in the cloud and witness a meteor shower.

The third letter

In his third letter, Schiaparelli enunciated his new theory of meteors in nine points. It took him almost eight weeks to complete the lengthy article.

1. Matter exists in all sizes in space, from stars to tiny particles. *(Schiaparelli was correct.)* The average particle that flashes through our sky as a falling star weighs one gram.[13] *(A shade heavy.)*
2. These tiny particles form by crystallization and fill the universe with "cosmic clouds." *(An anticipation of nebulae – interstellar matter?)*

3. Our Sun occasionally encounters a cosmic cloud and its gravity diverts the cloud onto a parabola-like orbit. *(No. Virtually all meteoroids travel closed elliptical orbits around the Sun and originated within the solar system.)*

4. As a cosmic cloud falls closer to the Sun, it is stretched out along its orbit so that the material takes years or even centuries to pass the Sun. We see these particles as meteors when they encounter the Earth.

5. There are many different meteor streams currently making loops around the Sun. They are of such low density that these streams can pass through one another without damage, but the planets do disturb these swarms with their gravity and can alter their paths into elliptical orbits. *(Planets do alter meteoroid orbits.)* The period of the November meteors, Schiaparelli says, is about 33 years. They are spectacular now because the particles are bunched up along one portion of their orbit. As time passes, they will spread out more evenly along their orbit, like the meteors of August.[14] *(Again, correct.)* To create the meteor showers we see, the particles in a meteor cloud are about 100 miles (160 kilometers) apart on the average.[15]

6. Cosmic clouds with short periods cannot exist permanently without violating the law of gravity – that is, without having their orbits warped and warped again by close-passing planets. *(True.)*

7. Most meteor particles return to interstellar space *(No)*, but encounters with planets may divert some particles onto special orbits so that they light the skies of Earth as sporadic meteors. *(Yes.)*

8. Meteors are related to and come from the realm of the stars, so their name *stelle cadenti* – falling stars – "expresses purely and precisely the truth of the matter." *(Meteor particles are little-changed remnants of the materials from which the Sun, planets, and comets formed, so they have significant scientific value. The elements beyond hydrogen and helium were all created inside exceptionally massive stars that expelled these elements when they died, so meteoroids are almost entirely stardust.)* Schiaparelli adds that asteroids are to Mars and Jupiter as meteors are to comets – small versions of larger but similar bodies. What they lack in size, they make up in number. *(Correct.)*

9. Falling stars, bolides (bright meteors), and aerolites (meteorites on the ground) differ only in size. *(A first approximation, but close examination allows a better distinction, as we shall see.)* "Material that has fallen from the sky is a sample of that which formed the universe." *(An important realization, although meteorites allow us to determine the age and initial composition of our solar system rather than the universe.)* No chemical element not already known on Earth has been found in meteorites or in meteors (by spectroscopy), so all the bodies in the uni-

verse are composed of the same elements. *(A breathtaking realization being reached during this period by observational astronomers in many fields who were using spectroscopy.)* The cosmic clouds that bring us meteor showers, Schiaparelli said, can contain comets in addition to small particles, thus meteors and comets could have similar orbits. Various astronomers, he noted, have seen double or triple comets close to one another. Because of their slightly different orbits, they now have significantly different periods. If meteors are found in the same orbits as comets, are they different classes of objects, or are they the same object differing only in size? Are comets just big meteors and meteors just tiny comets?

Schiaparelli concluded letter 3 with enthusiasm: he is looking forward to the return of the November meteors "next Tuesday."[16, 17]

The fourth letter

Letter 4, the decisive one, is one-third the length of the shortest of the previous three. Schiaparelli was in a hurry to publish. He knew now that Hubert Newton in the United States was about to reach the same conclusion that he had. Secchi had called Schiaparelli's attention to a letter Newton had written to Adolphe Quetelet in Belgium, which Quetelet had published in his annual observatory report for 1866. Newton had independently realized that meteors move in space with parabolic velocity – so they must be in parabolic orbits, just as comets seem to be. It was the same conclusion Schiaparelli had reached in letter 1.

Schiaparelli envisioned comets and meteor particles traveling the same orbit together as a "mixed system." He tended to think that meteors might aggregate to form a comet rather than that comet fragmentation would create meteors.[18] It would take him more than a year before he would adopt the correct view.

Meantime, he pushed forward into territory others had guessed at but no one before him could prove. He made the greatest single discovery about meteors in the 19th century.

In revealing that discovery, he introduced the terminology by which meteor showers are known today. Instead of referring to the falling stars of August, Schiaparelli said, he will for the sake of brevity from now on call them the *Perseids*, "based on the constellation [Perseus] from which they seem to come." That would make the November meteors that radiate from Leo the *Leonids*.

Having given the August meteors a name, Schiaparelli furnished his calculation of the orbit of the Perseids, using the position of the radiant and the time of maximum meteor fall (August 10, 18 hours) to derive the elements that uniquely define the path on which the meteors orbit the Sun. He then compared the elements of the Perseids' orbit to the elements that Theodor Oppolzer had derived for Comet 1862III.[19]

They were more than just *very* similar. "We have arrived therefore at this truly unexpected conclusion," Schiaparelli said, "that the great comet of 1862 is none other than one of the Perseids of August, and probably the most important of them all."[20]

Each August the meteors from the constellation Perseus rained down on Earth as the Earth cut through the orbit of this particle swarm. These particles were striking the Earth but were so tiny that they burned up in the Earth's atmosphere. Now Schiaparelli had identified a comet in essentially the same orbit as the meteoroids – just the "largest of them all." Every year the Earth was crossing the comet's orbit as well. Was there any danger that the comet could arrive at that intersection at the same time as Earth? Schiaparelli calculated that the comet could pass as close to the Earth as 439,000 miles (707,000 kilometers),[21] less than twice the distance of the

Schiaparelli's first meteor–comet connection

	Orbit of Perseid meteors (1866)	Orbit of Comet 1862III (Comet Swift–Tuttle)
Perihelion passage (closest to Sun)	July 23.63	August 22.9 (1862)
Date of passing descending node (where the object crosses the plane of the Earth's orbit moving southward)	August 10.75	
Longitude of perihelion	343°38′	344°41′
Longitude of ascending node	138.16°	137.27°
Inclination of orbit (to that of Earth)	64.3°	66.25°
Distance of the perihelion from Sun (in Earth-to-Sun distance units: 1 astronomical unit [AU] = 93 million miles; 150 million kilometers)	0.9643 AU	0.9626 AU
Direction of orbital movement (direct or retrograde)	retrograde	retrograde
Period of revolution	108 years?	123.4 years (Stämpfer says 113)

Comet Swift–Tuttle: parent of the Perseid meteors

Two Americans independently discovered a comet in 1862 within three days of one another, and the comet bears their names: Swift–Tuttle.

Lewis Swift discovered his first comet on July 16, 1862 with his 4½-inch refracting telescope. He didn't publish his discovery immediately because he thought he was looking at recently discovered Comet Schmidt (1862II). Three nights later, Harvard College Observatory astronomer Horace P. Tuttle noticed it and was first to publish. At first the comet was designated 1862III, but today it is known by both their names: Comet Swift–Tuttle. Both men went on to discover other comets.

After its discovery, Comet 1862III grew brighter until it was visible to the unaided eye, with a tail stretching 25 degrees across the sky, longer than the constellation Orion is tall.

Four years later, Giovanni Schiaparelli matched the orbit of Comet Swift–Tuttle to the orbit he had calculated for the Perseid meteors – the first proof that comets and meteors were intricately related.

Moon. Telescopes showed comets as extended nebulous objects, so that in the hours around noon on August 10, the Earth could indeed encounter the denser inner portions of a comet – "a real passage," said Schiaparelli ominously.

How often would we face this danger? The comet returns close to the Sun, and hence passes perilously close to Earth, about every 120 years.

Schiaparelli pictured the meteoroids as particles – minor Perseids – that had separated from the "main" Perseid – the comet.[22] These particles now traveled their own orbits so that, because of planet perturbations, they gradually developed orbits slightly different from the comet. The orbit for the Perseid meteoroids he had calculated, reinforced by historical records of better-than-usual displays of August meteors, pointed to an orbital period for an average Perseid particle of about 108 years.[23]

Having linked beyond doubt one meteor shower with one comet, Schiaparelli knew what to do next. He calculated the orbit of another meteor shower, the Leonids of November, and searched through catalogs of comets to find a match. He could not find one with the 33⅓-year period he sought. But he planned to keep looking.

Lewis Swift: self-trained comet seeker

Lewis Swift (1820–1913) grew up on a farm in upstate New York and might have remained there the rest of his life except for an accident. At the age of 13, a piece of machinery fell on him, fracturing his hip. The injury and ineffective surgery that followed left him permanently lame. So he dedicated himself to his studies, rocking on crutches two miles to school and back each day for the next three years. Farmwork was out of the question.

For eight years, he helped make horse hayforks, one of his father's inventions. When his father died, Swift left the farm and moved to Marathon, New York, where he set up a hardware business. He had a long-standing interest in science and technology, and made his living in part by giving public lectures on inventions and machinery. In 1855, he bought some astronomy books, which excited him so greatly that he built a 3-inch (7½-centimeter) telescope and embarked on an astronomical career at the age of 35.

Swift went on to discover or co-discover 10 comets, the last in 1899 when he was 79 years old. He also discovered about 1,200 previously unknown nebulae. While observing the 1878 total eclipse of the Sun in Colorado, Swift thought he discovered one or perhaps two small planets closer to the Sun than Mercury. He was wrong, but the controversy raged for years. Swift's son Edward discovered a comet of his own in 1894.

Lewis Swift

Reaction

Other astronomers also knew what to do when they read or heard about Schiaparelli's fourth letter.[24] The race was on to match comets and meteor showers. It was a time of the most feverish work in the field – and the period of most rapid discovery. The biggest remaining prize was the comet corresponding to the grandest meteor display of all, the Leonids.

Schiaparelli was right that many other astronomers were on the same or a parallel trail to his. Each in his way would contribute simultaneously to the search for or confirmation of the source of the Leonids.

To find a comet associated with the November meteors, astronomers needed to know the orbit of that Leonid swarm. Schiaparelli's initial inability to identify a comet with the 33.3-year orbit he derived for the Leonids indicated that a better orbit would be useful.

While Schiaparelli was writing long journal articles about meteors and comets and sending them as letters to Angelo Secchi in Italy, Urbain Le

Urbain Le Verrier. National Maritime Museum Greenwich

Verrier in France stepped in to solve the November meteor orbital problem. Neither knew of the other's work on the November meteors.

Le Verrier was the most formidable figure in European astronomy, and had been since the time 20 years earlier when he used irregularities in the motion of Uranus to calculate that an unknown planet must exist still farther from the Sun. When he could not persuade astronomers in France or England to trust his calculations and search for the planet, he wrote a letter to Johann Galle, an astronomer at the Berlin Observatory, provided the planet's expected position, and asked him to search. Galle and graduate student Heinrich d'Arrest found the new planet – Neptune – after less than an hour of searching on September 23, 1846. It was a triumph of mathematical astronomy: a planet discovered on a sheet of paper.[25] Le Verrier was an acknowledged master of orbital calculations.

In mid-January 1867, he presented to the Scientific Society of France and then published in the French journal *Comptes Rendus* an improved orbit for the November meteors, adjusting the orbital period to 33.25 years and revising other elements very slightly.[26] He concluded that the August meteors have had time to spread out fairly evenly along their orbit, but the November meteors have not, indicating that the November swarm has not existed for long in astronomical time – perhaps only since A.D. 126. The meteor stream returns close by Uranus, so maybe meteor particles were ejected from Uranus at that time. *(Wrong. But Le Verrier was correct that the meteor stream was debris shed by a larger body.)* There is no reason to suppose, said Le Verrier, that the meteoroids originated anywhere except inside our solar system. Perturbations of the particle stream by Jupiter and other planets then altered the abundance of November meteors annually and every 33.25 years. These planetary perturbations gradually scattered the meteoroids far from their original courses so that they trickled into our atmosphere as sporadic meteors.

Convergence

Schiaparelli had offered an orbit for the November meteors. Almost simultaneously, and without knowing of Schiaparelli's work, Le Verrier had offered a very similar one. Astronomers rushed to the catalogs of comets to match partners for this cosmic dance. The comet wasn't there...

Because at that moment the orbital calculation for that comet, spotted at the beginning of 1866, had just been finished – by Oppolzer again – and was being published in *Astronomische Nachrichten*.[27] Christian A. F. Peters, editor of *Astronomische Nachrichten*, had an advantage in putting one and one – meteor swarm and comet – together. He saw articles *before* they were published. He had a headstart in looking for correlations. If he were not a man of principle, he could also delay publication of someone else's idea or discovery by a week or two until he had a chance to publish. Le Verrier's orbit for the November meteors came out in mid-January 1867. Oppolzer's article on the orbit for Comet 1866I was dated January 7 and published by Peters on January 28.

Peters' 22-year-old son Carl was briefly home after graduating from college and before beginning doctoral studies in astronomy at Göttingen. Peters brought the opportunity to the attention of his son.

Carl F. W. Peters politely dated his article on the correlation between the comet and the November meteors January 29, the day after Oppolzer's orbital data appeared in *Astronomische Nachrichten*. The article reached print six days later. In this half-page, four-paragraph article, Peters printed Le Verrier's orbital elements for the November meteors and then Oppolzer's orbital elements for Comet 1866I. "At first glance," Peters wrote between the two, "we notice the great similarity of this orbit with that of Comet Tempel (1866I)."[28] Peters did not mention in his article Schiaparelli or any other scientist whose work might have alerted him to watch for a meteor-comet connection.

When the January 28, 1867 issue of *Astronomische Nachrichten* containing Oppolzer's orbital elements for Comet Tempel reached Schiaparelli in Milan, he instantly saw how his orbital calculation for the November meteors coincided with the orbit for the comet seen at the beginning of 1866. He sat down immediately – February 2, 1867, two days before Peters' article appeared – to identify a second meteor–comet connection and to claim credit for discovering the first.

"I have the honor," he began, "of sending you a copy of my letters to R. P. Secchi on the course and probable origin of shooting stars.[29] In these writings, I have taken on the task of gathering all the evidence for an analogy between these mysterious bodies and comets. . . . It is no longer to be doubted that certain comets, if not all, are part of the numerous meteor streams that occupy celestial space. Thus Comet 1862III is none other than one of the shooting stars of August and Tempel's most recent comet (1866I) is part of the [meteor] stream of November."[30]

In his article, Schiaparelli improved his orbit for the November meteors because the radiant was not precisely Gamma Leonis as he thought American observers believed. Observations "made with great care" by Alexander Herschel in England[31] have demonstrated that the American radiant "is faulty by several degrees," so his orbit for the November meteors can be regarded only as a "very gross approximation." Here, he said, "is a more exact calculation, compared with the elements of Comet 1866I given by Oppolzer . . ." The revision provided modest improvements,

Connection between Leonid (November) meteors and Comet Tempel–Tuttle (1866I)

	Leonid orbit by Schiaparelli[a]	Leonid orbit by Le Verrier	Comet 1866I orbit by Oppolzer
Longitude of perihelion	71°	(42°24′)[b]	
Longitude of ascending node	231°	(231°18′)[c]	(231°26′)[c]
Date of descending node	November 13.5		
Inclination to ecliptic	15°	14°41′	17°18′
Distance from Sun at perihelion	0.96 AU	0.989 AU	0.977 AU
Date of perihelion passage	October 30.5		
Semi-major axis (average distance from Sun)	10.4 AU	10.340 AU	10.325 AU
Maximum distance from Sun		19.691 AU (just beyond Uranus)	
Eccentricity		0.904	0.905
Direction of revolution	retrograde	retrograde	retrograde
Period of revolution	33.3 years	33.25 years	33.18 years

Notes: [a]Not shown in Peters' article.
[b]Provided by Carl F. W. Peters in *Astronomische Nachrichten*, number 1626, February 4, 1867; columns 287–288.

[c]The longitude of the ascending node is inferred from the longitude of the descending node given by Peters.

but Schiaparelli's original elements matched closely enough to establish a meteor–comet relationship. For his new calculation, Schiaparelli assumed that the period of the November meteors was 33.25 years, as proposed by Hubert Newton.

The striking similarities between the orbits of the November meteors and Tempel's Comet "require no commentary," said Schiaparelli. Instead, he asked a question: "Should we regard shooting stars as a swarm of tiny comets or, equally well, as the product of the disintegration of great comets? I don't dare answer such a question."

Schiaparelli's article was published in *Astronomische Nachrichten* on February 20, 1867. On the next page of the same issue of *Astronomische Nachrichten* was an article by Oppolzer, dated February 6, before the February 4 issue with Peters' announcement could have reached him. In a two-paragraph article occupying less than half a page, he thanked

Comet Tempel–Tuttle: parent of the Leonid meteors

Wilhelm Tempel at an observatory in Marseilles, France discovered a comet on December 19, 1865. Two and a half weeks later, on January 6, 1866, before news of the discovery had spread from Europe to the United States, Horace P. Tuttle at the Harvard College Observatory independently reported the same object. Comet 1866I became known as Comet Tempel and then gradually as Comet Tempel–Tuttle.

By the end of 1866, Theodor von Oppolzer had computed an orbit for the comet, and a period of 33.17 years. Oppolzer published his calculations for the comet's orbit at almost the same time as Urbain Le Verrier published his calculation for the orbit and 33.25-year period for the November meteors. Three astronomers working separately – Carl Peters in Germany, Giovanni Schiaparelli in Italy, and Oppolzer in Austria – instantly recognized that Comet Tempel–Tuttle and the Leonid meteors essentially shared the same orbit and thus must be causally connected to one another.

Schiaparelli for sending him an abstract of his letters to Secchi and he thanked Edmund Weiss for calling his attention to Le Verrier's computation of the orbit of the November meteors. Schiaparelli has shown, Oppolzer said, how the orbits of the August meteors and Comet 1862III nearly coincide. However Schiaparelli was not able to find a comet to fit the November meteors' orbit at that time. I have, says Oppolzer. Comet 1866I, for which I provided the orbital elements in *Astronomische Nachrichten* number 1624 (January 28, 1867), is the sought-after comet. He presented both Le Verrier's and Schiaparelli's original orbital data for the November meteors and his own elements for Comet 1866I, which were in "extraordinary agreement." The connection of the November meteors to Comet 1866I and the August meteors to Comet 1862III, said Oppolzer, "opens up a whole new perspective on the physical nature of comets and meteors." Comets, he said, are dust clouds, dense swarms of meteoroids, which disintegrate to give arcs and rings of meteors. Oppolzer then ended his very brief article dramatically: "A meteor shower therefore suggests how a collision of a comet with the Earth would take place."[32]

The history of science accords Peters, Schiaparelli, and Oppolzer equal honors for the recognition that the November meteors closely follow the path of Comet Tempel–Tuttle.[33]

The adventurous and checkered career of Horace P. Tuttle

The early career of Horace P. Tuttle was very promising. In 1857, at age 20, he became an assistant astronomer at the Harvard College Observatory and almost immediately discovered his first comet – actually an independent co-discovery.

Less than a year later, Tuttle was the sole discoverer of another comet. He co-discovered two more comets in 1858, for which, at the age of 22, he received the Lalande Prize in astronomy from the Academy of Sciences in Paris.

Just before he left for military duty in the Civil War in 1862, he and Lewis Swift, working separately, discovered a comet that four years later was found to be the source of the Perseid meteor shower each August.

During the Civil War, Tuttle served in the Union Army and then Navy from 1862 to 1865. As a sailor aboard an ironclad, he may have played a heroic role by going ashore to place a signal that deceived a British blockade runner into thinking it was safe to enter Charleston harbor. Tuttle's gunboat then captured British vessel.

After the war, Tuttle stayed in the Navy, returned to Harvard, and co-discovered Comet Tempel–Tuttle in 1866, which proved to be the parent of the Leonid meteors of November. By 1866, Tuttle had discovered four comets and co-discovered eight. He had also discovered two asteroids. Harvard awarded him an honorary master's degree in 1868.

But then, in his early 30s, Tuttle's career faltered. In 1869, the Navy found his records as paymaster deficient by four times his annual salary. No charges were filed at the time, but when Tuttle illegally cashed a Navy bill of exchange in 1873 and claimed that almost all of the money was stolen from him, the Navy charged him with embezzling a total of nearly $6,000. He was found guilty and dismissed from the Navy.

That blemish did not end his government career however. In 1875, three weeks after his dishonorable discharge, the U.S. Geographical and Geological Survey hired him as astronomer for the Rocky Mountain Region. In 1877 he helped mark the boundary between Wyoming and Dakota Territory.

In 1884, at age 47, the Navy either forgave or forgot his moral lapses as a paymaster and hired him to work at the U.S. Naval Observatory in Washington, D.C. He made an independent co-discovery of one more comet in 1888.

Tuttle lived to age 86. He left an estate of only $70 and was buried in an unmarked grave.*

* This vignette is based on Donald K. Yeomans: *Comets: A Chronological History of Observation, Science, Myth, and Folklore* (New York: John Wiley & Sons, 1991), pages 238–239. Yeomans credits Richard Schmidt of the U.S. Naval Observatory for the research upon which his story of Tuttle is based.

An elevated view

The spectacle of the 1866 Leonid meteors in Europe stimulated the first deliberate effort to observe an astronomical event from above the clouds. The experiment took place a year after the prime shower and a day after the prime night of November 13–14.

On the night of November 14–15, 1867, aeronauts identified as de Fonvielle, Jules Goddard, and A. V. Weyenberch took off from Paris in a hydrogen-filled balloon shortly before midnight to observe what might be left of the Leonid shower. With a telescope, barometer, star chart, and other instruments aboard, they reached an altitude of only a few thousand feet, but that was enough. Floating above the clouds, they recorded shooting stars not visible to observers on the ground. They landed safely on the coast of Belgium in the morning.

Astronomy had entered the era of the airborne observatory.*

* James Glaisher, Robert P. Greg, E. W. Brayley, Alexander S. Herschel, and Charles Brooke: "Report on Observations of Luminous Meteors, 1867–1868," *Report of the 1868 Meeting of the British Association for the Advancement of Science*, page 393 of pages 344–428.

Consolidation

Before more comet–meteor connections could be found, an article was published that enhanced confidence in the findings already made and another article provided a handbook for making new discoveries.

John Couch Adams showed that the 33.25-year period for the November meteors was the only orbit consistent with the way the date for the Leonids was slowly precessing to later in the year. When Adams spoke, scientists listened. It was not always so. As a student at the University of Cambridge, he had analyzed the aberrant motion of Uranus and concluded that a new planet still farther from the Sun must exist. He calculated accurately where that planet could be found, but neither England's astronomer royal nor the director of the Cambridge Observatory believed that such mathematics could be done, especially by such a young man. They did not begin a search until Le Verrier, well established as a scientist, contacted them and they realized that his independent calculations for a planet beyond Uranus closely matched Adams' earlier work. Their search however was slow and unenthusiastic because they didn't believe science could be done this way – theory leading observation – and because they lacked the best star maps, making the search more tedious. Before they recognized the new planet, Le Verrier grew tired of waiting, contacted the Berlin Observatory, and Neptune was found that very night within a degree of where it was supposed to be.

John Couch Adams. Master & Fellows of St. John's College, Cambridge

That was 20 years ago. Astronomers listened to Adams now.

The astronomical community had settled on 33.25 years as the orbital period for an average particle in the November meteor shower. It allowed the identification of Tempel's Comet with the Leonid meteors. But it was still an *assumption*.

Adams began his exceptionally clear article by summarizing Hubert Newton's work on the period between grand displays of November meteors. From October 12, 902 through November 13, 1866, historical records collected by Newton had shown 14 outbreaks surrounding dates that were 33.25 years apart on the average. If the meteors fell in large numbers every year, the meteors must be evenly distributed along their orbit. But for the meteors to fall in torrents every 33.25 years, they must be clumped up into about 1/10 or 1/15 of their orbit.[34] An orbit with a period of 33.25 years could, of course, produce this effect, but so could orbits of 180.0 days, 185.4 days, 354.6 days, and 376.6 days.[35] Newton, who pointed out clearly that the Leonid meteors tend to return in great strength every 33.25 years, nevertheless thought that the November meteors mostly likely traveled in a nearly circular orbit just inside the Earth's track that crossed the Earth's orbit every 354.6 days. Only every 33.25 years did the Earth and the meteor clump arrive at the same spot at the same time.

But Adams took special note of the way the date for the big Leonid displays was slowly shifting to later in the year, from October 12, 902 to November 14 in 1866.[36] That meant that the place where the meteor stream crossed the Earth's orbit (the node) was shifting forward along the Earth's path – approximately 29 minutes of arc (almost half a degree – about the size of the full moon in the sky) between visits every 33.25 years. That shift, said Adams, must be caused by gravitational disturbances inflicted on the particle stream by the planets they passed: primarily Venus, Earth, and Jupiter if the orbit were about half a year or a year; Jupiter, Saturn, and Uranus if the orbit were 33.25 years.

Adams began by calculating the perturbations imposed by Venus, Earth, and Jupiter on the meteor particles if their orbital period was 354.6 days, the one favored by Newton. For this orbit, he found, the planets would force a precession of about 12 minutes of arc in 33.25 years – far too little. The other half-year and one-year orbits would also be too slightly perturbed to fit reality.

But for the 33.25-year period, when the Leonid meteor stream would flow out beyond the orbits of Jupiter, Saturn, and Uranus, Adams' calculation showed that the planets would force the meteors' node forward along the Earth's orbit by these amounts in 33.25 years:

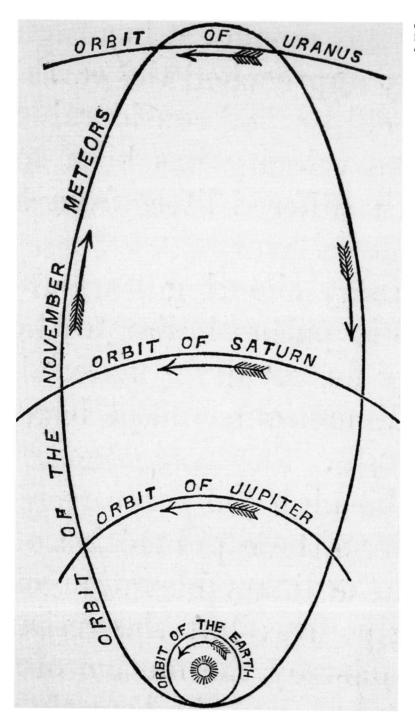

Orbit of Leonid meteors
Elias Loomis: *Elements of Astronomy*, 1869

Jupiter	20 minutes of arc
Saturn	7 minutes of arc
Uranus	1 minute of arc
all other planets	negligible

The total came to a forward displacement of the node of 28 minutes, effectively the same as the 29 minutes observed.

"This remarkable accordance," said Adams, "between the results of theory and observation appears to me to leave no doubt as to the correctness of the period of 33.25 years."[37]

Adams realized that his work on the orbit of the November meteors was a little late. Other astronomers, he said, had been led to the same conclusion about the 33.25-year period and other orbital elements for the November meteors on "totally independent grounds" from those upon which he was making his calculations. The importance of his work, although Adams was an exceptionally modest man and would never say such a thing, was that he *proved* rather than *assumed* that the meteors orbited on a 33.25-year period and that no other orbit was possible.[38]

As of February 1867, there was an undeniable similarity between the orbit of an annual meteor shower and a comet in two separate cases. One involved the best-known, most dependable annual meteors; the other involved the erratic but most spectacular meteors. That similarity could *not* be accidental, said Adams, but its significance was not yet clear.

More correlations

If there were two identifications established between meteor showers and comets, could there be more? Almost no one in the burgeoning new field of meteor astronomy doubted it. They went searching.

Edmund Weiss, an astronomer at the University of Vienna observatory, considered the meteor shower–comet connection and reached a conclusion opposite of Schiaparelli's: the rings of meteor particles do not coalesce to form comets. Instead, said Weiss, comets shed material as they round the Sun and it is that debris that forms the orbits of material that produce meteor showers if the Earth passes through or very near that comet's orbit. Therefore, other meteor showers must owe their existence to other comets. If comets round the Sun only once and never return, they can produce one shower, but not annual showers. Only comets with short orbital periods – a few years to a few centuries – can produce the annual meteors that we observe.

Weiss combed three lists of meteor displays[39] and decided that eight annual showers had been definitely established *(he was right)* and four more were probable but not certain *(he was right to be uncertain)*. Could comets be found to match these showers? The decisive clue would be the point at which the orbit of the comet cut through the orbital plane of the Earth – the node. Weiss provided a list of 32 short-period comets with orbits that crossed or passed close to the orbit of Earth for the benefit of astronomers seeking to make correlations. He himself then matched these comet nodes to meteor radiants. Where the comet's node would be seen in the sky when the Earth arrived at that intersection should be the point from which the meteors seemed to radiate. On that basis, Weiss offered[40] two more correct comet–meteor shower correlations:

Heinrich d'Arrest. Courtesy Copenhagen University Astronomical Observatory

Date of shower maximum	Radiant	Comet
April 20	Lyra	Comet 1861I
November 28	Andromeda	Comet Biela

Together again

The effort to establish the source of meteors briefly reunited on a single scientific project all the major contributors to the discovery of the planet Neptune: John Couch Adams, Urbain Le Verrier, Johann Galle, and Heinrich d'Arrest. And each made a contribution in confirming the comet–meteor connection.

Heinrich d'Arrest, now a professor of astronomy and director of the Copenhagen Observatory, had been fascinated by Biela's Comet and by its orbit that brought it close to Earth ever since he and Johann Encke, his boss at the Berlin Observatory, had observed it on its return in January 1846 and found that it had split into two parts.

The comet had first been observed in 1772 and then again in 1805, when Friedrich Bessel recognized it to be a return of the Comet of 1772. The first person to see the comet on its return in 1826 was Wilhelm von Biela, an Austrian infantry captain and amateur astronomer, who was one of three to accurately compute its orbit and its 6.7-year period of revolution – a comet with a very short period. Ever since, the comet has been known by his name.

On December 29, 1845, Edward Herrick at Yale College, co-discoverer of the annual return of the Perseids, and his friend Francis Bradley, a New Haven banker, were the first to observe that Comet Biela had fragmented,[41] a month before d'Arrest and Johann Encke, director of the Berlin Observatory, observed this anomaly.

It was now more than 21 years later, and d'Arrest was wondering if meteors accompanied Biela's Comet. "It is remarkable," he wrote, "that some extraordinary meteors around December 6–7 have not yet been connected to the orbit of Comet Biela, through which the Earth passes at that time."[42] To prove that the meteors have the same orbit as the comet is not straightforward, d'Arrest said, because there is too little data to allow for the accurate computation of the meteor orbit. Instead, he said, he will take the opposite approach and calculate what the radiant would be if meteors accompanied Biela's Comet. He found that the radiant would be at the border between Andromeda and Cassiopeia, the same as meteors to be seen in early December. Without knowing that Weiss had taken the same approach and arrived at the same conclusion about Comet Biela and its meteors, d'Arrest sent his article to *Astronomische Nachrichten* three days after Weiss sent his. Weiss and d'Arrest proposed that Comet Biela caused the Andromedid meteors. They both pictured the breakup of comets as the source of meteor particles, material left traveling in paths approximating the comet's orbit – and initially distributed unevenly. Through many revolutions, the particles would distribute themselves along their entire orbit.

If the early December meteors are caused by Biela's Comet, then, said d'Arrest tentatively, the early December meteors should rain down in great numbers when Biela's Comet returns in 1878. Weiss and Johann Galle predicted a major return of the Andromedids in 1872.[43] Remarkable outpourings of meteors did occur on November 27 in both 1872 and 1885 at times when Comet Biela should have returned. But after 1852 the comet was never seen again.

After 1885, the meteors from Comet Biela have been much less numerous, declining to the point that the Andromedids are scarcely detectable today and are scattered through most of November.

About two weeks after Weiss and d'Arrest finished their articles on meteor–comet linkages, Johann Galle, now professor of astronomy and director of the observatory at Breslau (then in Prussia, now in Poland), finished and mailed to *Astronomische Nachrichten* his identification of another meteor shower with a comet. The latest issue of *Astronomische Nachrichten* had just arrived, with the article by Weiss giving comet orbital data for correlation with meteor showers. Galle took Weiss' identification of Comet 1861I with the April meteors to be tentative. Galle offered his data to reconfirm Weiss' identification. The orbit of Comet 1861I, he said, crosses the Earth's path on April 20, when an annual shower of meteors occurs. These meteors radiate from Lyra, the Lyre, near the bright star Vega. Do these meteors seem to radiate from the position where the comet would be seen from Earth on April 20? Yes.[44]

Johann Galle. E. R. Phillips: *Splendour of the Heavens*, 1925

Verdict

It had been a whirlwind half year in astronomy: the period between the return of the Leonids in November 1866 as predicted and the recognition that specific comets cause all meteor showers.[45] Schiaparelli, quite appropriately, pronounced the benediction for this revolution:

> Not since the year 1759, when the predicted return of [Halley's] comet first took place, had the verified prediction of a periodic phenomenon made a greater impression than the magnificent spectacle of November 1866. The study of cosmic meteors gained thereby the dignity of a science, and took finally an honorable place among the other branches of astronomy.[46]

NOTES

1. J. Norman Lockyer: *The Meteorite Hypothesis* (London: Macmillan, 1890), page 147.

2. Mimas, the innermost large satellite of Saturn, is now recognized as the foremost culprit in the creation of the Cassini Division. This gap is not devoid of ring particles; it is a region of lower particle density.

3. D[aniel] K[irkwood]: "Cometary Astronomy," *Danville Quarterly Review*, volume 1, December 1861, pages 614–638.

4. Brian Marsden in his article on Kirkwood in the *Dictionary of Scientific Biography* says that Kirkwood offered the "first convincing demonstration of an association between meteors and comets."

5. Kirkwood, as he notes in his article, was improving on François Arago's estimate of 7 million. Today, astronomers estimate that there are more than a trillion (10^{12}) comets in orbit around the Sun, almost all located in the Kuiper Belt and Oort Cloud far beyond Pluto.

6. Thanks to Donald K. Yeomans who brought this contribution of Herrick to my attention. See his book *Comets: A Chronological History of Observation, Science, Myth, and Folklore* (New York: John Wiley & Sons, 1991), page 183.

7. D[aniel] K[irkwood]: "Cometary Astronomy," *Danville Quarterly Review*, volume 1, December 1861, page 637.

8. D[aniel] K[irkwood]: "Cometary Astronomy," *Danville Quarterly Review*, volume 1, December 1861, page 637.
   By 1867, when Edmund Weiss argued that comets shed material that becomes meteors, Kirkwood realized that his insight about the source of meteors had not reached the attention of the astronomical community and he republished an excerpt from his article as Appendix B: "Comets and Meteors" at the conclusion of his book *Meteoric Astronomy: A Treatise on Shooting-Stars, Fire-Balls, and Aerolites* (Philadelphia: J. B. Lippincott, 1867), pages 125–128. It was too late to earn Kirkwood more than independent discoverer status.

9. H[ubert] A[nson] Newton: "The original accounts of the displays in former times of the November Star-Showers; together with a determination of the length of its cycle, its annual period, and the probable orbit of the group of bodies around the sun," *American Journal of Science and Arts*, 2nd series; part 1 in volume 37, May 1864, pages 377–389; part 2 in volume 38, July 1864, pages 53–61.

10. H[ubert] A. Newton: "Abstract of a Memoir on Shooting Stars," *American Journal of Science and Arts*, 2nd series, volume 39, March 1865, pages 193–207.

11. Others before Newton, including Olmsted, Herrick, and Kirkwood, had noticed that the ellipticity of meteoroid orbits was similar to that of comets and that the diffuseness of meteor particles in a shower was reminiscent of the telescopic appearance of a comet's coma and tail. Newton was moving into new territory toward a rigorous mathematical demonstration of meteoroid and comet orbital similarities.

12. Schiaparelli had great facility in languages. He subsequently published many articles on meteors and other subjects in French and German.

13. Schiaparelli says this earlier in letter 3.

14. Schiaparelli says this earlier in letter 3.

15. Schiaparelli says this earlier in letter 3.

On the basis of the 1866 Leonid shower, Hubert Newton calculated that the particles were separated from one another by 30 to 40 miles (50 to 65 kilometers). H. A. Newton: "On certain recent contributions to Astro-Meteorology," *American Journal of Science and Arts*, volume 43, May 1867, pages 285–300.

16. Letter 3 contains no date, only this mention in closing that Schiaparelli is looking forward to the return of the November meteors "next Tuesday," so the date of the letter could be any time between November 7 and about the 12th.

17. Schiaparelli added a footnote to letter 3 that his first two letters were already published and his third one written before Secchi made him aware of Hubert Newton's work linking meteors to comets. Almost no one knew that Daniel Kirkwood had concluded five years earlier that comet debris causes meteor showers.

18. Schiaparelli probably favored the idea that comets form out of meteor particles because he and most astronomers of his day subscribed to the Kant–Laplace nebular hypothesis of the origin of the solar system: a cloud of gas and dust coalesced by gravity to form the Sun, planets, and all the bodies in the solar system. That basic idea is well substantiated today, but the process is better understood. Comets did form out of the same material as the Sun and planets, but the nebulosity remaining in the solar system today is far too tenuous to spawn comets.

19. Theodor Oppolzer: "Bahn-Bestimmung des Cometen II.1862" (Orbital Elements of Comet 1862II), *Astronomische Nachrichten*, number 1396, December 19, 1862; columns 49–58, article dated November 25, 1862, refined from an initial calculation published as Theodor Oppolzer: "Ueber die Bahn des Cometen II.1862" (Concerning the Orbit of Comet 1862II), *Astronomische Nachrichten*, number 1384, October 14, 1862; columns 249–250, article dated September 8, 1862.

Oppolzer refers to Comet Swift–Tuttle as Comet 1862II instead of Comet 1862III because, as was common at that time, he was not counting the return of short-period comet Encke.

20. Alternatively, ". . . and probably the largest of them all." Schiaparelli used the word *principalissima*, implying both biggest and most important, rather than either larger or most important, for which specific words were available.

21. 0.004 72 astronomical units, where 1 AU equals the mean distance between the Earth and the Sun.

22. Schiaparelli was inconsistent on this point. More often he conceived of meteoroids coalescing to form a comet.

23. In a footnote, Angelo Secchi called for annual observations to determine what year the Perseids are at a maximum.

24. Actually, more likely, other scientists began to learn of Schiaparelli's findings when they saw a French translation of his first letter to Secchi (as "Sur la marche et l'origine probable des étoiles météoriques") in the widely read French scientific journal *Les Mondes*, 2nd series, volume 12, December 13, 1866, pages 610–624, with the editor's promise that letter 2 would appear very shortly.

Before that happened, however, Secchi mentioned in his article "Sur les étoiles filantes de novembre 1866" (On the Shooting Stars of November 1866) for *Les Mondes*, 2nd series, volume 12, December 20, 1866; pages 645–648 that Schiaparelli would announce in the next issue of *Bullettino Meteorologico* that the August meteors come from a comet seen in 1862.

Schiaparelli's second letter to Secchi appeared as "Sur le mouvement et l'origine probable des étoiles météoriques" in *Les Mondes*, 2nd series, volume 13, January 24, 1867; pages 147–162. A summary of Schiaparelli's letter 4 ("Théorie des étoiles filantes") that identified the August meteors with Comet 1862III followed in the February 21, 1867 issue (pages 284–287). Letter 4 was accompanied by a French language version of a fifth letter that Schiaparelli wrote to Secchi (pages 287–289), dated February 2, 1867, that improved his orbital calculation for the November meteors and linked them to Comet 1866I – covering the same ground as his article "Sur la relation qui existe entre les comètes et les étoiles filantes" for *Astronomische Nachrichten*, number 1629 (combined issue with 1628), February 20, 1867; columns 331–332, article dated February 2, 1867, that made Schiaparelli one of the three independent discoverers of the source of the November meteors.

25. This story is told in detail, with a bibliography listing other accounts, in Mark Littmann: *Planets Beyond: Discovering the Outer Solar System* (New York: John Wiley & Sons, 1988, 1990).

26. [Urbain J. J.] Le Verrier: "Sur les étoiles filantes du 13 novembre et du 10 août" (On the Shooting Stars of November 13 and August 10), *Comptes Rendus*, volume 64, number 3, January 21, 1867, pages 94–99. On January 14, 1867, Le Verrier presented these findings in a talk to the Société Scientifique de France. Le Verrier's purpose in the article was to respond to John Herschel's recent article in *Monthly Notices of the Royal Astronomical Society* about how meteoroids in a retrograde orbit could fit in with the nebular hypothesis about the origin of the solar system.

27. Th[eodor] Oppolzer: "Bahnbestimmung des Cometen

I.1866 (*Tempel*)" (Orbital Calculation for Comet 1866I [Tempel]), *Astronomische Nachrichten*, number 1624, January 28, 1867, columns 241–250, article dated January 7, 1867.

28. C. F. W. Peters: "Bemerkung über den Sternschnuppenfall vom 13. November und 10. August 1866" (Remarks about the Meteor Shower of November 13 and August 10, 1866), *Astronomische Nachrichten*, number 1626, February 4, 1867; columns 287–288, article dated January 29, 1867.

29. The "R. P." before Pietro Angelo Secchi's name stands for Reverendus Pater – Reverend Father. Secchi collected Schaparelli's four letters and republished them immediately as "Intorno al Corso ed all'Origine Probabile delle Stelle Meteoriche, Lettere di G. V. Schiaparelli al P. A. Secchi, Estratte dal Bullettino Meteorologico dell'Osservatorio del Collegio Romano, Vol. V, N.[i] 8, 10, 11, 12" (About the Path and Probable Origin of Meteors . . .); 1866 and Vol. VI, N.[i] 2, 1867.

30. J. [Giovanni] V. Schiaparelli: "Sur la relation qui existe entre les comètes et les étoiles filantes" (On the Relation that Exists Between Comets and Shooting Stars), *Astronomische Nachrichten*, number 1629 (a combined issue with 1628), February 20, 1867; columns 331–332, article dated February 2, 1867.

31. A[lexander] S. Herschel: "Radiant Point of the November Meteors, 1866," *Monthly Notices of the Royal Astronomical Society*, volume 27, number 2, December 14, 1866, pages 17–19. Immediately following Alexander Herschel's article was one by his father, John Herschel, on the same subject with similar results (pages 19–21). John's article also examined the more general problem of how a radiant is determined in a meteor storm.

32. Th[eodor] Oppolzer: "Schreiben des Herrn Dr. *Th. Oppolzer* an den Herausgeber" (Letter from Dr. Th. Oppolzer to the Editor), *Astronomische Nachrichten*, number 1629 (combined issue with 1628), February 20, 1867, columns 333–334, article dated February 6, 1867.

33. Schiaparelli did not fault Peters for not acknowledging his work. In an article for *Comptes Rendus* dated March 12, 1867, Schiaparelli wrote:

Peters did not know of the orbit that I had calculated and published three weeks before in my fourth letter [to Secchi]. Thus, although I owe him nothing, he owes nothing to me, and he has, without doubt, priority of publication, his remarks having been published in no. 1626 of *Astronomical News* [Astronomische Nachrichten], dated February 4.

Schiaparelli goes on to praise Ernst Chladni and Hubert Newton:

It would be unjust not to add that the relationship between comets and shooting stars had already been predicted by Chladni in his book *Fiery Meteors* in 1819

and that the necessity of great eccentricities in the orbits of shooting stars had already been recognized by Newton in the last Reports of the British Association [for the Advancement of Science] and in the Brussels [Observatory] Annual for 1866, page 201. W. [actually Giovanni V.] Schiaparelli: "Sur les étoiles filantes, et spécialement sur l'identification des orbites des essaims d'août et de novembre avec celles des comètes de 1862 et de 1866" (Extrait d'une Lettre à M. Delaunay), *Comptes rendus*, volume 64, 1867, pages 598–599, article dated March 12, 1867.

34. 33.25 ÷ 360 ° = about 1/10.

35. Bodies in orbits that complete these multiples or fractions of a revolution per year will return every 33.25 years: 2 ± 1/33.25, 1 ± 1/33.25, 1/33.25. The Leonid meteors could not have shorter periods than the shortest of these because those orbits would not reach out as far from the Sun as the Earth.

36. Adams and Newton both allowed for the shift from the Julian to the Gregorian calendar in 1582, which dropped 10 days out of that year to bring Easter back to the beginning of spring. The historical record of Leonid meteors shows the calendar correction between October 27, 1602 and November 8, 1698, by which time the correction factor had expanded to 11 days.

37. J[ohn] C[ouch] Adams: "On the Orbit of the November Meteors," *Monthly Notices of the Royal Astronomical Society*, volume 27, April 1867, pages 247–252.

38. In his article on Adams for the *Dictionary of Scientific Biography*, Morton Grosser offers this assessment of Adams' achievement:

By dividing the orbit into small segments, he calculated an analysis of perturbations for the meteor group, resulting in improved values for its period and elements. This work provided another demonstration of Adams' extraordinary ability to manipulate equations of great length and complexity without error.

39. Weiss gleaned his list of meteor showers from Humboldt's *Cosmos*, Quetelet's "Catalog der Sternschnuppenerscheinugen" ("Nouveau catalogue des principales apparitions d'étoiles filantes," *Mémoires de l'Académie Royale des Sciences et Belles-Lettres de Bruxelles*, volume 15, 1841, pages 3–60), and Heis' "Abhandlung über Sternschnuppen."

40. Edmund Weiss: "Bemerkungen über den Zusammenhang zwischen Cometen und Sternschnuppen" (Remarks about the Connection Between Comets and Falling Stars), *Astronomische Nachrichten*, number 1632, March 9, 1867; columns 381–384, article dated February 22.

41. Donald K. Yeomans: *Comets: A Chronological History of Observation, Science, Myth, and Folklore* (New York: John Wiley & Sons, 1991), page 183. Herrick and Bradley

were the first to observe that Biela's Comet had split, but Matthew Fontaine Maury of the United States Naval Observatory in Washington, D.C. is often credited with the discovery because he published first, although he didn't see that the comet was in two pieces until 15 days later, January 13, 1846.

42. [Heinrich] d'Arrest: "Ueber einige merkwürdige Meteorfälle beim Durchgange der Erde durch die Bahn des *Biela*'schen Cometen" (Concerning a Remarkable Meteor Shower During Which the Earth Passed Through the Orbit of Biela's Comet), *Astronomische Nachrichten*, number 1633, March 19, 1867; columns 7–10, article dated February 25.

43. Edmund Weiss: "Beiträge zur Kenntniss der Sternschnuppen" (Contributions to the Understanding of Meteors), *Astronomische Nachrichten*, number 1710–1711, 1868, columns 81–102, article dated July 16, 1868. Weiss also predicted an Andromedid meteor storm in 1879. It did not occur.

44. J[ohann] G. Galle: "Ueber den muthmasslichen Zusammenhang der periodischen Sternschnuppen des 20. April mit dem erstan Cometen des Jahres 1861" (Concerning the Supposed Connection of the Periodic Falling Stars of April 20 with the First Comet of the Year 1861), *Astronomische Nachrichten*, number 1635, April 2, 1867, columns 33–36, article dated March 11.

45. Contemporary reviews of this burst of discovery appear in James Glaisher, Robert P. Greg, E. W. Brayley, Alexander S. Herschel, and Charles Brooke: "Report on Observations of Luminous Meteors, 1867–1868," pages 344–428 (especially pages 393–422) in *Report of the 1868 Meeting of the British Association for the Advancement of Science*; and H[ubert] A[nson] Newton: "On Certain Recent Contributions to Astro-Meteorology," *American Journal of Science and Arts*, volume 43, May 1867, pages 285–300.

46. G[iovanni] V. Schiaparelli: *Entwurf einer astronomischen Theorie der Sternschnuppen* (Outline of an Astronomical Theory of Falling Stars), translated and edited by Georg von Boguslawski (Stettin: Nahmer, 1871), page 55. The work was originally published as *Note e riflessioni intorno alla teoria astronomica delle stelle cadenti* (Firenze: Stamperia reale, 1867). This translation is from page 108 of J. Willard Gibbs: "Memoir of Hubert Anson Newton, 1830–1896," *Biographical Memoirs of the National Academy of Sciences*, volume 4, 1902, pages 99–124.

CHAPTER 9

Meters for meteors

We may measure the place of a comet or asteroid at our leisure, but a Leonid meteor, like the Irishman's pig, will not stand still to be counted.

G. D. SWEZEY (1901)[1]

The accuracy of results obtained from the photographs is unspeakably greater than those gotten by any other known method.   HUBERT A. NEWTON (1893)[2]

ERNST ÖPIK complained of a problem that had plagued meteor science for more than a century, from its inception through the discovery of the connection between meteors and comets to 1940, when he wrote:

> There can hardly be found another branch of astronomy where the mere accumulation of observational data has so small an influence upon the progress of our knowledge as it has in meteor research.

The chief difficulty, Öpik said, was that meteor observations rely on the observer's memory, which is apt to distort what was seen. Because meteors have only a momentary existence, it is impossible for observers to check their observations. For meteor science to progress, Öpik said, "duplicate simultaneous observations by two or more persons are a prerequisite . . ." Even so, visual observations could not help but include wide margins of error.

Meteor science clearly needed some way to catch and hold its elusive quarry for precise measurements.

In the last third of the 19th century, stimulated to a significant degree by the Leonid meteors, two of those essential tools made their appearance in meteor research: spectroscopy and photography. Öpik had invented the rocking mirror, the most intricate device for the visual determination of meteor velocities. But he recognized that it was not good enough. "The importance of accurate photographic observations cannot be overestimated," he said. And for "the study of meteor spectra, the photographic plate yields results which are unrivalled."[3]

The spectrum of meteors

The Leonid meteor storm of 1866 was the proving ground for the first of a wave of new techniques applied to meteors that would transform the field. The first of the new techniques was spectroscopy.

In 1814, Joseph Fraunhofer in Germany used a prism to diffract the Sun's light into a rainbow. He noticed hundreds of dark lines at specific places among the colors. Those lines, corresponding to specific wavelengths, allowed scientists to determine many features of the object that was glowing, even if it was completely out of reach. This light analysis was called spectroscopy. It was just what astronomers needed. The stars were hopelessly remote. Light was the only clue a star offered about itself. Without spectroscopy, astronomers could determine little more about objects in space than their position, brightness, transverse motion, and perhaps their distance (if close). Spectroscopy turned a heavenly body's light into a chemical and physical laboratory.

John Herschel. Courtesy Special Collections, San Diego State University

Spectroscopy used a prism (or, later, a diffraction grating) to intercept the light coming from an object. The prism spread the incoming light into a pattern of colors – either a rainbow with distinctively placed dark lines running across it or a succession of precisely positioned colored lines or bands. The spectral lines corresponded to wavelengths that identified the nature of the object that emitted light (and the material between the emitter and the observer on Earth). Spectroscopy quickly proved its value to astronomy by revealing the composition, temperature, motion toward or away from the observer, rotation, magnetic fields, and other characteristics of the planets, Sun, and other stars.

Could spectroscopy be applied to meteors? The first person to see a meteor spectrum was the famed English astronomer John Herschel, son of even more famous astronomer William Herschel. He saw the spectrum by accident in 1864 when he was using a spectroscope to examine the star Capella and a meteor zipped by.

The prediction that the Leonids would return in 1866 had the scientific community ready. Astronomer Alexander Herschel, one of John's sons, thought it might be possible to observe the spectra of meteors on purpose. He conceived of a hand-held spectroscope suitable for meteor observation and teamed up with John Browning, a renowned instrument maker, to build it. Their creation was about the size of a cigar box and was used like binoculars. Light entered through two openings in front, passed through a prism at each opening, and emerged from the prism and the box into the observer's eyes as a spectrum.[4] Instead of an image of the stars as dots or an occasional meteor as a streak, the observer saw a spectrum for each of the objects in the field of view.

Photography was also a new science of the time, but cameras and emulsions were not yet fast enough to record the blink of a meteor. For the time being, meteor spectroscopy required extraordinary patience.

Herschel and Browning had to wait for a meteor to pass across their fixed field of view – or spot a meteor as it flew by and, in that second or less, aim the spectroscope at the meteor. Then, in that fleeting second, they would try to note as many as possible of the features in the meteor's spec-

Alexander S. Herschel.
Courtesy Mary Lea Shane
Archives, Lick Observatory

trum. They decided to fix their spectroscopes in position and wait for a bright meteor to pass through their field of view.

Herschel and Browning warmed up by watching the Perseid meteors in August 1866 and saw spectra for 17 of them.[5] Browning did even better in November when the Leonids came through with a meteor storm over Europe.[6] They reported that meteor spectra varied, but that yellow tended to predominate. Herschel and Browning both noted that in a chemical laboratory a yellow spectrum would indicate sodium. Browning shied away from claiming to identify the chemical composition of meteoroids, but Herschel was confident that sodium was present. Sodium was also found in meteorites, and Herschel praised Ernst Chladni for his insight in 1794 that meteoroids and meteorites differed little except in size. Herschel and Browning continued their visual observations of meteor spectra for the next 20 years.

In 1896, the Harvard College Observatory began to compile a photographic full-sky catalog of stellar spectra. They used a prism placed over the large (objective) lens at the skyward end of the telescope so that every star was recorded not as a dot but as a small spectrum. Over the years, this project, the *Henry Draper Catalogue*, allowed the stars to be classified by similarities in their spectra. Accidentally, an occasional meteor burned its way into the emulsion and into the scientific archives. Thus the first photograph of a meteor spectrum was made in 1897.

If photographing a meteor was hard because of its brief existence, photographing a meteor's spectrum was much harder, because the momentary flash of light was spread by the prism over a wide area. By 1932, only eight photographs of meteor spectra existed. In that year, Peter Millman, a Canadian astronomer, began the first comprehensive program of meteor spectroscopy, taking and collecting spectrograms, greatly multiplying the supply. The elements and their proportions in meteoroids were very similar to those in meteorites, confirming the meteor spectroscopy that Alexander Herschel had begun, verifying what Chladni had suspected – and going far beyond.

Modern studies of meteors show their spectra to be composed of emission lines, indicating that the light comes from hotly glowing gases as the solid meteoroid vaporizes. The emission lines identify those gases as iron, sodium, magnesium, silicon, and other elements expected to form the solid particles in comets.[7]

A blind alley

The Leonids in 1866 were a proving ground for spectroscopy, but they were also a source of inspiration. In 1866, an 18-year-old English boy in Bristol was enchanted by the Leonid meteor storm and devoted most of the rest of his 83 years to the study of meteors. William Frederick Denning did not marry, did not attend college, did not have a professorship – or any regular job – to support himself.[8] He was truly an amateur astronomer – one who did it for love of the subject.

William Frederick Denning

Denning published relentlessly in scientific journals, averaging 23 articles and notes a year throughout the 1880s. He worked hard to bring rigor and reliability to the observation of meteors in the days before photography became a dependable tool of astronomers.

Beginning with the Leonid storm of 1833, observers had noticed that the radiant of a meteor shower rises and sets with the stars, moving westward due to the rotation of the Earth. In 1877, Denning demonstrated through the precision of his observations that the radiant of the Perseid meteors was not fixed among the stars but was slowly edging eastward because of changing perspective as the Earth revolves around the Sun.[9] Orbiting meteoroids move in parallel paths, sometimes (as with the Perseids) in a broad stream. As the Earth passes through the stream, it is constantly altering its direction of motion because it is circling the Sun. The Earth's change of direction makes the meteors in the stream appear to come from a radiant slightly east of where it was the previous night.

Denning realized that the Perseid radiant drifted during the course of the Earth's passage through the stream. Yet Denning's observational passion for meteors led him one year later, 1878, into one of the two great missteps in the history of meteor science – and into the leadership of that position, which became a stumbling block for the field for more than a third of a century.

He proposed that all but a few meteor shower radiants are fixed, that they do not move with respect to the background stars. What he was saying was not contrary to his discovery about the Perseids. He was content that the Perseids and perhaps a few other once-a-year meteor showers did have radiants that crept eastward while the shower was in progress and that these showers were created by particles traveling near the orbits of comets from which they were dislodged. What Denning claimed was that there was another, far larger category of meteors whose radiants remained stationary among the stars throughout the night and throughout the year. The meteors from these stationary radiants looked at first like sporadic meteors – from no radiant in particular. They were comparatively few in number compared to shower meteors at their peaks. But Denning insisted that his careful observations demonstrated that every sporadic meteor had a precise active radiant that produced a handful of meteors each night – at least 50 different radiants being active in the course of one evening. And each of these radiants was fixed, unmoving, among the background stars.

Denning carefully recorded what he saw and, on the eve of the expected return of the Leonids in 1899, produced a catalog of 4,367 radiants.[10]

Denning's claim of a huge class of meteors with stationary radiants was an arrow in the heart of existing meteor theory. The basis of meteor science was that comets shed debris which continues to orbit the Sun. When the Earth speeds through these orbiting swarms of particles, they burn up in the atmosphere. If a meteor shower caused by orbiting particles lasted several days, it was possible, as Denning himself demonstrated with the Perseids, to see the radiant of the shower slide eastward among the stars. However, if the radiant did not move, then these meteors could not be caused by comets in orbit around the Sun, or by any planet, moon, asteroid, or unknown body in orbit around the Sun. If the radiant did not move, it was as if these particles were coming from the distant stars.

Initially, most astronomers were sure that Denning's observations or analysis were somehow in error. A few astronomers offered theories that could support Denning's observations. One proposed that fast-moving, dust-shedding stars were passing close to the solar system so that their particles, raining down from a vast distance, spread across the entire orbit of Earth and would always seem to come from the same spot in the sky. Denning himself rejected that idea. He saw no nearby, dust-spewing, speedy stars. Another proponent suggested the existence of a dust ring around the Earth from which meteoroids steadily fell, but Denning rejected that idea too. Another suggested a combination of these ideas. Again Denning said no. He could not explain why all but a few of his meteor radiants should be fixed, but he trusted his observations and hoped one day theory would catch up. If, he said, fixed radiants are

> a false effect, further observations are capable of showing it in that character, but if it is a real feature, then the most elaborate arguments or refined mathematical objections cannot obliterate it from the sky.[11]

Several astronomers objected to the concept of fixed radiants because it violated the well-established theory that meteoroids are spawned by comets and because the idea of fixed radiants had no theoretical basis of its own – no mechanism to explain it. In response, Denning simply offered his observations. One of the few meteor specialists to challenge Denning's observational methods – actually his way of analyzing those observations – was George Tupman, author of an earlier catalog of meteor radiants with many fewer entries.[12] Tupman particularly objected to Denning assigning a radiant to meteors that were seen many nights apart. Tupman argued that radiants were valid only if established by one night's watch. Beyond one night, one could not be sure that the Earth had not moved on into another meteoroid stream. Denning replied that most of his radiants produced only one meteor every three or four hours, so he needed to rely on observations spread over many nights to "bring out" the radiant.

As the 19th century drew to a close, Denning, the amateur astronomer, was acknowledged by most professional astronomers as the leader in meteor observation. Most of them now thought that stationary radiants for meteors were a proven fact. In 1898, the Royal Astronomical Society awarded Denning its Gold Medal. That same year, H. G. Wells used Denning's stature to give authenticity to his *War of the Worlds*:

> Then came the night of the first falling star. It was seen early in the morning rushing over Winchester eastward, a line of flame, high in the atmosphere . . . Denning, our greatest authority on meteorites, stated that the height of its first appearance was about ninety to one hundred miles . . . [13]

Denning's reputation was further enhanced as the 20th century began. Two respected astronomers independently examined Denning's observational methods and praised them as capable of establishing radiants with accuracy. But the need for that examination revealed an increasing uneasiness the astronomical community was feeling about Denning's work. His claim of fixed radiants defied every other piece of evidence available. Something was wrong, but no one could put his finger on it.

The first real challenge to Denning's observational methods came from American astronomer Charles P. Olivier in 1912. Like Denning, Olivier had made thousands of meteor observations. Olivier's key point was that a comet could be responsible for two or more nearby streams of particles, so that one comet's meteor shower could have two or more radiants close to one another. The argument focused on the Orionid meteors of October. Today we know that the Orionids are created by Halley's Comet (which also provides the Eta Aquarid meteors in May – but that is a separate matter). The gravity of the large planets continually perturbs Comet Halley, changing its orbit slightly, so that each new dust outpouring from the comet travels a slightly different orbit from previously shed debris. These separate dust outpourings can create variations in the meteor numbers as the Earth crosses through successive swarms, which will seem to come from slightly different radiants of small areas. Each of these radiants would move slightly eastward each day. An observer like Denning who focused on a fixed radiant in the sky would see activity there on consecutive nights, but from different meteoroid streams from the same comet. There was no need here – or elsewhere – to conceive of fixed radiants that each, according to Denning, encompassed a region of 5 to 7 degrees.

Thereafter the battle went badly for Denning. Gradually his support withered away as his staunchest supporters died and a new generation of astronomers ignored him, basing their work on the straightforward theory that meteors have their origin in comets.

Denning's health failed him in 1906, just after the British government awarded him a modest pension in appreciation of his 40 years of fulltime volunteer service to astronomy. The passion for meteors never left him. Two days before his 66th birthday, he wrote a poem, "Falling Stars." Here are the opening and closing lines:

> Bright falling stars I greet you with a smile,
>   While you beguile,
> My loneliness, with pleasure pure and sweet
>   In moment's fleet. . . .
> Whene'er you come you bring a joyous thrill
>   My soul to fill.
> Oh messengers from distant worlds! I yearn
>   Your tale to learn,
> And I await, amid earth's frosted dews,
>   Celestial news.[14]

Observing meteors was perhaps Denning's one remaining pleasure. From 1906 on, with a small pension, in poor health, he lived his remaining 25 years fighting a rearguard action for his stationary radiants, sinking ever deeper into poverty, and cringing from the taunts of neighborhood boys. He died in 1931.

A schism between theory and observation like Denning and his generation of meteor astronomers experienced is always easier to analyze in retrospect. In his enthusiasm, Denning had tried to coax every sporadic meteor into a specific, if weak, meteor shower. For Denning, calling a meteor "sporadic" was just an admission that one was ignorant of that meteor's actual radiant. Olivier, Denning's most avid opponent, returned to the original perspective on sporadic meteors:

> . . . there is not the least reason to try to force every meteor, or perhaps
> even the majority seen on an average night, into radiants. Not all meteors
> are parts of existing streams. There must be vast numbers following
> entirely unique paths in space.[15]

By collecting meteors over a period of many nights and looking for a common radiant, Denning could not help but find many intersections that were only accidents of randomness, not true radiants.[16]

Eventually astronomical photography provided observational evidence to settle the dispute. Especially influential was the work of Ernst Öpik in Estonia who in 1921 began publishing radiant positions measured from photographs of meteors.[17] As the years passed, Denning's radiants were found to be void of true activity or, if real, then in motion. There were no stationary radiants.

Manning Prentice, a lawyer who was director of the Meteor Section of the British Astronomical Association, had championed Denning in 1922.

In 1948 he looked back with embarrassment on meteor science's fixed radiant blunder as "a source of so much wasted effort and unsound method."[18]

In the fallout from the controversy, meteor astronomers adopted some guidelines to avoid such confusion in the future. No meteor radiant could exist unless at least four meteors had paths that could be projected backwards to intersect within a circle whose diameter was no more than 4 degrees. Those meteors must all have been seen by one observer in the course of one night's observation, not to exceed 8 hours.

Meteor photography

The general public paid no attention to the confusion of meteor astronomers about whether meteor shower radiants were stationary or in motion against the background of stars. But they remembered what they saw of shooting stars in 1866, 1867, or 1868 and what they heard about 1833.

Helping to keep the memory vivid between the late 1860s and the expected return of the Leonids at the end of the century were the Andromedid meteors, which produced two meteor storms in this period only 13 years apart. The Andromedids traced their particles to Comet Biela, which was seen to break up as it passed the Sun and Earth in 1845. Both storms were as good or better than the Leonids as seen from Europe in 1866. Both occurred on November 27, one in 1872 and the other in 1885. Because these meteors traveled along and near the orbit of Comet Biela, they were frequently called the Bielids, although efforts started by Schiaparelli to standardize meteor nomenclature urged that meteor showers be named after their radiant. These meteors seemed to radiate from the constellation Andromeda; hence the Andromedids. That was tidy, orderly, useful. Astronomers of course went on calling them the Bielids or the Andromedids according to their whims.

On November 27, 1885, the Andromedid (Bielid) meteors were expected to return, as they had in great numbers in 1872. Ladislaus Weinek had a new idea for welcoming them. He set up two cameras, one in Prague and one 75 miles (120 kilometers) away in Dresden. He thought it might be possible to photograph meteors. He wanted to capture the same meteor with both cameras so that the altitude of the meteor could be calculated with precision.

Meteors are hard to photograph. A few may appear as bright or brighter than the brightest nighttime stars, but they last for only a second or less, making it hard for a camera, particularly with the slow emulsions of the 19th century, to record anything at all. The Andromedids provided a splendid show. The Dresden camera recorded nothing. Weinek's bold experi-

A segment of a fireball trail photographed by John E. Lewis on January 13, 1893. Courtesy of Dorrit Hoffleit, Yale University

ment had failed. But . . . not quite. His Prague camera captured one streak. He was the first person to photograph a meteor.

It was easier to photograph stars and planets. They might be faint but they shined from the time they rose until they set, allowing a time exposure to soak up the photons. Without intending to, early astronomical photographs recorded a number of meteors.

At Yale University, Lewis Elkin intentionally embarked on a program to photograph meteors. Elkin was Hubert Newton's successor as director of the Yale Observatory. During the Perseid meteor shower of 1893, Elkin pointed his camera at the radiant in Perseus and over four hours caught three trails, two of them Perseids. John E. Lewis, an amateur astronomer 10 miles (16 kilometers) away, also captured the brighter of the two Perseids on film.[19] The two photographs allowed Elkin to calculate that the meteor was visible to the cameras from a height of 68.0 down to 51.65 miles (109 to 83 kilometers) along a path of 29.3 miles (47 kilometers).

During the Geminid meteor shower in December that same year, Elkin photographed three Geminids. Tracing their paths backward revealed the Geminid radiant, only 1 minute of arc in diameter. He was the first to determine an exact meteor shower radiant by photography. Locating a meteor shower's radiant is a vital step in calculating the orbit of a meteor stream and identifying its parent comet. Hubert Newton was proud of his colleague: "The accuracy of results obtained from the photographs is unspeakably greater than those gotten by any other known method."[20]

The Harvard College Observatory made extensive observations of the Leonids from 1897 to 1904. On November 14, 1898, they obtained 34 photographs of 11 different meteors.[21]

W. Lewis Elkin. Courtesy of Dorrit Hoffleit, Yale University

Joseph Sykora in Russia followed Elkin's lead. He used five Perseid trails he photographed in 1901 to determine the Perseid radiant. In 1910, with cameras at three stations, he measured the descent of a bright Perseid from 70 miles to burnout at 50 miles (112 to 80 kilometers). By 1924, Sykora had photographic images of more than 30 meteor trails.

Having demonstrated that meteors could be photographed reliably, Elkin launched his great quest. He picked two separate sites and set up four to six cameras at each. He already knew that he could use photography to calculate meteor heights and radiants. But Elkin had a more challenging measurement in mind. He positioned a bicycle wheel in front of each camera array. Between half the spokes he placed opaque screens so that as the wheel turned, the light from a meteor headed for a camera would be alternatively cut off and allowed to pass in rapid succession. He used a motor to spin the wheel so that the light would be cut off 6 to 10 times a second. The result was that a photograph would show a meteor streak chopped up into a dashed line. Elkin knew how fast the wheel was turning, so he knew how much time elapsed between gaps in the meteor streak. In this way he could calculate how many degrees of the sky the meteoroid traversed per second – the meteor's angular velocity. If – if – a camera at the other station had also photographed that same meteor, Elkin could calculate the meteor's height and that would allow him to convert the meteor's angular velocity into linear velocity – miles per second.

In 1899, Elkin was ready, and he caught one Andromedid (Bielid) meteor with two cameras separated by 2 miles (3.2 kilometers). He could calculate that the meteoroid's speed was about the same as Comet Biela, its parent comet, now broken up and vanished, would have had at this point in its orbit. But he could not calculate a more precise speed. His baseline was too short for pinpointing the meteor's altitude.

In 1901, Elkin lengthened the distance between camera stations to 5 miles (8 kilometers). He photographed meteors and analyzed data, but

Model of W. Lewis Elkin's meteorograph. Courtesy of Dorrit Hoffleit, Yale University

never published them. His baseline was still too short.[22] In 1910, he resigned from Yale, perhaps out of frustration, and devoted his remaining 23 years mostly to music.

Elkin had failed at his grand task but he demonstrated to astronomers that meteor photography could be used to determine meteor velocities.[23]

Beginning in 1936, Fred Whipple at Harvard University used Elkin's basic strategy in a new effort to determine meteor speeds. Synchronous motors were now available for precise light chopping. The shutters interrupted the light 20 times a second. Most importantly, Whipple positioned his cameras 23 miles (37 kilometers) apart.

Meteor velocity can now be calculated to an accuracy of one-tenth of one percent.[24]

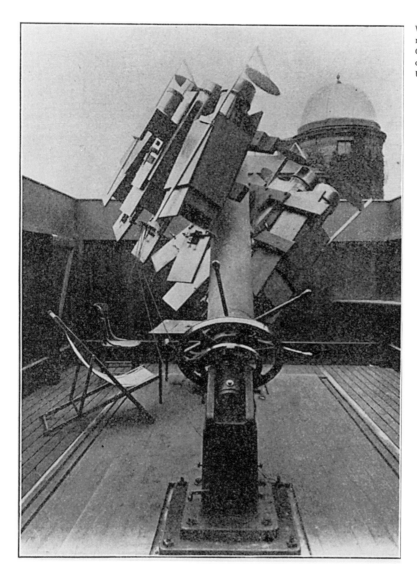

W. Lewis Elkin's
meteorograph at Yale
Observatory. Courtesy
of Dorrit Hoffleit, Yale
University

The fall of the Lost City Meteorite, January 3, 1970, captured by an automatic camera of the Smithsonian Astrophysical Observatory. This photograph and others from the Smithsonian's camera network in the midwestern United States enabled scientists to calculate that a meteorite probably fell near Lost City, Oklahoma. Searchers found a 26-pound (12-kilogram) meteorite fragment ten days later. Notice that this camera was equipped with a rotating shutter that created chop marks in the fireball's trail so that the meteorite's speed could be gauged Smithsonian Astrophysical Observatory

NOTES

1. G. D. Swezey: "The Present Status of Meteoric Astronomy," *Proceedings of the Nebraska Academy of Sciences*, volume 7, November 1901, pages 79–89.

2. H. A. N. [Hubert A. Newton]: "Photographs of August and December Meteors," *American Journal of Science*, volume 47, 1894, pages 154–155.

3. E[rnst Julius] Öpik: "Meteors," *Monthly Notices of the Royal Astronomical Society*, volume 100, February 1940, pages 315–326.

4. Spectroscopes usually require light to enter the instrument through a slit so that the spectral lines are distinct. Alexander Herschel gauged that a meteor streak was already a line and did not require a slit, especially because a meteor is so brief and faint that a narrowed opening would not admit enough light to form a visible spectrum.

5. A[lexander] S. Herschel: "Prismatic Spectra of the August Meteors," *Intellectual Observer*, volume 10, October 1866, pages 161–170. For a diagram of Herschel's meteor spectroscope, see [no author given]: "The

Coming Meteor Shower. – The Spectra of Meteors," *Intellectual Observer*, volume 10, August 1866, pages 38–40.

6. John Browning: "On the Spectra of the Meteors of Nov. 13–14, 1866," *Monthly Notices of the Royal Astronomical Society*, volume 26, December 1866, pages 77–79.

7. This section on meteor spectroscopy was based primarily on Martin Beech: "A Simple Meteor Spectroscope," *Sky & Telescope*, volume 80, November 1990, pages 554–556; D[onald] W. R. McKinley: *Meteor Science and Engineering* (New York: McGraw-Hill, 1961); and David W. Hughes: "The History of Meteors and Meteor Showers," *Vistas in Astronomy*, volume 26, 1982, pages 325–345.

8. For a fuller presentation of Denning's work and life, see Martin Beech: "The Stationary Radiant Debate Revisited," *Quarterly Journal of the Royal Astronomical Society*, volume 32, 1991, pages 245–264, and Martin Beech: "William Frederick Denning: In Quest of Meteors," *Journal of the Royal Astronomical Society of*

Canada, volume 84, number 6, 1990, pages 383–396. Much of the information in this section on Denning was based on these articles.

9. William F. Denning: "The Radiant Centre of the Perseids," *Nature*, volume 16, August 30, 1877; page 362.

Alexander C. Twining thought he had discovered the motion of the Perseid meteor radiant 15 years earlier when he observed the phenomenon in separate years. However, his measurement of an eastward shift of the radiant by about 12 degrees in little more than 24 hours is too great. The actual shift is about 1 degree a day. Hubert A. Newton doubted that Twining was correct. H. A. Newton: "Summary of observations of shooting stars during the August period, 1863," *American Journal of Science and Arts*, volume 36, September 1863, pages 302–307.

10. W[illiam] F. Denning: "General Catalogue of the Radiant Points of Meteoric Showers and of Fireballs and Shooting Stars observed at more than one Station," *Memoirs of the Royal Astronomical Society*, volume 53, 1899, pages 203–292 and plate 5.

11. William F. Denning: "The Stationary Meteor Showers," *Sidereal Messenger*, volume 5, 1886, pages 167–174.

12. George L. Tupman: "Results of Observations of Shooting Stars, Made in the Mediterranean in the Years 1869, 1870, and 1871," *Monthly Notices of the Royal Astronomical Society*, volume 33, 1873, pages 298–312.

13. H. G. Wells: *War of the Worlds*, 1898, first paragraph of chapter 2.

14. William F. Denning: "Falling Stars," poem (dated November 27, 1914) at the end of his "The Claims of Meteoric Astronomy," *Journal of the Royal Astronomical Society of Canada*, volume 9, 1915, page 60 of pages 57–60.

15. Charles P. Olivier: *Meteors* (Baltimore: Williams & Wilkins, 1925), page 122.

16. Martin Beech has demonstrated very cogently how Denning's methods led to this error in "The Stationary Radiant Debate Revisited," *Quarterly Journal of the Royal Astronomical Society*, volume 32, 1991, pages 245–264.

Denning recognized this problem to a minor degree, but thought he and other experienced observers could recognize characteristics of meteor families. "The photography of meteors, in an ample and effective manner, has failed . . . ," he said. W[illiam] F. Denning: "The Claims of Meteoric Astronomy," *Journal of the Royal Astronomical Society of Canada*, volume 9, 1915, pages 57–60.

17. E[rnst J.] Öpik: "Teleskopische Beobachtungen der Perseiden," *Astronomische Nachrichten*, volume 217, 1922, columns 41–46.

18. J. P. M[anning] Prentice: "Meteor Section" ("Brief History"), *Memoirs of the British Astronomical Association*, volume 36, 1948, pages 104–110.

19. Dorrit Hoffleit points out that John Lewis photographed his first meteor by accident earlier that year when he was trying to photograph Comet Holmes. He brought the picture to Hubert Newton, who analyzed it and encouraged Elkin's two-station meteor photography effort with Lewis. H. A. Newton: "Fireball of January 13th, 1893," *American Journal of Science*, volume 46, September 1893, pages 161–172.

20. H. A. N. [Hubert A. Newton]: "Photographs of August and December Meteors," *American Journal of Science*, volume 47, 1894, pages 154–155.

21. E[dward] C[harles] Pickering: "The November Meteors of 1899," *Scientific American*, volume 81, October 28, 1899, page 279; and Charles P. Olivier: *Meteors* (Baltimore: Williams & Wilkins, 1925), page 15.

22. Dorrit Hoffleit points out that Elkin had made his reputation with his parallax measurements, determining the distances to nearby stars by measuring extremely small angles. David Gill and W. L. Elkin's parallax for Alpha Centauri was 0.75 *seconds* of arc. A meteor 50 miles high seen from stations 2 miles apart would be displaced 2 *degrees*. She suggests that Elkin was confident that he could work with a shorter, more convenient baseline. "Unfortunately in the case of a meteor varying smoothly in brightness along its path," she writes, "it is not always possible to match corresponding points on the two photographs. Hence, even increasing the base line to five miles did not yield the desired accuracy." Dorrit Hoffleit: "Yale Contributions to Meteoric Astronomy," *Vistas in Astronomy*, volume 32, 1988, pages 117–143.

23. This section on meteor photography was based primarily on Dorrit Hoffleit: "Yale Contributions to Meteoric Astronomy," *Vistas in Astronomy*, volume 32, 1988, pages 117–143; and *Astronomy at Yale: 1701–1968* (New Haven: Connecticut Academy of Arts and Sciences, 1992).

24. David W. Hughes: "The History of Meteors and Meteor Showers," *Vistas in Astronomy*, volume 26, 1982, pages 325–345.

CHAPTER 10

The coming fire shower – 1899

The expected return of the main body of the November meteors dwarfs every other astronomical event this fall.   GARRETT P. SERVISS (1899)[1]

Our world is assured of the most magnificent and imposing exhibit of celestial fireworks...

CHICAGO TRIBUNE, *ATLANTA CONSTITUTION*, and other newspapers,
November 12, 1899

INSPIRED BY the American centennial celebration in 1876, Richard Devens gathered together what he considered to be the 100 greatest and most memorable events in the first hundred years of the United States. The thick volume had a 57-word title that began *Our First Century*. Devens wrote chapters about each of the 100 historic events, including the signing of the Declaration of Independence, the ratification of the Constitution, the freeing of the slaves, the completion of the transcontinental railroad, various battles . . . and the "Sublime Meteoric Shower All Over the United States – 1833." It was, Devens said, "The most grand and brilliant celestial phenomenon ever beheld and recorded by man."[2]

Rising expectations

On November 14, 1898, places in the United States received a decent Leonid shower, as good or a little better than the annual Perseid meteors of August. In his historical account of this period, meteor astronomer Charles Olivier says that this 1898 display was "enough of an increase over 1897 to raise the hopes of everyone."[3]

Finally it was November 1899. The astronomical community was full of good will toward the citizenry and wanted them to enjoy the rare phenomenon with them. As astronomical discoveries relied more and more on larger telescopes, sophisticated photography and spectroscopy, and mathematics, it was harder and harder for the public to participate in astronomical discoveries, or even many astronomical events. The "canals" on Mars fascinated the public, but these purported markings could only be glimpsed through the best telescopes under ideal conditions. To view the return of the Leonids required no telescopes or equipment at all. With a little training, amateurs could even help with the meteor counts and recording their paths.

Newspapers all over the United States began receiving nationally distributed essays on the Leonids from astronomers trying to be educational and helpful. A quick look by newspaper editors at their back issues from 33 and 66 years ago told of a stirring sight and a stirred-up public.

A contribution from Columbia University professor John K. Rees appeared in many papers around the country, explaining what meteors are, their origin in comets, how they orbit, even why meteors are more numerous after midnight – because we are then on the front side of the Earth as it travels around the Sun, plowing through meteoroid streams. Rees tried to make the concept understandable by calling on knowledge shared by everyone:

> Riding on the cowcatcher of a locomotive engine in the midst of a down-pour of rain, we are certain to get wetter than if we sat on the platform of the rear car.[4]

Some newspapers got out their purple prose pens in their attempts to prepare the populace. A reporter for the *San Francisco Chronicle* wrote:

> An army of meteors is advancing toward the earth at the rate of twenty-six miles a second, while the earth is moving toward these visitors at a speed of eighteen miles a second. Already the advance guard of the meteoric army has been seen and in the early morning of to-morrow, sometime between midnight and dawn, the main body of meteors will be sighted. Unless the full moon or clouds intervene San Francisco will have the opportunity of witnessing the famous November shower of Leonids which occurs every thirty-three years. It is expected that the display on this occasion will be one of the grandest in the history of meteoric showers . . .[5]

Many papers recounted an Arabian myth about meteors. Evil genies flew up to the barriers of heaven each night to eavesdrop on the angels. The angels were on guard, however, and their bowmen shot fire-tipped arrows at their foes across the nighttime sky. Their aim was so perfect that every shooting star meant one less evil spirit to vex the world.[6]

Scientific, historical, and mythological background was one thing. But most newspapers went far beyond. On Saturday, November 11, 1899, the *Seattle Post-Intelligencer* announced in large type: "November's meteors will be the crowning glory of the heavens . . . one of the most magnificent sights of a lifetime." They urged readers to buy their Sunday edition, which would contain an entire page of information from eminent scientists.

The *Chicago Daily Tribune* and the *Atlanta Constitution* on Sunday, November 12, 1899 both carried an unsigned article with numerous errors that proclaimed: "our world is assured of the most magnificent and imposing exhibit of celestial fireworks presented since the great performance of November 14, 1866." It promised plenty of meteors: "Millions upon millions of them will be seen to dart from the quarter of Leo on Wednesday night if the sky is unclouded."

Northwestern University astronomy professor Elias Colbert was incensed by the hype and the errors in his Sunday *Chicago Daily Tribune*. He wrote an article that appeared the next day in the *Tribune*. He tried to lower expectations by explaining the difficulty in making accurate predictions about meteor showers; by pointing out that light from the nearly full moon would interfere with viewing; and by emphasizing that the Leonids posed no threat to Earth.

Voices such as Colbert's were seldom heard. Not many scientists were emphasizing uncertainties and those that did kept their volume low. Why not? It was pleasant to have the public excited about astronomy, particularly their field of meteor science. By all accounts, a meteor storm equaled or exceeded a total solar eclipse in grandeur. Besides, the Leonids had never failed; they had already returned as expected – once. Well, it was a little dicey in 1866 when the Western Hemisphere got only a trickle, but next day Americans heard from Europe that the storm had blanketed them, so the Leonids were off the hook, and in 1867 America got its share, and everyone was reasonably satisfied. Now 33 years had passed and it was Leonid time again.

Worries

In March of 1899 two English astronomers had sounded a cautionary note, but softly. As the Leonids approached, G. Johnstone Stoney and Arthur M. W. Downing went back to John Couch Adams' computation of the orbit of the Leonid meteors in 1867 and carried it forward to 1900. They focused on the position of the swarm that caused the meteor storm in 1866, followed it along its orbit past the orbits of three giant planets and back to where it would again be closest to Earth. In 1867, Adams had done all the calculations himself. Stoney and Downing used a crew of human computers from the Royal Observatory at Greenwich.

What they found was disturbing. Outbound in 1870, the swarm passed close enough to Saturn to be yanked a little off course by its gravity. Worse yet, in 1898, on its way back to the Sun and Earth, the swarm had passed even closer to Jupiter – about 84 million miles (135 million kilometers). The result was that the orbit of that storm-producing swarm was shifted so that the center of the swarm would no longer cross the orbit of the Earth. Instead, it would pass to the sunward side of Earth by more than a million miles (almost 2 million kilometers). Still, said Stoney and Downing, a shower could be expected if the meteor swarm was wide enough to stretch all the way to Earth. A shower could also occur if there were other unknown dense swarms traveling near the one that provided the memorable 1866 display. The perturbations that carried the 1866 storm-producing particles closer to the Sun might have placed one of the other swarms on a collision course with Earth. In either case, the Earth would be closest to the orbit of the known Leonids on November 15, 1899, so the shower should occur then, with the time of passage favoring a display for Europe and the eastern Americas.[7]

Just a few days before the Leonids were scheduled to arrive, Stoney and Downing looked over their calculations and conclusions one last time . . . and had an anxiety attack. In a short article published in *Nature*[8] on November 9 and in a speech before the Royal Astronomical Society[9] on November 10, Stoney and Downing refined their figures. They reemphasized the extent of the perturbations which the Leonid stream had suffered: "the stream through which the earth passed in 1866 will on its return pass at some distance from the earth." In November 1899, the orbit of the Leonids would be no closer to Earth than 0.0141 AU (1.3 million miles, 2.1 million kilometers). The abnormal perturbations experienced by the Leonids in the last 30 years meant, Stoney and Downing said, that the date they had given for the meteors' arrival was less certain; the shower should be watched for on both November 15 and 16. Nevertheless, Stoney and Downing were confident that, despite its displacement, the Leonid stream was wide enough to engulf our planet. They predicted that "the earth is likely to receive one of the great showers this year."

The reckoning

Reviewing the public fervor of 1899, Charles Olivier found that "the average person as fully expected to see a great meteoric shower as he did to see the sun rise next morning."[10] Astronomers and the general public tried their best. They were outside watching in cold (and most cloudy) weather in numbers never seen before for any astronomical event.

French astronomer Jules Janssen was in no mood to trust French weather in November. At the age of 75, Janssen used a balloon to ascend to 3,300 feet (1,000 meters) to have a better look at the Leonids. In several hours aloft he saw about 100 meteors.[11] Thus Janssen, a pioneer in astronomical (especially solar) spectroscopy, became a pioneer in high-altitude astronomy. As usual, he took the concept one step further. He called attention to the advantages of observing the Sun, planets, and stars from space.[12]

England was almost completely covered with clouds on the night of November 15–16. That didn't stop an English trio intent on seeing the Leonids. They also sailed skyward in a balloon. Aeronaut Percival Spencer lifted off from west of London with the Reverend J. M. Bacon and his daughter aboard. They brought with them a device "to catch some of the

Jules Janssen

fiery vapor" of the meteors. They saw only five shooting stars, but were confident they had collected the meteor gases. Experimental astronomy can be hazardous however. They found that the wind was blowing them toward the coast and they had to make a sudden descent. Bacon was badly shaken up and his daughter's arm was broken. There was no word about secrets gleaned from the meteor vapors they thought they had trapped.[13]

New York City entrepreneurs knew an opportunity when they saw it. They set up portable telescopes in the streets and charged people for an optically enhanced view of the – nonexistent – meteors.

And what would a Leonid storm season be without a story of the fear and bizarre behavior that the Leonids generated? "Leonids Caused a Panic in Russia" blared the headline in *Seattle Post-Intelligencer* on November 17, 1899. The story appeared in the *San Francisco Chronicle* and the *New York Times* that day as well.

> In Russia the Leonid displays caused a panic in many places. It was believed that the end of the world had come. Churches were opened all night and hundreds of thousands spent three nights in the open air, fearing earthquakes and a general cataclysm. There are even rumors that in some villages Russian parents murdered their children to save them from an expected worse fate. There was a brilliant display between 2 and 5 o'clock Thursday morning at Berlin.

No confirmation of this story has been found, and nowhere were the Leonids numerous.

When the Leonids failed to appear in the early morning of November 16, the *Atlanta Constitution* invited readers to telephone the newspaper and leave their names. If the meteor shower finally came before dawn the next morning, the newspaper staff would call everyone to alert them. The paper was overwhelmed. Hundreds of people registered. Alas, no calls went out.

Nothing doing in Denver

CONDENSED FROM THE *DENVER POST*, NOVEMBER 17, 1899

The meteors didn't meet. At least not enough of them to make a quorum. Saturn and Jupiter played policemen and kept all the disturbers of the peace away from Earth.*

The Denver University Observatory people were very much disappointed. The observatory looked like a great toadstool in the moonlight. Running in and out and around the toadstool were an excited lot of little black ants. The ants were people – most of them students of the university – come to count the meteors.

* Note that the journalist was aware, presumably from Professor Howe, that there had been a warning (from G. Johnstone Stoney and A. M. W. Downing) that the perturbations of Saturn and Jupiter had diverted the Leonids, casting some doubt on a meteor storm.

They were tremendously scientific – the students. They gathered in groups and talked about orbits and the Sickle and all kinds of astronomical things.

At the stroke of 1 on Thursday morning an agile and cheerful little professor called all the students together with a loud and urgent whistle. He divided them up into relays, and each relay had some particular part of the compass to watch. Each watcher counted the shooting stars in his little corner of the heavens. Every fifteen minutes the whistle blew and a new person came to relieve the watchers. It was all very methodical and precise. The students wandered out into the moonlight and sat down in the prairie and counted.

It was easy enough to count. There were so few meteors. One, two, three – the long, cold hours wore away. A rooster arose in some far-off barnyard and hailed the shivering wind that presaged the coming of the cold gray dawn.

The students were getting discouraged. Up all night and nothing to see – nothing but the splendid pageant of the night – the Moon and the stars, and the shadows on the great peaks creeping, creeping – a pageant that is worth more than all the human efforts at grandeur and dignified parade on Earth.

And it was cold. And the students were hungry. Dean Howe was not cold. He was not hungry, and he didn't know it was lonely and scary up there in the dome so near the frosty sky. The great telescope was busy. It was taking pictures.

Four o'clock, 5 o'clock – a faint, faint blush in the east – a smile of promise above the gray horizon: the Moon was gone now. "Oh-ho-ho-ho!" said a tall student with a jaw-dislocating yawn. "Uh. Hish. I wish I was in bed."

"Go-o-d morn-ing, How do you like the view?" called the rooster from somewhere across the fields. A dog barked and from some lonely corral a melancholy cow lifted up her voice and bewailed the Earth and the leanness thereof.

Half past 5 – the gray was turning to lavender – the pink in the east was scarlet – no, deep rose – no, crimson – and above it rolled a majesty of golden mist. The splendor of the night was gone. Clang, clang, clang, a student bell rang somewhere. "All aboard," shouted the tall student. The students piled upon the car, tired, cross, cold, and sleepy. "I'm going to sleep till doomsday," moaned a little blonde girl with a face like a kitten.

"Nothing to see," wailed all the students. The glory of the moonlight and the splendor of the Sun rising was as nothing to these young men and young women, who would walk a mile to see a circus procession with a gilt chariot driven by a lot of painted women.

This *Denver Post* cartoon showing the Earth threatened by meteors accompanied the article about the failure of the 1899 Leonids. Courtesy *Denver Post* (artwork restored by Mark Maxwell)

"No meteors this time," said Dean Howe, climbing down from his ladder. "Saturn and Jupiter must have scattered them. We'll have to wait another thirty-three years."

And so the tired, dizzy old world will whirl on a little longer. We talk a great deal of the weariness of it. I wonder how many of us are sorry to find ourselves still alive this bright hopeful day?

The Leonids were conspicuous by their absence.

In England, William Denning, most prominent of meteor observers, wrote:

> The firmament, with its glittering stars and silver moon, was just as still as on an ordinary mid-November night. Only now and then, indeed, a shooting star rapidly streaked along the sky to prove that the Leonids were present in a weak and scattered shower in place of the dense and brilliant display that had been awaited.[14]

Newspapers and their readers razzed the astronomers. The *Chicago Daily Tribune* offered an taunting article and a bit of doggerel, concluding with a dreadful pun:

> "I watched all night," he said,
> "Despite my years,
> But saw no shooting stars –
> Hence these me-teors!"[15]

Writing in 1923, meteor astronomer Charles Olivier was still reeling from the experience: "The twenty-four years that have since elapsed have not been enough to lessen the memory of the bitter disappointment felt when dawn finally broke and the last hope of a shower faded away."[16]

Professional astronomers were in despair. Oxford University astronomer Herbert Hall Turner opened his short report on the Leonids to the Royal Astronomical Society by saying: "Preparations were made to observe the expected shower, but came to nothing."[17]

Miss Arden sees the Leonids in the afternoon

For thousands of astronomers and millions of other people around the world, the Leonids failed to appear in 1899. But Mima Arden in Liverpool, England saw them because she watched during the day, while everyone else tried to see them at night. She wrote the *Liverpool Echo* to explain:

> Not having seen any account in the papers concerning the arrival of the meteoric showers, I beg to state that I saw them on Thursday afternoon, the 16th. I first noticed them at 12:15; they were shooting in all directions and kept on until about 4 o'clock. Then on Friday, the 17th, I again saw them at the same time. I called the attention of several people, with the result that they could also see them. Owing to the bright sky, one had to stare for a few seconds before perceiving the stars, as they were very dazzling to the eyes.

William Denning, the world's most famous meteor observer, answered Arden in the scientific journal *Nature*: "There is not the slightest doubt that

Olivier looked back a quarter of a century and pronounced an assessment of the fiasco that continues to resonate:

> ... the failure of the Leonids to return in 1899 was the worst blow ever suffered by astronomy in the eyes of the public, and has indirectly done immense harm to the spread of the science among our citizens.[18]

Bernard Lovell, a pioneer in the use of radar to study meteors, looked back half a century to the 1899 debacle and agreed, "The failure of the shower to manifest itself undoubtedly led to a serious diminution of interest in meteor astronomy."[19]

William Denning summarized the grief:

> No meteoric event ever before aroused such an intense and widespread interest, or so grievously disappointed anticipation. The scientific journals and newspapers all contained references to the subject, and the occurrence was predicted in such confident terms to take place that the public became enthusiastic, and looked forward to its appearance as a certainty. Many people regard the prescience of the astronomer as something marvellous, he can foretell the moment of an eclipse that will occur generations hence, and no thought of questioning either his accuracy or veracity ever enters their heads. Thus everyone expected that when they looked up to the sky on the night of November 14–15 they would see it full of meteors. But the fiery storm did not appear. . . . The finest celestial sight of a generation had failed to come at its appointed time, and the disappointment was all the keener in some quarters from the impression which prevailed that another chance of witnessing it would not occur until 1933.[20]

the objects were illusory and had nothing whatever to do with the November meteors." He explained that the radiant of the Leonids set at 2:30 p.m. and "we cannot have a shower of Leonids with the radiant below the horizon."

Denning drew on his many years of observing with the unaided eye to provide an explanation of what Arden had seen as shooting stars:

> ... they may be easily produced by bending the neck and gazing intently for a few minutes at a bright sky. I have observed many of these spectral meteor showers on occasions when I have been looking for Venus or some other object in bright daylight.

But what about all the people who saw the meteors when Arden pointed them out? "It is astonishing," wrote Denning, "that if one calls the attention of people to imaginary phenomena of this kind and asks them to look, they will, in ninety-nine cases out of one hundred, see the same thing and encourage similarly mistaken ideas!" Inexperienced observers are especially at risk, Denning added.*

*W[illiam] F. Denning: "Supposed Daylight Leonids," *Nature*, volume 61; December 14, 1899; page 152.

Puzzlement . . . and hope for 1900

Two weeks after he predicted a great shower of meteors in 1899, Johnstone Stoney published a new article in *Nature* to examine what went wrong with his and Downing's prediction. The problem, he said, was that a meteor swarm is not visible in space, so its exact location is uncertain. Its momentary position can only be known when it creates a meteor display, as a portion of the Leonid stream did in 1866. That point in the orbit of the Leonids can then be projected forward in time to see what gravitational forces will act upon it, and that was what he and Downing had done. The absence of more than a trickle of Leonids in 1899 proved, Stoney said, that the Leonid stream is not wide enough to reach across a distance of 1.3 million miles (2.1 million kilometers) – about five times the distance to the Moon – to brush the Earth.

Already, however, in the ashes of expectations dashed and opportunity lost, there were flickers of hope. As daytime wiped out the last chance of seeing a storm of Leonids on November 16, 1899, astronomer Edward Barnard at the University of Chicago's Yerkes Observatory told readers of the *Chicago Tribune* not to give up: that there was reason to expect the Leonids in 1900. Northwestern University astronomer Elias Colbert was willing to grant the Leonids one more chance. He thought a display might be seen in the western United States in 1900, but "after next year we shall have no more imposing exhibitions of meteor downfall from the Leonid radiant. The famous thirty-three-year showers will pass into history."[21]

In England, Denning held out hope for 1900 or 1901: "When we remember that there were brilliant displays in 1866, 1867, and 1868 we need not despair of seeing the Leonids at their best until after 1901 is past."[22]

In November 1899, newspapers typically devoted much or all of one page of a Sunday edition to prepare readers for the Leonids, then offered extra articles daily leading up to the observing nights, then provided daily coverage of the sad results.

When November 1900 came, few newspapers devoted any space to the Leonids or encouraged readers to watch for them. The *New York Times* published just one preparatory article marked "Special to the New York Times," meaning that it was not written by their staff. "Leonids Due on Tuesday" was satisfactory in style and popular appeal, but it contained mistakes that demonstrated that the author was not well informed about astronomy, among them: *comets* caused the perturbations experienced by the Leonid meteoroid stream; meteoroids are *perhaps* debris shed by a comet. It was quite a come-down from the commentary by experts the *Times* published in 1899. Except for "Leonids Due on Tuesday" and a letter to the editor about the Leonids immediately following it, that was all the

New York Times offered. It was very little, but it was more than most papers.

In England, Stoney and Downing were back with a new forecast for 1900. This time they had analyzed the portion of the Leonid stream that had furnished the meteor storm over the Western Hemisphere in 1867. The projected it forward to 1900. It too had undergone considerable perturbation from Saturn and Jupiter. At its closest, it would be 0.018 AU (1.7 million miles; 2.7 million kilometers) from the Earth, farther than the 1866 swarm had been in 1899, when it produced nothing. Stoney and Downing had been burned; now they were being extra careful to cover all possibilities. If the Leonid swarm is wide enough, they wrote, "we shall have a great meteoric shower this year." Or if there is a sinuosity in the stream, it may bring a portion of the meteoroids close enough to Earth to provide a shower. But, they concluded, neither the stream's great width nor its sinuosity is likely and there is little hope of a great shower this year. Still, Stoney and Downing added, if something does happen, it will be on November 15, timed for visibility over the Atlantic and the eastern Americas. Be prepared wherever you are, they urged. There is much uncertainty.

Stoney and Downing closed with an especially poignant and prescient observation:

> The perturbations during the last revolution, which have for the present carried the ortho-stream [meteor-storm-causing-stream] of Leonids so far from the earth's orbit, belong to the class of perturbations which act at different times with equal effect in opposite directions; so that there is reasonable ground for expecting that further perturbations must at some future time bring this remarkable stream back to the earth's orbit.[23]

Just carry forward our computations, they said, and see what the planets will do to the Leonids' orbit in the future.

The 1900 Leonids disappointed no one. No one was expecting anything. It was just an ordinary annual Leonid shower night in November, with the Leonids outnumbering the sporadic meteors, but not by much. The meteor century closed with a whimper. If anything, 1900 was even weaker than 1899.

Unnoticed

Stoney and Downing made no analysis or prediction for 1901. There was no lead up in the papers. In England and Ireland, the Leonids were stronger than in 1899 or 1900, but about 20 meteors an hour was a paltry show.

Then it happened. Two hundred an hour over the eastern United States – more than three a minute. Four hundred an hour in the west – averaging one every five seconds. Observers in Claremont, California claimed a peak of 800 an hour. It wasn't a storm, but it was many times more abundant than the impressive annual Perseids. The Leonids had finally put in an appearance.

But almost no one saw it. The public remembered only the disappointment of 1899.

The Leonids through the generations

The Leonids dashed expectations in 1899. They failed to use 1900 to redeem themselves. Little hope was held for them in 1901 – except by "Stargazer," an anonymous woman from Peekskill, New York. She wrote the *New York Times* before the 1900 no-show to explain that she saw a torrent of Leonids in 1868, that the Leonids return every 33 years, and therefore she expected to see the Leonids in November 1901.

"Will you pardon here a personal reminiscence of the last shower?" she asked.

> My husband, myself, and two children had been spending the evening of November 13, 1868 with some friends. On our return ride about midnight, suddenly there shot up from the eastern horizon what appeared to be a very large and brilliant skyrocket. A friend was with us, and we all said, "What is that?" In a moment we remembered the date, and we cried out, "The meteors!" The show had begun.
>
> We stopped a short time at my parents' home and with them viewed the thrilling sight, while they told us the story so often told before of the visitation they had witnessed years earlier, when "the stars fell" in 1833. During the remainder of our ride home the sky was filled constantly with the most brilliant Roman candles of every color, shooting in all directions. It was the most magnificent spectacle that I have ever witnessed. My children, though young then, will never forget it. Soon after our return to our home the family retired, excepting myself. I watched the ravishing display all night, going from one window to another on all sides of the house, in spellbound admiration.
>
> The next night we looked for them again, but the shower had passed over.*

In 1901, a strong shower by the Leonids surprised everyone … except one woman in upstate New York.

* [Stargazer]: "An Astronomer's Error?" *New York Times*, November 11, 1900. The reminiscence has been edited slightly.

NOTES

1. Garrett P. Serviss: "The Heavens in November," *Scientific American*, volume 81, October 28, 1899. Serviss was careful to mention that the Leonids might be disappointing because of Jupiter's perturbations, the scattering of the stream along its orbit, and bright moonlight.

2. R[ichard] M[iller] Devens: *Our First Century: Being a Popular Descriptive Portraiture of the One Hundred Great and Memorable Events of Perpetual Interest in the History of Our Country* ... (Springfield, Massachusetts: C. A. Nichols, 1877). Devens was a slovenly researcher. He mentions the return of the Leonids in 1867, but gives no information about meteors past what was known and conjectured in 1834. Devens included in his book three other astronomical events: the total solar eclipse of 1806, an impressive aurora in 1837, and the great comet of 1843.

3. Charles P. Olivier: *Meteors* (Baltimore: Williams & Wilkins, 1925), page 37.

4. *Seattle Post-Intelligencer*, November 12, 1899, page 27 – and other newspapers.

5. *San Francisco Chronicle*, November 15, 1899, page 14.

6. *Chicago Tribune* and *Atlanta Constitution*, November 12, 1899, among others.

7. G[eorge] Johnstone Stoney and A[rthur] M[atthew] W[eld] Downing: "Perturbations of the Leonids," *Proceedings of the Royal Society of London*, volume 64, March 2, 1899; pages 403–409.

8. G[eorge] Johnstone Stoney and A. M. W. Downing: "Next Week's Leonid Shower," *Nature*, volume 61, November 9, 1899; pages 28–29. The quotations that follow are from this article.

9. Charles P. Olivier: *Meteors* (Baltimore: Williams & Wilkins, 1925), page 37.

10. Charles P. Olivier: *Meteors* (Baltimore: Williams & Wilkins, 1925), page 37.

11. *Denver Post*, November 17, 1899, page 1, was one of several newspapers to report this story. Several newspapers, including the *New York Times* (November 16, 1899), mention an astronomer named de la Vaulx who also made a flight.

12. For a sketch of Jules Janssen's contributions as a solar spectroscopist, inventor, and balloon pioneer, see Mark Littmann and Ken Willcox: *Totality: Eclipses of the Sun* (Honolulu: University of Hawaii Press, 1991), pages 68–72.

13. *New York Times*, November 17, 1899, page 4.

14. W[illiam] F. Denning: "Report of the [Meteor] Section, 1899," *Memoirs of the British Astronomical Association*, volume 9, 1900, pages 123–136 (pages 1–14 within report).

15. *Chicago Tribune*, November 19, 1899, page 15.

16. Charles P. Olivier: *Meteors* (Baltimore: Williams & Wilkins, 1925), page 38.

17. H[erbert] H[all] Turner: "Observations of the Leonids of 1899, made at the University Observatory, Oxford," *Monthly Notices of the Royal Astronomical Society*, volume 60, December 1899, pages 164–165.

18. Charles P. Olivier: *Meteors* (Baltimore: Williams & Wilkins, 1925), page 38.

19. A. C. B[ernard] Lovell: *Meteor Astronomy* (Oxford: Clarendon Press, 1954), pages 337–338.

20. W[illiam] F. Denning: "Report of the [Meteor] Section, 1899," *Memoirs of the British Astronomical Association*, volume 9, 1900, pages 123–136 (pages 1–14 within report).

21. *Chicago Tribune*, November 17, 1899.

22. W[illiam] F. Denning: "Report of the [Meteor] Section, 1899," *Memoirs of the British Astronomical Association*, volume 9, 1900, pages 123–136 (pages 1–14 within report).

23. G[eorge] Johnstone Stoney and A[rthur] M[atthew] W[eld] Downing: "The Leonids: A Forecast," *Nature*, volume 63, November 1, 1900; page 6.

CHAPTER 11

The world's safest fireworks display – 1932

The world's safest fireworks display, the Leonid meteors, are due … In more than 1,000 years, during which the Leonids have been startling mankind about three times a century with veritable rains of fire, there is no record of any person having been hit. CHARLES P. OLIVIER (1931)[1]

THERE WAS a surprising buildup in meteor numbers during Leonid season in 1930. Charles C. Wylie and his astronomy students at the University of Iowa reported a single observer peak rate of 120 meteors an hour. Meteor astronomers began to twinkle inside while trying to keep their exteriors composed. They remembered or had read or had been reminded by meteor astronomer Charles Olivier what it was like in 1899 to pump up the public's expectations and then have a bust. On the eve of the 1931 return, the *Chicago Daily Tribune* reported "Astronomers are not predicting the intensity of this year's display."[2]

Without much publicity, not many people were watching, but those who were came away pleased. Not dazzled but pleased. John A. Theobald, a Catholic priest and astronomer at Columbia College in Dubuque, Iowa, had his students well organized for the event. In the early morning of November 16, 1931, the appointed solo observer counted 90 meteors an hour at maximum. Even better, a Mount Wilson Observatory astronomer, returning to California by airplane, noticed the meteors during a stop at the Las Vegas airport. He recorded 120 an hour. That was much better than the 25 an hour that Manning Prentice and other English observers saw the next morning.[3] Olivier did better than anyone. He traveled to the Catskill Mountains in upstate New York where he counted 180 meteors an hour at their peak.[4] The 1931 Leonids certainly favored the United States.

The expected return of the Leonids in 1931 was the inspiration for another innovation in meteor studies. Iowa professor Wylie reported in *Popular Astronomy* the strange experiment and its curious result:

> Mr. A. M. Skellett observed the effect of the meteors on the transmission of wireless waves: he says, "A cloud of electrified particles accompanies every meteor in its descent. When the meteor enters the Kennelly–Heaviside conducting layer this disturbs the condition of balance, so that the effective height of the layer is temporarily lowered. A fogging of radio-signals results." This is a point that should be examined if the expected rich shower occurs next November.[5]

The young field of radio astronomy was beginning to detect and study meteors. The new technique would have profound consequences for meteor science.

Preparing for 1932

Astronomers weren't predicting much publicly, but they were thinking – and hoping – lots. A definite, strong return of the Leonids in 1930. A shade better in 1931. It was just the pattern they were looking for: as if the Leonids were building for 1932, the time when Comet Tempel–Tuttle would return and the stream of Leonid particles would be its richest.

In February 1932 Wylie confided to readers of *Popular Astronomy*: "The recent excellent displays of the Leonids [in 1930 and 1931], the investigation of the orbit of Tempel's Comet by Crommelin, and the study of the period of the Leonids by Maltzev all encourage us to hope for a spectacular display in November 1932."[6] British astronomer Andrew C. D. Crommelin had continued what Stoney and Downing had begun in 1899. He projected Comet Tempel–Tuttle, source of the Leonid particles, through another 33-year circuit to see what gravitational perturbations by the planets had done to its orbit. There was good news. The comet's node had shifted half a million miles (about 800,000 kilometers) closer to the Earth than it was in 1899. Closer meant a better chance that the meteoroid stream would be wide enough to sweep over the Earth. Yes, the Moon would be a nuisance for Leonid watching in 1932 – three days past full, near maximum brightness for the entire meteor shower. And the timing of the Earth's near encounter with the comet's orbit suggested that the best meteor fall would be in the Far East. But why grumble? Overall, these were the best circumstances since 1866.

Charles Olivier, as president of the American Meteor Society that he had founded in 1911, decided that this time around public information about the Leonids would be done right: lots of information, no promises, no fear, and plenty of public participation. He would personally run a public information campaign that the newspapers could not resist. He would even make use of that new medium radio. Everywhere one turned in October and November 1932, there with information about the Leonids was Charles P. Olivier, Director of the Flower Observatory, Professor of Astronomy at the University of Pennsylvania, President of the American Meteor Society, President of the Meteor Commission of the International Astronomical Union.

He spoke for 15 minutes nationally on the CBS radio network. His speech was printed as an article in *Science News Letter* prior to the expected shower. He wrote a newspaper article that was distributed throughout the country by the Associated Press. In every forum, Olivier stressed the same themes: that meteor predictions are uncertain, that meteor showers pose no danger, and that grand meteor displays are well worth seeing.

Near the beginning of his radio address, Olivier told his audience that he must "be very frank about one point."

> As to eclipses of the sun, there is absolutely no doubt of the day, hour, – even the approximate second can be computed for each phase. Such predictions do not fail; they never can or will. Eclipses depend upon the positions of the sun and moon, both of which have been observed daily for centuries. But with meteor showers the case is different.

We are dealing not with a single large body like a planet, he explained, but with a loose aggregation of small particles whose very existence we infer only by the meteors we see. We cannot follow a meteor stream with a telescope because its particles are too small and too far apart. Olivier's point was that astronomers can compute the orbit of a meteor if it is carefully observed, but that does not provide exact information on the rest of the meteoroid swarm and its return. "The very fact that a meteor, popularly known as a 'shooting star,' is seen means that that particular body is destroyed forever; it has entered our atmosphere and has been burned up; it cannot come back."

"Nevertheless," said Olivier, "having given all this warning, we feel reasonably confident that this November will witness another notable meteor shower furnished by the Leonid stream . . ."[7]

Olivier had done what he could. How would the media and the citizenry react? What would the Leonids bring?

The *Seattle Post-Intelligencer* plunged itself into meteor science by volunteering to the American Meteor Society to serve as the clearinghouse for reports from volunteer observers throughout western Washington. It would help recruit and train observers. "Get ready to be an official observer for the society. . . Just read the Sunday Post-Intelligencer and act."[8] Hey, meteors could sell newspapers.

The article recalled the sight in 1833 when the meteors "fell almost as thick as snowflakes but with almost lightning speed, giving the impression of a brilliant waterfall of fire." Based on the 1931 display, the newspaper said "Another increase is expected by astronomers this year."

The Sunday, November 13, 1932 *Seattle Post-Intelligencer* offered an article with background on the Leonids (competent), an article on how to observe meteors (competent), a simple observation report form (of dubious usefulness), and a diagram of the Jupiter-perturbed meteor stream (utterly botched: the stream never crossed the orbit of Earth). The articles were the work of J. Hugh Pruett, a professor at the University of Oregon and western representative of the American Meteor Society. Pruett quoted Olivier liberally. He also tried to toe the Olivier line on avoiding the impression of certainty: "Astronomers are quite certain that those who watch the sky during the night starting November 15 (or a night or two before and after this date) will be rewarded, especially after midnight, by the appearance of many more meteors than usual, although nothing very unusual may develop." It would not be as good as in 1833, he said, but it would "probably" match 1866. (Oops. But he recovered.) "Let us keep a 'general watch' and enjoy what the Leonids have for us. It is quite certain to be something, even though not a noted display."[9]

The *San Francisco Chronicle* published Olivier's article, although his name became Oliver. The headline said "Leonid Meteor Fall Expected to Be Heavy." Olivier told readers that in 1928 there were "a fair number of Leonids." In 1929 there were "very few." In 1930, "many more than in 1929." And in 1931, "an excellent shower in some places, rates of over 100 an hour being reported..." "The gradual increase," Olivier said, "is what would be expected if the major shower were coming this year or in 1933."

In his article, Olivier modified history just a tad in his effort to restrain optimism. He said that the people in 1899 had failed to heed the warning that "careful computations, published shortly before the Leonid date, had indicated that the meteor stream had been moved out of its path by the powerful gravitational pull of Jupiter." That was indeed what Stoney and Downing had found, but Olivier did not include that they persisted in predicting a major Leonid shower. Olivier explained that the meteoroid stream had now been shifted to a more favorable position, but not back to its position a century earlier, so "We should not have so good a show as in 1833." (Caution, restraint.) With the Moon so bright, the dimmer meteors would be invisible.[10]

Not all newspapers caught the spirit. The *Denver Post* ignored Olivier in favor of a different Associated Press story out of New York. The first paragraph could have set Olivier trembling: "A celestial show, playing a return engagement after an absence of thirty-three years, will dazzle sky-watchers before daybreak Wednesday morning." The story quoted Clyde Fisher, astronomer at the American Museum of Natural History, as saying that "At the height of the display in 1833, 1,000 meteors a second flashed across the sky." Whether the astronomer or the reporter slipped up, 1,000 meteors a *second* was too many. Even 100 a second exceeded reliable reports. Probably a maximum of 30 a second in 1833 was closer to the truth. The *Denver Post* article, in its phrasing and its fanning of expectations, indicated that the reporter knew little about astronomy and had not bothered to check his information.

The *Chicago Daily Tribune* looked to George Van Biesbroeck, professor at the University of Chicago's Yerkes Observatory, for advice. He was in no mood for hype. The Leonids are "the residue of some degenerated comet," the article said. "The exact time of their coming cannot be foretold," nor where they will be seen. The reporter continued: "Prof. Van Biesbroeck, a specialist on comets, has been attempting, without success so far, to photograph the Tempel comet, which is also due at this time. The comet was discovered and last seen in 1866." It was not observed in 1899. "It is believed doubtful," the article said, "that the comet will ever be seen again."[11]

The *Atlanta Constitution* published Olivier's article in its Sunday edition and then, two days later, just before the expected Leonid peak, it offered an article distributed by a different news service by Willem J. Luyten, assistant professor at the University of Minnesota. He too emphasized the uncertainties: "It is out of the question to predict accurately what is going to happen." The shower wouldn't match 1799 or 1833. "On the other hand, it could hardly be as bad as the failure of 1899." But Luyten began his article with how impressive a great shower could be.

> Among the four truly great spectacles of nature – a total eclipse of the sun, a beautiful manifestation of the northern lights, a great comet with a long tail, and a real 'shower' of shooting stars – the last occupies a peculiar position. It is of exceedingly rare occurrence, perhaps not more than once or twice a century – and there is just a chance 1932 may bring us one of these.

Too much moonlight would be a problem, Luyten said, but citizens should take advantage of "an opportunity to see what may possibly be the sight of a lifetime."[12]

Albert M. Skellett of Bell Telephone Laboratories was back in 1932 for another look at – or, rather, listen to the Leonids. His experiment with radio signals bouncing off the ionosphere during the Leonid shower of 1931 had allowed him to recognize that vaporizing meteoroids create ionization of their own that disturbs the ionosphere's reflection of radio waves. The November meteors were perfect for his purpose: "The Leonids are the swiftest of recurring meteors and therefore have the most energy for ionization." Moonlight and clouds didn't bother radio scientists. Skellett saw nothing but good omens and bubbled over with excitement in the unemotional scientific journal *Science*: "the excellent return exhibited last November and recent calculations based on their orbit and period lead to the hope of a truly great shower this year."[13]

The outcome

"Seattle Observers See Fiery Meteors Falling Like Hail" screamed the headline in the *Seattle Post-Intelligencer* on November 16, 1932. The newspaper was proud of itself – and reminded readers why in the second paragraph of the story. It had appointed its readers as official observers for the American Meteor Society, giving them this chance: "they witnessed an event of cosmic importance – the apparent swinging back of the orbit of the Leonid meteors into the path of the earth." Picturesque descriptions flowed freely: "A brilliant meteor that actually broke through the overhanging cloud blanket and tore a path of flame from east to west was reported shortly before midnight by Oscar Jensen . . ." No matter that meteors are not visible below the clouds because they burn up 40 miles (65 kilometers) or more above the cloudtops.

Despite reports of only rare holes in the clouds, the newspaper reported a "tremendous increase in the Leonids." Minnie D. Lassen watched through a window from her home in the city on the "cloudy and misty night." She watched for 10 minutes just after midnight and counted 38 meteors. But a meteor about every 15 seconds was apparently not enough to keep her watching, even though she saw so many meteors despite clouds obscuring the sky, despite city lights dimming meteor brightness, and despite a window limiting and distorting her view.

C. A. Nurell, a night engineer at a hospital, said the sight of all the Leonids was

> so amazing – so magnificent – that I ran back to the hospital to get others to watch with me. . . . We could not count or estimate the number of meteors we saw. They fell like hail, coming from one point in the heavens where the sun rises. It was the most thrilling thing I ever saw. I wanted to count, as the meteor society and the Post-Intelligencer had asked us, but there were too many – you couldn't.

Interesting that the name of the newspaper just happened to appear in the quotation. More suspicious was Nurell's testimony that the Leonid radiant was on the eastern horizon. In the pre-dawn hours of mid November, the constellation Leo was high in the southeast and hence the Leonids would have been radiating from near the top of the sky. But the *Post-Intelligencer* had invested in this meteor shower and it was determined to see a return on that investment.

The next morning should have been Leonid maximum, but now the newspaper could only report rain, low clouds, and mist. Nevertheless, it urged meteor observers to file their reports with the *Post-Intelligencer* because it was the regional clearinghouse for the American Meteor

Society. Professor Pruett's report would be published soon, they promised. It wasn't. There was nothing to report.

The *Denver Post*, which had ignored the cautions issued by most astronomers in favor of flagrantly exaggerating hopes for the meteor shower, responded to the poor showing of the meteors by treating their performance as an opening night at the theatre, to be critiqued . . . by bashing the astronomers, of course:

> A highly ballyhooed revival of a centuries-old starring vehicle, the 1932 appearance of the Leonid meteors, scheduled to brighten the heavens early Wednesday, 'flopped' dismally . . .

Caught up in its play review format, the *Post* reported that the director of Philadelphia's Flower Observatory, "Lawrence P. Oliver," was clouded out at 3 a.m., having seen only a few meteors till then.[14] The next day the *Post* continued with an Associated Press story from California in which unnamed astronomers at Mount Wilson Observatory said they believed that the Leonid meteors were "scattering and will not again be the spectacle it was in past centuries."[15] The next day, there was even juicier pickings.

A University of California, Berkeley, astronomer, William F. Meyer, made some offhand remarks and his words radiated around the country by news service. They played into the hands of newspapers like the *Denver Post* that were eager to taunt astronomers for their unfulfilled but usually carefully qualified predictions about the Leonids. Meyer may have (should have) cringed when he read the first sentence of the article about him: "Astronomers who pride themselves on split second predictions of eclipses and such may have slipped up in the matter of the Leonid meteor shower in the opinion of Dr. W. F. Meyer." The reporter went on to say that Meyer "thinks a whole year may have been lost somewhere. He's sure something went wrong in the calculations on account of his neck. It's stiff." In three nights of watching meteors, the article explained, Meyer had seen a total of three. (Tsk, tsk. Poor Professor Meyer.) And still more misinformation was coming: "The last time the Leonid shower appeared was in 1899." (The Leonids were dismal that year.) Therefore, said the article, "as computed by mathematical experts of astronomy, the next display of fireworks was due in thirty-three years, or 1932." (Snicker, snicker.) Meyer recalled that 1931 had brought "a really big shower." The article concluded: "There is just the possibility, he thinks, that the Leonids may have made up a year on a detour somewhere."[16] A few more articles like that could stunt a professional career.

Much of the country had poor weather. Washington, D.C., Chicago, and San Francisco saw no meteors because of clouds.

The *New York Times* fell victim to hoaxes and bad reporting. On November 16, the paper reported that New Yorkers saw the Leonids pay "their scheduled visit with the regularity of clockwork" and that they "furnished a brilliant display in the heavens." Homeward-bound night workers, the article said, thought the meteors were only a few hundred yards away. The brief story concluded: "the show was furnished by celestial particles of iron or rock burning up as they came in contact with the upper atmosphere, but where they came from scientists do not know."[17] Actually, astronomers had proved more than half a century earlier that the meteoroids in annual meteor showers come from comets in orbit around the Sun.

It was time for the scientists to pool their reports and try to establish what actually had happened. In England, only Manning Prentice won his

My first scientific research

JAMES A. VAN ALLEN, PH.D., Regent Distinguished Professor, University of Iowa

My introduction to scientific research goes back to my freshman year (1931–32) at Iowa Wesleyan College. Thomas C. Poulter, *the* physics professor there, gave me a part-time job as an undergraduate teaching and research assistant at 35 cents an hour.

In August 1932, I was enlisted by Poulter to help observe the Perseid meteors, checking out the equipment and procedure he had devised for his meteor studies in Antarctica. Poulter had been chosen as chief scientist for Admiral Byrd's second Antarctic expedition, 1934–35. The observing equipment was a reticle made of welding rods. It consisted of a set of several concentric circles and six diametric rods corresponding to hourly divisions on a clock face. The outer diameter of the reticle was about 35 inches. Three straight rods connected this outer circle to a one-inch diameter circular eyepiece to provide a conical field of view of about 50°. The reticle was supported on a transverse horizontal axis by two ring stands.

Two stations were set up 48 miles apart on a north–south line, one at Mt. Pleasant, Iowa (which I manned) and the other at Iowa City (which graduate student Ray Crilley manned), for observing meteor trails in the upper atmosphere. The central axes of our lines of sight were tilted so as to intersect at a height of about 60 miles (100 kilometers) above the midpoint between the two stations.

On the night of 9–10 August 1932 Crilley and I observed the Perseids for about two hours, recording the time of occurrence of each meteor that we saw within the field of the reticle and, to the best of our abilities, the polar battle with the clouds and saw 20 Leonids an hour on November 16.[18] In the United States, Olivier saw 13 Leonids an hour, which, he said, would correspond to 30 an hour if the Moon had not interfered. Worried about clouds, Willem Luyten at the University of Minnesota persuaded the Navy to fly him to an altitude of 17,000 feet (about 5,000 meters), where he saw 30 meteors an hour. Yale reported 48 an hour. James Van Allen, a sophomore at Iowa Wesleyan College, saw 18 an hour, hampered by moonlight.[19] John Theobald and his students in Dubuque, Iowa saw meteors at a single observer rate of 92 per hour.[20] Overall, it was another disappointment.

A handful of astronomers held out hope that the big show might come in 1933 or 1934, even 1935.

coordinates of the beginning and ending point of the luminous trail in the upper atmosphere.

The object of the work was to determine by geometric triangulation the heights of appearance and disappearance of each trail that was observed simultaneously by both observers. We each recorded about two dozen meteors, five of which were clearly simultaneous sightings of the same object.

Our data were turned over to Professor Charles C. Wylie of the University of Iowa who analyzed them and published a short paper (with our names in it!) in *Popular Astronomy* (volume 40, October 1932, pages 500–501). Wylie found that the heights of appearance ranged from 71 to 92 miles and of disappearance from 55 to 65 miles. *Science News Letter* for November 12, 1932 also carried the story.

These Perseid meteor observations were my first participation in an original astronomical/geophysical project.

In the early morning hours of November 16, 1932, I was out again, trying to observe the return of the Leonid meteors. As in 1899, they failed to appear in significant numbers. *Science News Letter* for November 26, 1932 mourned that the shower "was far inferior to the display of last year, which it was hoped might be the forerunner of a brilliant shower this month, possibly rivalling that of 1866."

However, *Science News Letter* continued, "Professor James Van Allen at Iowa Wesleyan College" saw one fireball that "left a trace that lasted fifteen seconds." Alas, I have no recollection of that fireball . . . nor of my momentary promotion to professor. I was an 18-year-old college sophomore.*

* In 1958, James Van Allen used *Explorer 1*, the first American satellite in orbit, to discover belts of high-energy ionized particles surrounding the Earth, trapped by its magnetic field. He is therefore credited with discovering the Earth's magnetosphere and with founding the field of space plasma physics. The populations of energetic particles around the Earth are called the Van Allen Belts.

The other depression of 1933

November 1933 rolled around and the *Seattle Post-Intelligencer* was doing things a little differently. As before, there was an article in which Oregon professor Pruett, as regional director for the American Meteor Society, asked citizens to watch for the Leonids and file reports. "The Post-Intelligencer is cooperating with Pruett in interesting observers in making reports," the newspaper added. But that was a far as the newspaper was going. It no longer touted itself as appointer and trainer of observers and the regional clearinghouse for meteor reports. Its article about the possible return of the Leonids was half-hearted and slipshod. "Make a count of meteors seen to get the hourly rate," it advised, but "Do not bother with faint meteors." The article was illustrated with a photograph of a University of Washington coed with her eye to the eyepiece of a large telescope and with her hand wrapped tightly around the eyepiece holder. The headline above the picture read "Hurry, Meteors, I'm All Ready."[21]

There was a note of weariness in the *San Francisco Chronicle*'s brief article in preparation for the "long overdue rain of fire of leonid shooting stars." "Astronomers calculated last year as the real time," the paper reported, but "'Our hopes for an excellent shower in 1933 are still high,' said Dr. Charles P. Olivier, president of the American Meteor Society."[22]

The *New York Times*' one preparatory article was a "Special to *The New York Times*." It reported that 24 cameras and telescopes at Harvard were ready for the meteors, with spectroscopy to concentrate on the red portion of a meteor's spectrum to identify elements in addition to iron and calcium. Scientists hoped, the article said, to determine more than the meteors' height and rate. They wanted also to study the drift of fireball trails (trains) to measure wind velocities in the far upper atmosphere "in the hope that such knowledge may be of great value to aeronauts planning plane flights through the stratosphere."[23] The stratosphere lies between 9 and 30 miles (15 and 50 kilometers) up. Meteors leave trails from about 75 to 50 miles (120 to 80 kilometers) high. Only rockets could carry passengers into or through the meteor layer.

The Leonids suffered from lack of media and public interest in 1933. There were a few upbeat words from meteor scientists – moonlight would not be a hindrance this year – but inwardly the optimism was gone.

The Leonids of 1933 sagged like astronomers' hopes. Bad weather again in England. No better than 64 meteors an hour in Iowa.[24] In Phoenix, local amateur astronomer V. O. Wallingford told the *Phoenix Gazette* that the display had been hijacked by two of the giant planets. "Mr. Wallingford said scientists have decided that the tremendous magnetic powers exerted by Jupiter or Uranus may have drawn the meteors from their path."[25] He may have meant gravitational, not magnetic, forces.

Regardless of how few Leonids were visible, the *Seattle Post-Intelligencer* had a poem ready and was determined to publish it:

> Those meteors that fell last night
> Were wonderful to see,
> It thrilled our Scotchmen with delight
> Because the show was free.[26]

From Hyderabad, India, M. A. R. Khan summed it up: "This year's Leonids were a poor show."[27]

Meteor astronomy by radar

In the wake of 1932 and 1933, meteor astronomers were miserable. Most of the public forgot about meteors and the remainder thought they were a joke. The Leonids had not provided the show or the conventional data expected of them. But almost unnoticed by meteor astronomers, a new technique was born beneath the 1931 and 1932 Leonids. These meteors, far from plentiful, were plenty adequate for this new tool of investigation. Charles Olivier had used radio to spread the word about the Leonids. Now the Leonids were using radio to spread a message of their own.

The first two commercial radio stations went on the air in 1920. By 1929, much of the world was tuning in to their favorite radio programs. Radio could reach people miles beyond the transmitting tower's line of sight because radio wavelengths bounce off the ionosphere, an electrically charged layer of the uppermost atmosphere.

The Sun creates the ionosphere. When its ultraviolet and x-ray radiation strikes the Earth's upper atmosphere, it removes negatively charged electrons from neutral atoms, making them positively charged ions. At night, without the energy of the Sun impinging on them, the ionospheric layers tend to rise and the electrical charge of the lowest layer decreases almost to zero. With the reflective layer of the ionosphere higher, radio signals can bounce greater distances to antennae and receivers. During the day, a radio station can be picked up perhaps 50 miles (80 kilometers) away. At night, that radio station with the same power might be heard hundreds or even thousands of miles away.

Scientists couldn't resist exploring this phenomenon. One of the first discoveries they made was that the ionosphere at night had occasional sudden flashes of extra reflectivity. The Sun wasn't around to create these bursts of ionospheric strength. In 1929, Hantaro Nagaoka of Japan was the first to attribute these fluctuations to meteors, but he thought that

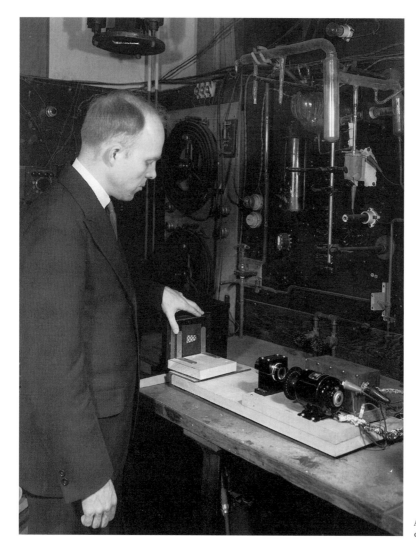

Albert M. Skellett. Courtesy
of Lucent Technologies

meteors swept electrons out of the ionosphere.[28] Albert M. Skellett of Bell
Telephone Laboratories in the United States got it right in 1931: meteors
were adding to the ionization. He and his colleagues J. Peter Schafer and
William M. Goodall used the sprinkling of Leonid meteors in 1931 and 1932
to show that visible meteors corresponded exactly to radio signals echoing
off the ionosphere.[29]

So far, however, meteors tampering with radio transmissions were a
curiosity explored only by radio scientists and engineers. Astronomers

J. Stanley Hey. Courtesy of J. S. Hey and the Royal Radar Establishment, Malvern

had not yet realized that bouncing radio signals – radar – off meteors could provide them with answers they desperately wanted. Radar could not detect the meteor particle itself, which was typically only the size of a grain of sand. But that speeding particle could by its friction in the atmosphere generate so high a temperature as it vaporized that it would strip electrons from atoms in the air as it passed, creating a long trail of ionization. Radio waves reflected strongly off that column of ionization, which remained suspended in the atmosphere for seconds or even minutes after the meteoroid had disappeared into ash and atoms.

Radar developed rapidly during World War II. J. S. Hey, a British scientist working for the government to improve radar, found that radar often warned of approaching enemy aircraft when there were none. In the last year of the war, German V-2 rockets descending on England warned of their approach by strong radar echoes from their disturbance of the ionosphere. Yet, Hey found, radar operators were often confused by "spasmodic transient echoes at heights around 60 miles [100 kilometers]. These echoes, appearing at an average rate of 5 to 10 per hour, were a maddening nuisance and caused frequent false alarms."[30] Hey and Gordon Stewart accused meteors of providing the misleading signal and used those signals to determine the radiant of the Delta Aquarid meteor shower.[31]

After the war, the British army allowed Hey and Stewart to use military radar to continue their study of meteors. They discovered the first two daytime meteor showers, June 6–13, 1945, and located the radiants from which meteors were plunging into the atmosphere and burning up unseen.[32]

Radar observation of meteors made its greatest leap from engineering nuisance to astronomical tool soon after World War II when the Draconid (Giacobinid) meteors returned as expected on October 9–10, 1946. The Draconids came from Comet Giacobini–Zinner, which has a 6½-year period. The Draconids had delivered a memorable meteor monsoon in 1933 when the Earth passed close to the orbit of the comet just 80 days after Giacobini–Zinner had passed. There was no shower in 1939 when the Earth passed near the comet's path 136 days ahead of the comet. In 1946, the Earth would pass the comet's orbit only 15 days after the comet – and the meteor science communities in the United States and Europe equipped themselves for visual, photographic, spectroscopic, and, for astronomical purposes for the first time, radar observation of a major meteor event.

The Draconids were so numerous that they created a temporary ionosphere of their own for a few hours. From their radar observations of meteor distances and how those distances changed with time, Hey, Stewart, and John Parsons determined that the Draconids were traveling at an average speed of 14.2 miles per second (22.9 kilometers per second), the first meteor velocity measurements ever made by radar.[33]

Edward Appleton and Robert Naismith realized in 1946 that the radio wavelengths returned by meteor trails changed as the meteoroids flew by, just as, for a person at a railroad crossing, the whistle of a locomotive seems to change from a higher to a lower pitch as the engine passes – the Doppler Effect. The change in wavelength of the meteor trails in motion picked up by radar could allow scientists to measure the speed of meteoroids, much as the police now use a radar gun to check the speed of cars. Radar had suddenly provided meteor science with several new and better ways to calculate meteor velocities.

Bernard Lovell. Courtesy of Sir Bernard Lovell

With radar's ability to detect meteors every night and even during the day, regardless of clouds; with its ability to detect meteors too faint for the eye to see; and with its ability to determine meteor velocities, radiants, and orbits, radar astronomy became a prime source of progress for meteor science.

When Bernard Lovell led the development of the first huge steerable radio telescope, the 250-foot-diameter (76-meter-diameter) dish at Jodrell Bank near Manchester, England, his primary purpose was the study of meteors, although the telescope immediately demonstrated its importance in all areas of radio astronomy.[34]

Going hyperbolic

The 1931–1932 Leonids saw the birth of the new technique of radar to study meteors. This Leonid return also provided the greatest battleground in an ongoing meteor science war, and perhaps its turning point.

No sooner had the stationary radiant dispute flickered out than a new controversy had ignited. Once again, the subject in question was sporadic meteors – this time their velocities. Astronomer and historian David Hughes views this misstep as the second of the two serious blind alleys that meteor science stumbled into.[35]

In 1926, Gustav Niessl von Mayendorf and Cuno Hoffmeister published a catalog of 611 fireballs.[36] They found that 79 percent of these bright meteors had been traveling along paths with hyperbolic velocities. Hyperbolic. Throughout the astronomical community, eyebrows rose and jaws dropped.

Meteoroids came – didn't they? – from comets, living or dead, in orbit around the Sun. To orbit the Sun, the velocity of a comet or meteoroid could not be too high or it would escape from the solar system. A body at the Earth's distance from the Sun traveling more than 26.1 miles per second (42.1 kilometers per second) would leave the solar system unless it hit something on the way. Similarly, a sunbound object falling past the Earth could not be traveling at more than 26 miles per second unless it came from beyond the solar system, beyond the gravitational hold of the Sun.

At first the Leonid meteoroids might seem to have hyperbolic velocities: 44 miles per second (71 kilometers per second). But their high speed when they blaze through our atmosphere is produced because they hit the Earth head on, adding their orbital speed to the orbital speed of the Earth. The orbital speed of the Leonids when they cross the Earth's orbit is 25.5 miles per second (41 kilometers per second).

Factor out the Earth's orbital velocity, its gravitational pull, and the angle of the collision, and if the meteoroids are traveling more than 26 miles per second, they are aliens intruding on the solar system. The annual meteor showers, associated with comets tied by gravity to the Sun, all have velocities that confirm their orbits to be ellipses.

But, said Niessl and Hoffmeister, the fireball reports they had collected showed that almost four out of five bright meteors had speeds that placed them on hyperbolic paths that would never return to the Sun. If the fireball observations were correct, that meant that most meteoroids came from interstellar space, from an unknown source.

Several astronomers voiced doubts about the accuracy of the fireball data. Although the fireball reports came from skilled observers, fireballs are unexpected brief visual sightings for which accurate measurements are extremely difficult. A small mistake in timing or direction of flight or the position of the radiant could easily transform elliptical velocities into hyperbolic.

Ernst Öpik from Estonia doubted that the fireball reports were reliable, but he thought that comets and meteoroids came from interstellar space. He hypothesized a great cloud of comets and meteoroids surrounding the solar system out to a distance of 1 parsec (3.26 light years), although most lay, he thought, within 10,000 AU of the Sun, 250 times farther from the Sun than the outermost planet. Thus Öpik in 1932 was the first to propose a vast comet reservoir at the outskirts of the solar system. But Öpik did not conceive of this comet-and-meteoroid cloud as the actual source of the long-period comets that visit the Earth. He conceived the mechanism somewhat differently. If the Sun carried around a cloud of debris with it, other stars must have their own comet-and-meteoroid clouds too. When these stars passed close to our solar system, they would lose some of their outlying comets and meteoroids to the Sun's gravity, so that they would fall into our solar system with hyperbolic velocities that would cause them to loop around the Sun and then escape from the solar system completely.

To discover whether any meteoroids really did have hyperbolic velocities, Harvard and Cornell universities together sent expeditions for three seasons (1931–1933) to the usually clear skies at and near the Lowell

Observatory in Flagstaff, Arizona. There astronomers attempted to observe meteors visually to plot their radiants and measure their velocities. Öpik, as a visiting lecturer at Harvard, went along on the expedition with his invention: a motor-driven rocking mirror that would convert a meteor streak into a series of loops or waves, the shape of which would provide the angular velocity of the meteor. Öpik continued his rocking mirror observations of meteors in Estonia from 1934 to 1938 and came to the conclusion that 66 percent of meteoroids have hyperbolic velocities – down from Niessl and Hoffmeister's 79 percent, but still confounding to most astronomers.[37]

Fred Whipple at Harvard launched a new attack on the problem of meteor velocities in 1936. His program used photography. Before World War II interrupted his work, his measurements were showing that meteoroids traveled in elliptical (closed cometlike) orbits around the Sun. Whipple resumed meteor work in 1951 with the new Baker Super-Schmidt telescopic cameras. Previous cameras could photograph meteors only if they rivaled or exceeded Sirius and the very brightest stars in brightness (magnitude −1). Whipple could photograph meteors that were 60 times fainter, down to magnitude +3.5. The Baker Super-Schmidts caught 260 meteors in 100 hours of exposure. Allowing for uncertainties, Whipple concluded that all the meteor trajectories could be explained with elliptical (cometary) orbits. No meteor showed a definite hyperbolic velocity.

In the meantime, 1948, Erik Sinding of Denmark analyzed comet orbits and how Jupiter (by far the most massive planet) perturbs them. He found that Jupiter could give comets hyperbolic orbits and send them flying out of the solar system – but that every comet he studied had an elliptical orbit before it encountered the planets.[38]

Even before Whipple had his answer, the possibility that streams of meteors had hyperbolic velocities had been undermined by a new technique that emerged from the war – radar. Radar beams reflected off the columns of ionized air that vaporizing meteoroids generated as they dashed through the atmosphere. A new breed of meteor specialists, radio meteor astronomers, used radar to measure meteor velocities much as a radar "gun" at a baseball game measures the speed of a pitcher's fastball. From 1948 to 1951, radar teams, especially one led by Bernard Lovell at Jodrell Bank in England, analyzed a series of meteor showers, including those for which claims had been made that their velocities were too high for them to remain in the solar system. They weren't.

By 1969, Öpik was convinced that meteoroids with hyperbolic velocities were rare and could be explained by gravitational assists from the planets.[39]

NOTES

1. *Chicago Daily Tribune* [no author given]: "Leonid 'Fireballs' Will Rain From Sky Beginning Tomorrow," November 13, 1931.

2. *Chicago Tribune* [no author given]: "Meteor Display May Greet Early Risers Tomorrow," November 14, 1931.

3. *Nature* [no author given]: "The Leonid Meteors," volume 128, December 5, 1931, page 972. See also John A. Theobald: "Dubuque Counts of the 1931 Leonids," *Popular Astronomy*, volume 40, January 1932, pages 54–56.

4. *Nature* [no author given]: "The Leonid Meteors," volume 128, December 12, 1931, page 1007.

5. *Nature* [no author given]: "The Leonid Meteors," volume 128, December 12, 1931, page 1007.

6. C[harles] C. Wylie: "The 1932 Return of the Leonid Meteors," *Popular Astronomy*, volume 40, February 1932, pages 97–99.

7. Charles P. Olivier: "Will the Great Shower Return?" *Science News Letter*, volume 22, November 5, 1932; pages 290–291, 297–298.

8. *Seattle Post-Intelligencer* [no author given]: "Meteor Rain Observers to Be Instructed by Seattle Post-Intelligencer," November 12, 1932, page 9.

9. J. Hugh Pruett: "Experts Tell Why Meteors Are Expected," *Seattle Post-Intelligencer*, November 13, 1932, page 11.

10. Charles P. Oliver [Olivier]: "Leonid Meteor Fall Expected to Be Heavy," *San Francisco Chronicle*, November 13, 1932, page 6.

11. *Chicago Daily Tribune* [no author given]: "Leonid Group of Shooting Stars Due This Week," November 13, 1932. Crommelin's orbit for Comet Tempel–Tuttle was not accurate enough to allow the comet to be found.

12. Willem J. Luyten: "Flashing Show of Leonid Meteors Expected to Be at Height Tonight," *Atlanta Constitution*, November 15, 1932, page 11.

13. A[lbert] M. Skellett: "Radio Studies During the Leonid Meteor Shower of November 16, 1932," *Science*, volume 76, November 11, 1932, page 434.

14. *Denver Post* [no author given]: "Leonid Meteor Show Falls Short of Being Spectacular Display," November 16, 1932, page 1.

15. *Denver Post* [no author given]: "Leonide [*sic*] Meteors Believed Scattering," November 17, 1932.

16. *Denver Post* [no author given]: "Miscalculation Blamed as Meteor Shower Fails," November 18, 1932.

17. *New York Times* [no author given]: "Leonids Illuminate the Sky; Pay Visit to City on Schedule," November 16, 1932, page 16.

18. *Nature* [no author given]: "The Leonid Meteors," volume 130, November 26, 1932, page 817.

19. C[harles] C. Wylie: "Preliminary Report on the 1932 Leonids," *Popular Astronomy*, volume 40, December 1932, pages 649–650; and *Science News Letter* [no author given]: "Earth Apparently Missed Main Leonid Meteor Swarm," volume 22, November 26, 1932, page 345.

20. John A. Theobald: "Dubuque Counts of the 1932 Leonids," *Popular Astronomy*, volume 41, January 1933, pages 56–59. The journal *Nature* incorrectly reported that Dubuque had recorded a single observer rate of 240 an hour, mistaking the group observing rate for the solo observing rate; *Nature* [no author given]: "The Leonid Meteors," volume 130, December 24, 1932; page 970. In his *Popular Astronomy* article, Theobald shows 240 meteors an hour as his group-observing rate. However the erroneous 240 meteors an hour solo observer rate for 1932 continues to appear in books and articles, such as in A. C. B. Lovell's *Meteor Science*, page 339.

21. *Seattle Post-Intelligencer* [no author given]: "Meteor Fall Due to Start Tuesday Night," November 13, 1933, page 11.

22. *San Francisco Chronicle* [no author given]: "Colorful Leonid Shooting Stars Due for Appearance This Week," November 13, 1933, page 4.

23. *New York Times* [no author given]: "Cameras to Record Leonid Shower; Data May Aid Stratosphere Fliers," November 14, 1933, page 21.

24. C[harles] C. Wylie: "Preliminary Report on the 1933 Leonids," *Popular Astronomy*, volume 41, December 1933, page 581–582; and John A. Theobald: "Dubuque Counts of the 1933 Leonids," *Popular Astronomy*, volume 41, December 1933, page 582–583. Wylie at Iowa City recorded 64 an hour. Theobald recorded 40 an hour.

25. *Phoenix Gazette* [no author given]: "Jupiter Uranus 'Steal Show' of Shooting Stars," November 16, 1933, pages 1, 8.

26. *Seattle Post-Intelligencer* [no author given]: in "'Round About" column, November 15, 1933, page 6.

27. *Nature* [no author listed]: "The Leonids, 1933," volume 132, December 16, 1933, page 928.

28. Hantaro Nagaoka: "Possibility of the Radio Transmission being disturbed by Meteoric Showers," *Proceedings of the Imperial Academy of Tokyo*, volume 5, 1929, pages 233–236.

29. A[lbert] M[elvin] Skellett: "The Effect of Meteors on Radio Transmission Through the Kennelly–Heaviside

Layer," *Physical Review*, volume 37, June 15, 1931, page 1668; A. M. Skellett: "The Ionizing Effect of Meteors in Relation to Radio Propagation," *Proceedings of the Institute of Radio Engineers*, volume 20, December 1932, pages 1933–1940; J[ohn] P[eter] Schafer and W[illiam] M[cHenry] Goodall: "Observations of Kennelly–Heaviside Layer Heights During the Leonid Meteor Shower of November, 1931," *Proceedings of the Institute of Radio Engineers*, volume 20, December 1932, pages 1941–1945; and A. M. Skellett: "The Ionizing Effects of Meteors," *Proceedings of the Institute of Radio Engineers*, volume 23, February 1935, pages 132–149.

30. J[ames] S[tanley] Hey: *The Evolution of Radio Astronomy* (New York: Neale Watson Academic Publications [Science History Publications], 1973), page 20.

31. J[ames] S[tanley] Hey and G[ordon] S. Stewart: "Derivation of Meteor Shower Radiants by Radio Reflexion Methods," *Nature*, volume 158, October 5, 1946, pages 481–483. Because Hey sent signals to astronomical bodies to gather data rather than waiting passively for electromagnetic radiation to reach him, when the Royal Astronomical Society awarded Hey its Eddington Medal in 1959, the citation described him as "the first experimental astronomer of all time." J. S. Hey, personal communication, October 21 and December 15, 1997.

32. J[ames] S[tanley] Hey and G[ordon] S. Stewart: "Radar Observations of Meteors," *Proceedings of the Physical Society*, volume 59, 1947, pages 858–883. They could specify the location of the radiants only to within 10 degrees.

33. J[ames] S[tanley] Hey, S. J[ohn] Parsons, and G[ordon] S. Stewart: "Radio Observations of the Giacobinid Meteor Shower, 1946," *Monthly Notices of the Royal Astronomical Society*, volume 107, 1947, pages 176–183. The true geocentric velocity of the Draconid (Giacobinid) meteors is 14.7 miles per second (23.7 kilo-meters per second). See also J. S. Hey: *The Radio Universe*, 3rd edition (Oxford: Pergamon Press, 1983), pages 118–122.

34. This section on meteor radio astronomy is indebted to D[onald] W. R. McKinley: *Meteor Science and Engineering* (New York: McGraw-Hill, 1961); David W. Hughes: "The History of Meteors and Meteor Showers," *Vistas in Astronomy*, volume 26, 1982, pages 325–345; and Bernard Lovell: *Astronomer by Chance* (New York: Basic Books, 1990), pages 117–137.

35. David W. Hughes: "The History of Meteors and Meteor Science," *Vistas in Astronomy*, volume 26, 1982, pages 325–345.

36. G[ustav] Niessl [von Mayendorf] and C[uno] Hoffmeister: "Katalog der Bestimmungsgrössen für 611 Bahnen grosser Meteore," *Denkschriften. [Österreichis-che] Akademie der Wissenschaften in Wien. Mathematisch-Naturwissenschaftliche Klasse*, volume 100, 1926, pages 1–70. Niessl died in 1919. His name appears on the article as G. v. Niessl. Hoffmeister completed the work in 1925 and it was published in 1926.

37. E[rnst J.] Öpik: "Meteors," *Monthly Notices of the Royal Astronomical Society*, volume 100, February 1940, pages 315–326.

38. Erik Sinding: "On the Systematic Changes of the Eccentricities of Near Parabolic Orbits," *Matematisk-Fysiske Meddelelser udgivet af Det Kgl. Danske Videnskabernes Selskab*, volume 24, number 16, 1948, paginated 1–8.

39. This section on suspected hyperbolic velocities for meteoroids was based primarily on David W. Hughes: "The History of Meteors and Meteor Science," *Vistas in Astronomy*, volume 26, 1982, pages 325–345; A. C. B[ernard] Lovell: *Meteor Astronomy* (Oxford: Clarendon Press, 1954); Bernard Lovell: *Astronomer by Chance* (New York: Basic Books, 1990), pages 137–148; and J[ohn] G[uy] Porter: *Comets and Meteor Streams* (New York: John Wiley & Sons, 1952), chapter 7.

CHAPTER 12

New horizons in meteor science

The early progress of meteoric astronomy is so intimately connected with the appearances of the great Leonid showers that it can best be traced through a study of these phenomena themselves and of the stream that caused them.

CHARLES P. OLIVIER (1925)[1]

THE YEAR 1950 was a watershed for meteor science because of two pivotal discoveries about comets, the parent bodies of meteors. The first breakthrough was made by Fred Whipple of Harvard College Observatory, who was already a leader in meteor research by demonstrating with photographic techniques that very few meteors had hyperbolic velocities – so few that it was best to consider all meteors to be members of the solar system. Whipple now offered a theory about comets: they were dirty snowballs. That phrasing was a bit informal for articles in the *Astrophysical Journal*. There he described comets as conglomerates of ices, mostly water, combined in a conglomerate with meteoric materials, all initially at extremely low temperatures. Dirty snowballs.

It was a breathtaking article because in one bold conceptualization, Whipple explained virtually every observed feature of comets.[2] Until Whipple published his article, astronomers lived uneasily with a theory that visualized a comet as a flying sandbank. The downpour of Leonid meteors in 1866 encouraged astronomers to continue to think of a comet as an oversized meteor swarm of icy rock particles traveling together, with somehow enough gravity to stay in a clump but not enough to coalesce into a solid body.[3] Through telescopes, comets looked fuzzy, with a coma and tail through which the distant stars shined undimmed. After 1866, meteors were recognized as debris from comets, *individual* particles traveling together. It was a short step to think of comets as just larger and denser meteoroid swarms.

But the flying sandbank was an uncomfortable theory. It could not explain how a loose agglomeration of particles could retain enough ice to form a comet's coma and tail each time it returned to the Sun. Nor could it explain how the sandbank could hold together through a near encounter with the tidal forces of the Sun or a large planet. A flying sandbank barely held together by gravity would be easily disrupted, but comets had been observed to pass near these massive bodies and most survived.

Whipple came to the rescue. Instead of a gravel pile that extended over hundreds or thousands of miles, Whipple proposed that a comet was in essence a small solid nucleus, only a few miles in diameter. At heart, a comet was a . . . dirty snowball. When this nucleus neared the Sun, its surface ices sublimated (changed directly from ice to vapor) to form a diffuse coma around the nucleus, veiling it from view. The light and particles of the Sun then propelled those particles outward from the Sun to form the comet's tail. Comets could gradually lose their ability to outgas as their surface ices vaporized, leaving behind some of the dusty material as an insulating layer. Whipple suggested that some asteroids were actually defunct comets whose surface ices had been vaporized and whose interior ices were so covered with insulating dust that they could no longer develop comas and tails.

On he went, matching observed characteristics of a comet to the natural results of sunlight shining on a dirty snowball. Comets were sometimes late and sometimes early in their returns to the Sun, but not because the law of gravity was flawed and not because astrodynamicists had mis-

What led me to the icy comet model of 1950

FRED L. WHIPPLE, PH.D., Harvard-Smithsonian Center for Astrophysics

In 1950 I published the icy conglomerate model of cometary nuclei. For a decade I had realized that the gravel-bank theory of comets was quite impossible and that comets had discrete nuclei of ices and small solid particles. This conclusion was based on my studies of the Taurid meteor streams from meteors doubly photographed by cameras with rotating shutters. These studies were summarized in my 1940 paper* showing that the Taurid meteor showers come from Encke's Comet and that the comet had been spewing out meteoroids for hundreds of revolutions about the Sun. Its age was estimated at 1,400 years or more, the comet having a period of 3.3 years.

Fred L. Whipple. Courtesy of Smithsonian Astrophysical Observatory

Because of the effect of light pressure and the light momentum from solar radiation on small particles that were supposed to emit gas to form cometary comas, any loosely bound gravel bank would be dissipated long before 100 revolutions. I considered this idea so obvious that I said nothing about it, assuming that sensible astronomers know it anyway. Also I had no simple positive way to *prove* the hypothesis.

It had long been known that Encke's Comet was returning sooner than it should be by Newtonian theory. The period changes averaged more than 2 hours each period. This had been ascribed to a *resisting medium* in space, perhaps supporting the idea of loose small particles in cometary nuclei. But Halley's famous comet had returned about 4 days *late* in its 1910

* Fred L. Whipple: "Photographic Meteor Studies. III. The Taurid Shower," *Proceedings of the American Philosophical Society*, volume 83, number 5, October 1940, pages 711–745.

computed their orbits. The reason, Whipple said, was that comets were contributing a small amount to their own motion by the rocket effect of their outgassing as they rotated. Depending on which way they rotated, a comet could expand or contract its orbit by the jets of gas pouring off of it as it rounded the Sun. Whipple could even use his model to estimate what fraction of a comet's mass was lost on each visit to the Sun. Short-period comets could be expected to survive several hundred perihelion passes.

Whipple's comet model elegantly explained how comets could spawn meteoroids as their surfaces sublimated. The vaporizing ices released trapped dust and grit and carried them away from the comet's nucleus, whose gravity was too weak to hold them. Each passage of a comet near the Sun would add more particles to the meteoroid litter along and near its orbit. Those outcasts from the comet would follow almost the same path as the comet and be subject to the same perturbations from the planets.

return and one or two other comets were late. No resisting medium could increase the period of comets, so the facts were ignored. It is not intuitively obvious why a resisting medium should reduce the period of a comet. The effect is to reduce the energy of the Sun–comet system. This drops the comet into a smaller orbit where it moves faster. The effects of both reduced orbit size and increased speed shorten the comet's period.

As I was trying to improve the theory of atmospheric drag on meteors to improve by determinations of atmospheric density above about 60 kilometers (about 40 miles), I realized that the vaporization of the meteoroid might produce a jet action that would increase the drag coefficient. Then, all of a sudden, *the light dawned* on me! The same thing could be happening in comets! The solar radiation is sublimating (evaporating without melting) the ices in a comet, producing a jet force on the sunny side. If the comet were rotating, the lag in heating the surface could cause the jet force to accelerate or decelerate the nucleus, depending on how the comet was spinning. Every body in free space is turning one way or the other, so why not?

Thus I had a reason to publish my idea of an icy conglomerate for comets.[§] In fact, the major effect of the jet action on cometary nuclei is now known to be produced mostly by the radial force towards the Sun depending on whether an individual comet is more active coming in toward perihelion, or in receding from the Sun after perihelion. But the jet action is real and well verified by comet orbits, and, of course, the spinning effect also occurs but is harder to observe.

[§] Fred L. Whipple: "A Comet Model. I. The Acceleration of Comet Encke," *Astrophysical Journal*, volume 111, number 2, February 1950, pages 375–394.

Gases vaporizing from a comet nucleus create a rocket effect that alters the comet's orbit. Because a brief time passes between the moment the nucleus receives sunlight and vaporization takes place, the rotation of the nucleus determines how that thrust is directed. (a) If the nucleus is not rotating, the rocket effect pushes the comet outward from the Sun. (b) If the comet is rotating the same direction as it is revolving, the rocket effect increases the comet's speed and boosts it into an orbit farther from the Sun. (c) If the comet is rotating in the direction opposite to its revolution, the rocket effect decelerates the comet, dropping it into an orbit closer to the Sun. Diagram by Will Fontanez and Tom Wallin, University of Tennessee Cartographic Services Laboratory.

Almost from the moment Whipple's article appeared, the flying sandbank model of comets was out; Whipple's dusty iceball model was in. It was a revolution swift as a meteor. There were very few holdouts among the scientists.[4] The field went on without them.

Sentinels of the solar system

The second meteor science revolution was contributed primarily by Jan Oort of the Netherlands.[5] In 1950 he published an analysis of comets with approximately parabolic orbits, the borderline between bodies with very elongated elliptical orbits around the Sun and those with truly hyperbolic orbits that would pass the Sun once and never return.

His analysis showed that comets that passed anywhere near Jupiter, Saturn, Uranus, and Neptune would be as likely as not to be given a boost of energy that would send them out of the solar system forever. Because of the giant planets, the solar system must be continuously losing comets. Yet there were always more comets making their appearance. Where did they come from? Were they being captured from interstellar space? If some comets truly came from beyond the solar system, why weren't there any velocities comparable to those of the stars – several or many times the comet velocities observed? Instead, the very highest comet velocities were just barely hyperbolic, with a probable error in the measurements that could reduce every velocity to elliptical. So, Oort concluded, the comets with extremely long orbits were not coming from interstellar space. They were members of our solar system.

Was it possible that our solar system had its own distant unseen reservoir of comets which could steadily replace those that passed too near the Sun and planets and were destroyed or ejected? Oort calculated that if

Jan Oort. Courtesy of Mary Lea Shane Archives of the Lick Observatory

The Oort Cloud is a spherical distribution of comets that surrounds our solar system out to a distance of about 100,000 astronomical units (AU) from the Sun. The Kuiper Belt of comets extends outward from Pluto, gradually broadening into the inner portions of the Oort Cloud. Trillions of comets orbit the Sun within the Oort Cloud and the Kuiper Belt. When disturbed by gravitational forces beyond our solar system, some of these comets take up new orbits which carry them close to the Sun and Earth. In this diagram, each increment of distance outward is 10 times greater than the previous increment. The result is that the outer regions of this diagram are greatly compressed. Otherwise, the size of the Sun and planet orbits would be an invisible dot at the center of the diagram. Diagram by Will Fontanez and Tom Wallin, University of Tennessee Cartographic Services Laboratory.

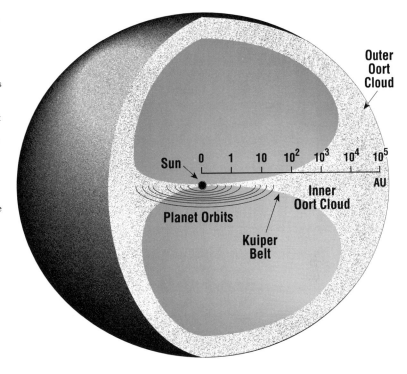

Ernst Öpik. *Annual Review of Astronomy and Astrophysics*, 1977

there were about 190 billion (1.9×10^{11}) comets scattered in orbits beyond those of the planets, extending out to perhaps 100,000 AU,[6] those comets could be affected by occasional close-passing stars whose gravity would divert some of the comets onto new orbits, a few of which would drop the comets in close to the Sun. With nearly 200 billion comets, such a comet cloud could last billions of years, since the solar system began, and still be supplying comets for earthlings to see.

Oort was not the first to suggest that a cloud of comets occupied the outer reaches of the solar system. In 1932, Ernst Öpik of Estonia was investigating the hyperbolic velocities attributed to some comets and meteoroids. To solve the problem, he proposed a comet-and-meteoroid cloud mostly confined within 10,000 AU of the Sun, but with outliers up to a million AU away. If our Sun had a comet cloud, other stars should too. The comets and meteoroids most distant from the Sun would be easily lost to the gravity of passing stars. Our Sun in turn would swipe a few of the outermost comets and meteoroids from the passing star's comet cloud and they would fall toward the Sun at speeds that marked them as being not of this solar system.[7]

Gerard P. Kuiper. Courtesy of Lunar and Planetary Laboratory, University of Arizona

In a footnote added just as his seminal article went to press, Whipple called Oort's paper an "extremely important" independent expansion of Öpik's work, and in a book he wrote about comets for the general public, Whipple referred to this comet reservoir as the Öpik–Oort Cloud. But almost everywhere else, in technical papers, in textbooks, in popular articles, the astronomical community refers to the outermost component of the solar system as the Oort Cloud, feeling perhaps that Oort conceived the comet cloud and its mechanisms more correctly.[8]

Expanding the reservoir

1950 was the year of Whipple and Oort. An important corollary to the Oort Cloud emerged into astronomers' awareness the next year, 1951, proposed by Gerard Kuiper at the University of Chicago's Yerkes Observatory. Kuiper was seeking to explain the origin of comets. Conditions were right at the beginning of planet formation, he calculated, for icy bodies to form in the outer planetary system, but the density was too low, especially beyond Neptune and Pluto, for large bodies to accrete.

Based on these conditions, Kuiper suggested that there must be a belt of billions of comets in the plane of the solar system starting just beyond the orbits of Neptune and Pluto and continuing out to the Oort Cloud.[9] As with the Oort Cloud, there was no direct proof of the existence of this belt of comets between the planetary part of the solar system and the innermost part of the Oort Cloud. Yet its existence was considered highly probable because this belt explained where the comets in the Oort Cloud came from and how they got there – by planetary perturbations. Gradually astronomers began calling it the Kuiper Belt.

As time passed, astronomers liked the concept of the Kuiper Belt more and more because it explained the behavior of short-period comets so well. Almost all short-period comets had direct orbits, traveling the same direction as the planets around the Sun. Almost all short-period comets had orbits only slightly inclined to the plane of the solar system. It was as if they all came from a belt moving in the same direction and in the same plane as the planets. The Kuiper Belt seemed to be where short-period comets came from.

Long-period comets, on the other hand, were often traveling in the opposite direction of the planets and with orbits tilted at every angle to the plane of the solar system. These traits suggested that they came from much farther away, from the Oort Cloud at the fringe of the solar system, where the gravitational effects of the galactic plane and passing stars and interstellar clouds could more easily perturb the dirty snowballs and send them in new directions, some to the center of the solar system, passing close to the Sun and Earth.

Long-period comets from the Oort Cloud, with their orbits tilted and perhaps reversed, could rarely be captured into short-period orbits. The Kuiper Belt seemed to be the source of Biela's Comet, Comet Giacobini–Zinner, and others with direct orbits and low inclinations that completed their circuits in under 100 years or so.

Recently, some astronomers have begun referring to the Kuiper Belt as the Edgeworth–Kuiper Belt in honor of Kenneth E. Edgeworth, an Irish astronomer who suggested a zone of comets beyond the outer planets in one sentence each in his 1943 and 1949 articles on the origin and evolution of the solar system. In 1943 he wrote: "It may be inferred that the outer regions of the solar system, beyond the orbits of the planets, is occupied by a very large number of comparatively small bodies." Edgeworth viewed comets as "heaps of gravel without any cohesion."[10] In 1949, he wrote: "It is not unreasonable to suppose that this outer region [beyond the orbit of Neptune] is now occupied by a large number of comparatively small clusters, and that it is in fact a vast reservoir of potential comets."[11]

Burning up

After World War II, Öpik, Whipple, and others worked out the details of a meteoroid's fiery plunge to oblivion in our atmosphere.[12] An incoming meteoroid begins to feel drag from the atmosphere of Earth at an altitude of about 90 miles (145 kilometers). That's why all spacecraft and artificial satellites must be placed 100 miles (160 kilometers) or higher above the Earth to stay in orbit more than a few days. Even 300 miles (500 kilometers) up, where the Hubble Space Telescope orbits and the Space Shuttle sometimes flies, there are enough atoms and molecules escaping upward from the air below to cause spacecraft to gradually fall to lower altitude. Left alone, the Hubble Space Telescope and other instruments in low Earth orbit would eventually reenter the atmosphere and become fireballs and meteorites.

When a meteoroid reaches 75 miles (120 kilometers) above the ground, the Earth's atmosphere is only one ten-millionth of its density at sea level. Yet the air is dense enough so that the friction between the meteoroid and the molecules in the atmosphere cause the meteoroid to begin to melt.

Depending on the meteoroid's size, composition, density, speed, and angle of entry, its surface begins to boil off at an altitude of about 60 miles (100 kilometers). The meteoroid begins to have a tail.

Depending on speed, the surface of the meteoroid reaches tempera-

New horizons for meteor research

IWAN P. WILLIAMS, PH.D., Professor of Mathematics and Astronomy, Queen Mary and Westfield College, University of London

Meteor science enjoyed its first golden age from about 1794 through 1867 as meteoroids were recognized to be of celestial, not meteorological, origin and as meteoroids were identified as particles lost by comets.

A second golden age began following the Second World War, as the use of radio waves in astronomy started to develop and one of its first uses was in detecting meteor trails. At this time the electronic computer also made its appearance, allowing the evolution of meteoroid orbits to be traced. By the end of this period, computer models were showing remarkably good agreement with observations. Our general understanding of meteoroid streams seemed to be complete.

Iwan P. Williams. Courtesy Iwan P. Williams

This agreement prompted the belief that virtually all that could usefully be known about meteors was already known. In consequence, especially in the West, many asked the question why we should waste more effort on meteor showers. Professional astronomers virtually abandoned the study of meteor streams, while theoretical development declined to almost nothing. The field was kept alive by studies in Soviet Bloc countries and by the observations of amateur astronomers.

Should we let this field die peacefully? Emphatically no. I believe that we are at the beginning of new period of high interest and activity.

tures of 5,500 to 18,000 °F (3,000 to 10,000 °C). A typical temperature is 12,000 °F (6,500 °C), hotter than the surface of the Sun – 10,000 °F (5,500 °C). Yet the meteoroid – or part of it – would survive this intense scorching during its brief plunge through the atmosphere if it could pass the heating of its surface along to the molecules of the Earth's atmosphere. It does so, enough to cause the molecules in the air to fluoresce brightly along and around the path of the meteoroid. But air, especially thin air, is a good insulator and the heat cannot be radiated or conducted away fast enough. Most of the heat of friction remains on the surface of the meteoroid.

Even so, the meteoroid, especially a larger one, might survive if it could quickly conduct the heat of friction to its core, thereby spreading the heat through its volume to prevent the surface from becoming so hot that it would melt and vaporize. But meteoroids, especially rocky ones, are also poor conductors of heat, so the temperature from friction builds up on the surface of the meteoroid. The surface vaporizes, exposing a new surface, which also ablates – and a meteor blazes briefly across the sky.

Computers allowed us to model meteor streams. Newer, more powerful computers have capabilities in data handling that have allowed us to identify a large number of discrepancies between simple meteor theories and the observations. For example, by studying the orbits of individual Quadrantid (Boötid) meteors rather than considering only their mean orbit, it was possible to understand why no Quadrantids were seen before the beginning of the 19th century. Computer simulations also showed that Comet Biela passed through the denser part of the Leonid meteor stream shortly before breakup.

Resolving these conflicts will allow us to use a meteor shower as a remote – and cheap – probe of the conditions on and in a comet's nucleus. We can deduce what the ejection speed of meteoroids must be, hence we can refine our models for outgassing from the nucleus. We can determine the densities of meteoroids, hence get information about the densities of the nuclei of comets. Further, since some meteor streams may be associated with asteroids, we can use those stream observations to characterize differences between asteroids and comets.

And yet another opportunity beckons us. Meteors are the remnants of interplanetary matter that has collided with the Earth. In understanding the dynamics of meteor motion, we may also obtain an understanding of the motion of potentially far more dangerous bodies, the near-Earth asteroids and comets.

The intense heat of friction from the streaking meteoroid instantly ionizes the air for a few yards (meters) around it. The thin, hot column of ionization that the meteoroid has left behind expands to a half mile (a kilometer) or more in diameter over the next second or two. It is the ionization of this column that reflects radio signals, allowing the study of meteors by radar. It is the light given off by this column that we see as a meteor's train.

By the time it reaches 45 miles (75 kilometers), there is almost nothing left of the meteoroid. It has melted, boiled, evaporated away into gases, now mingling with the Earth's atmosphere. Its flight to extinction through the upper atmosphere of Earth has taken two to three seconds.[13] It was visible as a glowing meteor for part of one second.

Sampling meteoroids

Not all meteoroids burn up, however. Large ones, especially from metallic asteroids, will hit the ground. Ironically, the smallest meteoroids survive too. Particles smaller than 0.1 millimeter (100 microns; 0.004 inch) will decelerate gradually without melting in the first faint wisps of the uppermost atmosphere.

Beginning with the expected return of the Leonids in 1966, scientists attempted to capture some of these particles – samples of comets, samples of nearby material least changed over the 4.6 billion years of solar system history. The United States Air Force launched an Aerobee rocket from White Sands Missile Range in New Mexico. Its nose cone had four arms that would extend, exposing surfaces to collect micrometeoroids – a Venus flytrap. On the way down, the petals would close into the nose cone to preserve the catch; then the rocket would parachute back to Earth. But the Venus flytrap failed to open.

Success in capturing meteoric specimens was finally achieved in 1973 by Donald Brownlee of the University of Washington and his colleagues by using a U-2 high-altitude, long-range reconnaissance (spy) aircraft converted by NASA for upper atmospheric research.[14] The jet carried a container on one wing that opened to expose an oil-coated surface to gently cushion and hold the impacting particles. To keep the fragile spacedust from vaporizing as it slammed into the container, the U-2 flew at 400 miles per hour (660 kilometers per hour), as slowly as it could in the thin air at 66,000 feet (12½ miles, 20 kilometers). Before the U-2 dropped down for landing, the micrometeoroid box was closed and remained so until it reached the laboratory. Brownlee and his team found they had caught minute rocky particles that often looked like clusters of grapes. The particles averaged about 10 microns (10 millionths of a meter) across and were

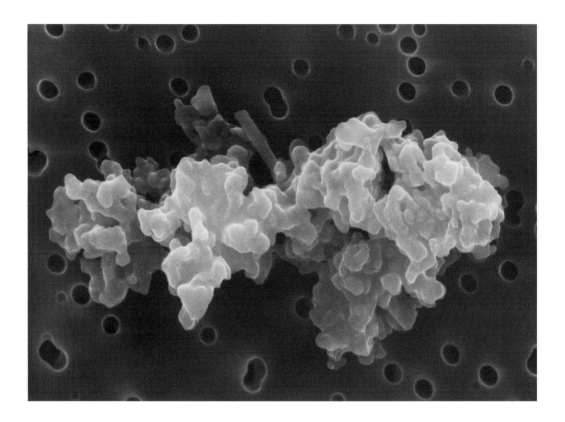

Interplanetary dust (Brownlee particle) collected in the atmosphere by a U-2 aircraft. Width of micrograph ~ 10 μm. Courtesy of Donald Brownlee, University of Washington

typically formed of oxides of such elements as silicon, iron, and magnesium, similar to a type of stony asteroid called a carbonaceous chondrite. Such asteroids might be defunct comets. Electron microscope photographs made the particles look white. Actually they were black as soot. Mixed with the ices in a comet, these dust particles turn a comet's surface black or dark in color.

Because the mineral composition of these particles is far closer to stony meteorites of the carbonaceous chondrite class than to dust on Earth, and because these particles crumble readily, like the meteoroids from comets that burn up so easily in the Earth's atmosphere, these particles are considered to be actual cometary dust – micrometeoroids. Their diminutive size allows the highest and most rarefied level of the Earth's atmosphere to gently decelerate them so that they do not incinerate themselves, but gradually float down toward the ground.

Technically, formally, they are interplanetary dust particles. Talking among themselves, astronomers often call them Brownlee particles.

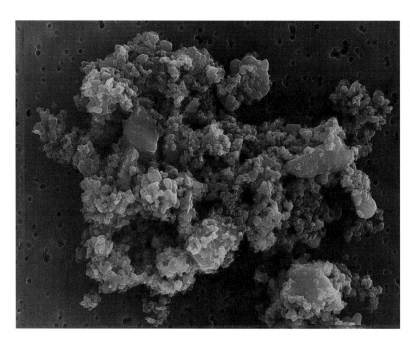

Another example of a particle of interplanetary dust (Brownlee particle). Width of micrograph ~ 10 μm. Courtesy of Donald Brownlee, University of Washington

NOTES

1. Charles P. Olivier: *Meteors* (Baltimore: Williams & Wilkins, 1925), page 23.
2. Fred L. Whipple: "A Comet Model. I. The Acceleration of Comet Encke," *Astrophysical Journal*, volume 111, number 2, February 1950, pages 375–394. See also Fred L. Whipple: "A Comet Model. II. Physical Relations for Comets and Meteors," *Astrophysical Journal*, volume 113, number 3, March 1951, pages 464–474.
3. England astronomer Richard A. Proctor (1837–1888) is frequently credited with this theory. Richard Arthur Lyttleton, who disagreed with Whipple's icy conglomerate model and tried to refine the flying sand-bank theory as an alternative, cited Proctor as the pro-genitor. See R. A. Lyttleton: *The Comets and Their Origin* (Cambridge: Cambridge University Press, 1953). But neither he nor others who credit Proctor cite his work on this subject. For the titles of Proctor's articles from 1868 to 1873, see Royal Society of London: *Catalogue of Scientific Papers (1864–1873)* (London: [Royal Society of London], 1879), pages 666–668. For the titles of most of Proctor's books, see Agnes M. Clerke's biographical note on Proctor in the *Dictionary of National Biography*. None of these books or articles by Proctor have titles that suggest he was proposing such a theory. The biographical entries on Proctor by Clerke in the *Dictionary of National Biography* and by J. D. North in the *Dictionary of Scientific Biography* make no mention of the flying sandbank. None of the familiar histories of 19th century astronomy credits Proctor for the flying sandbank model of a comet.

   Olmsted in the 1830s (chapter 2), Herrick in the 1830s (see chapter 6), and Oppolzer in the 1860s (chapter 8) are just a few of the astronomers who assumed that comets are dense swarms of meteoroids. The essence of the flying sandbank theory prevailed for more than a century.
4. One holdout for the flying sandbank theory was Raymond Arthur Lyttleton. See his *The Comets and Their Origin* (Cambridge: Cambridge University Press, 1953) and "Does a Continuous Solid Nucleus Exist in Comets?" *Astrophysics and Space Science*, volume 15, 1972, pages 175–184.

5. Jan H. Oort: "The Structure of the Cloud of Comets Surrounding the Solar System and a Hypothesis Concerning Its Origin," *Bulletin of the Astronomical Institutes of the Netherlands*, volume 11, 1950, pages 91–110.

6. Oort actually proposed 50,000 to 150,000 AU. Subsequent considerations, especially the perturbations on the outermost comets caused by the tidal effect of the galactic plane and occasional near passages of interstellar molecular clouds (nebulae), have reduced the proposed radius of the Oort Cloud to about 100,000 AU. The estimated number of comets in the outer Oort Cloud has grown to a trillion (10^{12}), with 5 trillion more in the inner Oort Cloud and Kuiper Belt.

7. E[rnst J.] Öpik: "Note on Stellar Perturbations of Nearly Parabolic Orbits," *Daedalus* (Proceedings of the American Academy of Arts and Sciences), volume 67, 1932, pages 169–183.

8. For a fuller assessment of Whipple's and Oort's work and credit for other major contributors, see Donald K. Yeomans: *Comets: A Chronological History of Observation, Science, Myth, and Folklore* (New York: John Wiley & Sons, 1991), chapters 9 and 11.

9. Gerard P. Kuiper: "On the Origin of the Solar System,"
pages 357–424 in J. A[llen] Hynek, editor: *Astrophysics: A Topical Symposium* (New York: McGraw-Hill, 1951).

10. K[enneth] E. Edgeworth: "The Evolution of our Planetary System," *Journal of the British Astronomical Association*, volume 53, July 1943, page 186 of pages 181–188.

11. K[enneth] E. Edgeworth: "The Origin and Evolution of the Solar System," *Monthly Notices of the Royal Astronomical Society*, volume 109, number 5, October 1949, pages 600–609.

12. See D[onald] W. R. McKinley: *Meteor Science and Engineering* (New York: McGraw-Hill, 1961), chapter 7.

13. Duncan Steel: *Rogue Asteroids and Doomsday Comets: The Search for the Million Megaton Menace that Threatens Life on Earth* (New York: John Wiley & Sons, 1995), page 5.

14. D[onald] E. Brownlee, F. Horz, D. A. Tomandl, and P[aul] W. Hodge: "Physical Properties of Interplanetary Grains," pages 962–982 in B[ertram D.] Donn, M[ichael J.] Mumma, W. Jackson, M[ichael F.] A'Hearn, and R[obert S.] Harrington, editors: *The Study of Comets* (Washington, D. C.: National Aeronautics and Space Administration [NASA SP–393], 1976).

Surprise – 1966

Hardly anything in 1899. Nothing much in 1932. Bernard Lovell, a pioneer in using radio telescopes to study meteors, had little hope that the Leonid meteors would ever again be noteworthy. The Leonids' parent comet, Tempel–Tuttle, had not been seen since 1866. Despite orbital calculations to figure out where to search for the comet and then energetic efforts to find it in 1899–1900 and 1932–1933, Tempel–Tuttle was not found. Most astronomers considered it to be lost, probably broken up. Without a continuing supply of particles, the Leonid meteor showers were doomed to fade away. In 1954, Lovell wrote a book devoted to the science of meteors and carved an epitaph for the Leonids: "It now seems certain that the main part of the Leonid orbit has been removed from the earth's orbit by successive perturbations and the recurrence of the tremendous meteoric storms of the Leonids in the future seems unlikely."[3]

It looked like Lovell was right. There was a flicker of hope in 1961 when the Leonids produced an unexpected burst of 50 an hour, scarcely as good as the Perseids of August typically are, but about four times the normal off-year Leonid rate. After 1961, nothing. Back to the usual annual Leonid drearies in 1962 and 1963 – about 10 to 15 per hour. In 1964 the Leonids doubled that rate. Ho hum. With Leonid season approaching in 1965, *Sky and Telescope* magazine saw no reason to alert readers. Observers on the American mainland saw up to 50 Leonids an hour on November 17, 1965. But observers in Hawaii and Australia saw meteors falling at the rate of perhaps 100 per hour, better than 1932 at its best.[4]

Embers of hope for the Leonids in 1966 began to glow. Harold B. Ridley made what many considered a somewhat enthusiastic prediction about the Leonids in the *Handbook* of the British Astronomical Association: "a strong maximum may be confidently expected in 1966, when the hourly rate is unlikely to be less than 100."[5] After all, that was about as good a display as the Leonids had managed in the 20th century.

There was additional good news. German astronomer Joachim Schubart used a new speedy computer to recalculate the orbit of Comet Tempel–Tuttle, which allowed astronomers to find the comet – for the first time in a century. With the comet still intact, it should be continuing to supply debris for future meteor showers on each visit to the Sun. *Science News* was waxing optimistic:

> The heavenly show produced by the Leonids is expected to be spectacular this year since it comes close to the time of the return of Comet Tempel–Tuttle. The shower could well rival the one in 1833, where so many shooting stars fell that they were reported to resemble a storm of snowflakes falling from the sky.[6]

The path of Comet Tempel–Tuttle would soon be so well known, *Science News* continued, "that this comet should never be lost again."

Charles Olivier, founder and still president of the American Meteor Society, recalled the public relations disaster in 1899 and had worked hard to educate the public about the Leonids in 1932, only to have them trickle down the sky. "No certain prediction is possible," he warned. Brian Marsden of the Smithsonian Astrophysical Observatory warned that the perturbations the comet experienced might be different from those affecting the meteoroid swarm. Everyone was hedging his bet with very careful phrasing. Thomas Nicholson, chairman of the American Museum of Natural History and its Hayden Planetarium, wrote with elegant ambiguity: "It could well be that we may see, this November, another interesting, if not spectacular, Leonid display." And, he continued, "If you are looking, you should certainly see at least several dozen meteors per hour in the eastern sky."[7] That kind of prediction could not technically be wrong. For meteor astronomers, two-thirds of a century of having their faces slapped by the press, the public – and the meteors – was enough. They were being extra careful this time.

Still, observers should be alert. The Moon would be near new phase, rising and setting with the Sun, and therefore completely absent from the nighttime sky. Olivier urged readers to get far from city lights and to make single observer counts. "Combined counts by two or more observers cannot be used." *Sky and Telescope* reminded readers that few Leonids would be visible until their radiant near Gamma Leonis was well up in the sky, about 2 a.m. local time.[8] (For some reason, that time would be misinterpreted by some news sources and observers.)

The newspapers were never sloppier. Poor George Van Biesbroeck at the University of Arizona's Lunar and Planetary Laboratory. The *Phoenix Gazette* said he was advising observers that "the best view should be from the east" – whatever that meant – and that this "display is the first in 33 years" – as if 1933 was a significant Leonid shower.[9] The next day the *Gazette* printed an Associated Press story from Harvard that advised the public that the meteors would be seen "about midnight (Phoenix time) toward the east and a little south."[10] A Leonid display can't really start until midnight, when its radiant in the constellation Leo rises in mid-November. A Leonid shower then typically builds toward dawn as the radiant is higher in the sky and as the Earth spins the observer into a position facing forward along the Earth's orbital path as it plunges through the meteoroid stream. Furthermore, the meteors would be visible everywhere in the sky.

Walter Sullivan, science reporter for the *New York Times*, wrote a major story about the Leonids. He reported that Kenneth L. Franklin of New York's Hayden Planetarium was optimistic: "the position of the parent comet, Tempel–Tuttle, is now roughly what it was in 1833. This had led him to suspect that a dramatic shower, like the one of [1833], may be in the offing." But Sullivan stumbled on some of the information. On the origin of meteoroids, he wrote: "Presumably the material was torn from the comet by a close encounter with the gravity of the sun or a large planet." (It is the Sun's radiation, not gravity, which vaporizes ices on the comet, releasing dust and rocky bits trapped within the ice.) "Only the Leonids," Sullivan wrote, "have produced dramatic displays such as the one that may occur tonight."[11] (Certainly the Leonids have produced the most consistent and probably the most torrential meteor storms, but the Lyrids in 1803, the Bielids (Andromedids) in 1872 and 1885, and the Giacobinids (Draconids) in 1933 and 1946 were among the meteoroid streams that had wowed the public in the years between Leonid outbursts.)

Based in part on Franklin's confident prediction, New York City decided to throw a meteor party, a family festival of shooting stars. The *New York Times* reported:

> Parks Commissioner Thomas P. F. Hoving invited New Yorkers to gather in Sheep Meadow in Central Park at midnight tomorrow for his biggest "happening" to date – the expected 3-hour shower of Leonids that occurs only once in a generation.
>
> He urged parents to put their youngsters to bed early and then bring them to the park – together with warm blankets and jugs of hot chocolate or coffee – "because of the magnificence of this shower of stars, which may not repeat itself for another 33 years."
>
> A team of astronomers from Hayden Planetarium, headed by Dr. Thomas D. Nicholson, will describe over a public address system just what is happening. They will use an electric beam, stabbing 300 feet into the sky, to point out various star formations.[12]

Several thousand observers showed up with mugs of coffee and soup, which "gave the evening the appearance of a nocturnal picnic."[13] Alas, "they saw only a gloomy, yellowish cloud cover that reflected the lights of the city below." Shortly after midnight, Nicholson told the crowd "that the clouds extended up to 24,000 feet [7,300 meters] and were not expected to break until daylight." With their blankets, sleeping bags, and lawn chairs, most of the amateur astronomers resolved to wait for a break in the clouds. "The first rule in watching meteor showers," Nicholson cheerfully told the crowd, "is to pick a clear night." Half the crowd booed while the other half laughed. He then gave the audience a 25-minute talk on what the Leonid shower might have looked like. "This year's shower," the reporter wrote, "had been expected to be the most brilliant since 1866, when the meteors were so bright that New Yorkers were startled out of their beds." Actually, with good publicity in 1866, many New Yorkers were outside watching, just as they were a century later – and seeing nothing. Europe was the lucky longitude in 1866. It is curious the way so many Leonid no-show years later get referred to as wondrous events. Parks Commissioner Hoving, later director of the Metropolitan Museum, wasn't present at his party in Central Park. He was in Japan, where the visibility was excellent.

"The crowd, obviously desperate for amusement, was easily pleased," the reporter wrote. "A plane passing overhead with its running lights blinking drew a lusty cheer." Gradually the people drifted away. "Rotten deal," said one man. Said another, "I'll be around 33 years from now. I'll catch it then." A woman observed, "It's marvelous. All these people in the park after midnight and no one is getting mugged." The article concluded, "Perhaps to stay warm, a group of young people began to sing: 'When you wish upon a star...'"

Trails of 1966 Leonid meteors, photographed by James W. Young. Courtesy Jet Propulsion Laboratory/NASA, Table Mountain Observatory; photos, by James W. Young

(a)

(b)

The reports begin to arrive

The next day, Walter Sullivan of the *New York Times* began his article: "The most spectacular celestial fireworks of this century were observed early yesterday morning over the southwestern United States."[14] Nathan A. Fain, a night assistant at the University of Texas' McDonald Observatory,

James W. Young photographed these 1966 Leonid meteor trails showing the radiant near the top of the picture just left of center. Note the point meteor at that position. Courtesy Jet Propulsion Laboratory/NASA, Table Mountain Observatory; photo by James W. Young

estimated that the Leonids had fallen at a rate of perhaps 50,000 an hour. A rate almost three times higher was reported from Arizona.

In Atlanta, hundreds of stargazers had assembled in hopes of seeing a shower of stars. A hundred members of the Atlanta Astronomy Club met on a rifle range 25 miles from the city. But according to the article, a local astronomer was skeptical.

> Agnes Scott College professor of physics and astronomy William Alexander Calder said, "If a really red-hot shower were coming, we would have seen evidences of it the night before . . . I'm going to continue watching and if anything happens I will rout out the students. Frankly, I think it's going to be a big disappointment."[15]

The president of the Atlanta Astronomy Club was an airline pilot on duty that night, flying from Columbia, South Carolina back to Atlanta just before dawn. In about one hour, he and his copilot counted 350 meteors through the side windows in the cockpit. Bob Fried thought that 350 Leonids an hour was what a single observer probably would have tallied if his view had not been restricted. " 'I've never seen anything like it in my life,' said Fried. 'They went swoosh, swoosh, one after the other, and almost every one left a smoky trail behind it.' "[16]

As for the Georgians on the ground, the newspaper reported, most of them expected the meteor display between midnight and 2 a.m. and "went to bed disappointed." The story continued: "Dr. William Calder, Agnes

Scott astronomy professor, said he gave up the vigil about 2 a.m."[17] Bad information about astronomical happenings can be worse than no information at all.

Phoenix was rejoicing. At least those who saw it were. It was a page 1 story-with-photograph for the *Phoenix Gazette*, happy as always to play cheerleader for the city's clear skies, but lax on accuracy, insisting on referring to meteors as meteorites despite the expert advice available to them. "Most Arizonans just weren't 'up for it.' Those who were out late or up early today saw a spectacular meteorite display – a brilliant shower of shooting meteorites. Dennis Milton [Milon] of the University of Arizona viewed the display from Kitt Peak Observatory and called it a 'snowfall of meteorites.'"

The *Gazette* had lots of anecdotes to report. The Lowell Observatory in northern Arizona estimated about 1,000 meteors a minute near 5 a.m. "Dr. Keith Parizek, Paradise Valley dentist and president of the Phoenix Observatory Association, said his peak count reached about 198 every two minutes at about 4:45 a.m. He said he 'tried to count so fast at one point that I broke my wife's grocery-shopper calculator.'" With good journalistic instincts, the reporter made a phone call and found that "Phoenix police said only one woman called to say the sky was falling."[18] Public understanding of meteors had apparently progressed in a century and a third.

Meanwhile the eastern United States newspapers were having a hard time grasping the intensity of the meteor storm that the southwestern states had seen. The *Washington Post* published an Associated Press-distributed photograph that carried a caption with nary a trace of accuracy: "Despite some reports of as many as 1,500 meteor sightings an hour in the Southwest, the shower – expected to be the greatest star show in more than a century – didn't live up to expectations. Unusual activity on the planets Jupiter and Saturn and cloudy skies were blamed for the less-than-expected show."[19] There was no accompanying story.

Science News headlined their story "Leonids Fulfill Promise" and reported rates of 200 meteors an hour along the East Coast to 40 meteors a second over Arizona. They repeated, as did several publications, Hayden Planetarium astronomer Ken Franklin's forecast that "In 1966, the position of the earth in relationship to Comet Tempel–Tuttle ... was the same as it was when the awe-inspiring Leonids showered 'like snowflakes' 133 years ago."[20] That was not quite true. (And in the *New York Times* article, Franklin had qualified the similarity as "roughly.") It was true that the comet came close to the Earth's orbit on the sunward side and that it crossed the Earth's plane before the Earth arrived. But the Earth in 1966 had passed more than two times farther away from the comet's orbit than in 1833 and had reached that plane-crossing point 8½ months later. Those Earth–comet-encounter distances and times would be important to correlate with what happened in the skies: meteor storm, meteor shower, or meteor miss, so that it might be possible to better predict Leonid tempests in the future.

Sky and Telescope published six pages of reports and photographs, noting that "the 1965 zenithal hourly rate for Leonids was only about 50, and few astronomers shared H. B. Ridley's confident expectation of at least 100 Leonids an hour in 1966. It turned out that this prediction was about 1,000 times too small!"

The great Leonid meteor shower of 1966

JAMES W. YOUNG, Resident Astronomer, Table Mountain Observatory, Jet Propulsion Laboratory

Late in the afternoon of November 16, 1966, Dale Etheridge of Mt. San Antonio College, along with his wife, came to Wrightwood for dinner with my wife Frances and me. We made extensive preparations for the expected meteor shower the following morning. Unfortunately, the weather was cloudy, with a heavy alto-cirrus overcast. We waited up till after midnight with no change in the cloud cover. We went to bed.

But I forgot to visit the bathroom before I went to sleep. I awoke at 2:30 a.m., finding that I needed to take care of that matter. As I stood in the dark, I peered out the bathroom window to the south, and thought I saw a streak of light in the sky. As I looked more carefully, I began to see additional streaks – meteors just above a now thinning overcast. I rushed outside in my pajamas and saw a few holes in the clouds, with meteor streaks everywhere I looked. I woke everyone. We got dressed and rushed up to Table Mountain Observatory, which is operated by the Jet Propulsion Laboratory. We woke up Charles (Chick) Capen, the resident astronomer, along with his wife Virginia and son Mars.

Dale set up his camera along with a timed rotating shutter placed in front of the lens to determine meteor speeds, a project in conjunction with other observers in Southern California. Chick and I set up 35mm cameras with 35mm and 50mm f/2 lenses. We used Kodak Plus-X film.

It was just after 3 a.m. local time. Meteors were everywhere. We couldn't count them, but 1 or 2 per second was a good guess. They were extremely fast, and tended to be white. Every now and then a bright fireball would appear, leaving a long, persistent train. As I tried to photograph some of the trains, I would catch other fireballs in the same field of the camera.

By 3:30 the intensity of the shower had grown to 5 to 10 meteors per

James W. Young and Charles Capen watched the show from the Jet Propulsion Laboratory's Table Mountain Observatory in Wrightwood, California. Capen said "we saw a rain of meteors turn into a hail of meteors and finally become a storm of meteors too numerous to count . . . " Young estimated 10 meteors a second for 30 minutes, the equivalent of 36,000 meteors an hour. Some fireballs left trains that lasted up to 20 minutes, Young said.

second. We felt like it was "hard-hat" time. Some of the fireballs were in the −8 to −12 magnitude range – at brightest rivaling the full moon. The earlier clouds had completely disappeared and the sky was absolutely crowded with meteors, radiating from the "sickle" of Leo. The sight was utterly breathtaking.

Sometime just after 4 a.m., the shower peaked at what we estimated to be upward of 50 meteors per second! Opening our eyes for the same duration as we normally blink our eyes, the sky was full of meteor streaks in every direction. One fireball of magnitude −10 came directly toward us from the radiant point. Awesome! It was a spectacle none of us had ever expected or ever witnessed before. We were numbed by the sight. We no longer yelled to the others where a bright fireball was. Each of us was watching his own show and didn't want to be interrupted.

At 4:30, the shower was still intense, 10 to 20 meteors per second. We took frame after frame on various cameras, averaging 3- to 6-minute exposures. It was obvious that the predicted point on Earth for this event – Europe – was incorrect. The maximum was where we were, in the southwestern United States.

As the intensity of the shower continued to taper off and twilight began, we all talked together about what we had seen, an event unprecedented in modern times. The films were developed, washed, and dried. After some of the better frames were printed, each negative was inspected. We had recorded over 1,100 meteors in about 90 minutes using only slow-speed film. We had seen 22 fireballs and photographed 12 of them, along with a few of their trains.

As I look back some 30 years at Table Mountain, I cannot think of any other event that came close to the awesome experience of that meteor shower. I still retain the visual impressions I had that morning of November 17, 1966 when I blinked my eyes open for only an instant and saw a sky full of meteor streaks. I can still see it vividly, but it is difficult to describe.

September 3, 1997

H. Gordon Solberg, Jr. of Las Cruces, New Mexico reported a single observer rate of at least 30,000 Leonids an hour. Jose Olivarez at Mission, Texas and many observers elsewhere were struck by the number of brilliant fireballs they saw. Olivarez saw ten, three of which cast shadows. He also found a striking image to describe the event: "As the shower reached its climax, the radiant was very high in the southeastern sky, and the meteors falling in all directions gave the impression of a gigantic umbrella."

Leonid fireball, November 17, 1966 – magnitude about −10; photograph by James W. Young. Courtesy Jet Propulsion Laboratory/ NASA, Table Mountain Observatory; photo by James W. Young

In Boulder, Colorado, astronomer Dana K. Bailey remembered watching for the Leonids that never came in 1933. He went out looking again in 1966, but his skies were mostly overcast. But through occasional breaks in the clouds, he saw numerous meteors. He jumped in his car and drove up Boulder Canyon until he found clearer skies. "I was simply stunned by the awesome display," he said. "There were far too many meteors for any direct count . . . . I had the feeling that I should be hearing something." He tried to devise some means of estimating the meteor falls.

The 1966 Leonid meteor trails demonstrate their radiant. Note also the two point meteors near the radiant in the sickle of Leo. Photographed by Dennis Milon in a 3½-minute exposure from Kitt Peak, Arizona. Copyright 1966 Dennis Milon

Counting "one thousand and one" over and over again, I tried to guess how many new meteors appeared before me in a one-second interval. I also tried opening my eyes for one-second periods and keeping them shut afterward while I visualized what I had seen. After several minutes of such observing, I decided that not fewer than 10 new meteors were appearing each second, for many minutes. Sometimes the rate was double or triple that, and a one-second interval with fewer than 10 new meteors in view was rare.[21]

Dennis Milon, leader of a group of 13 amateurs who watched the spectacle from Kitt Peak National Observatory in Arizona, reported "A rate equal to around 150,000 per hour for a single observer was seen for about 20 minutes. This is perhaps twice the rate of the 1833 Leonids, when the peak rate for a single observer was 20 per second." Milon's estimate was the highest rate ever offered as a systematic meteor count.[22] As great as the 1966 display was, it is doubtful if it was twice that of 1833 or perhaps even equal to it. Other competent observers in good locations in 1966 estimated between 30,000 and 60,000 meteors an hour. Even then, they – and Milon and his colleagues – were having great difficulty estimating meteor numbers past 10 per second. Given the techniques available then and even now, there is no way to resolve which display – 1833 or 1966 – was greater, nor is there a need for rivalry. The two certainly stand as the most awesome Leonid showers in the past two centuries. They were beyond counting . . . and beyond the words that strove to tell of them.

After his condensed account appeared in *Sky and Telescope*, Milon published a fuller account in the *Journal of the British Astronomical Association*. Six members of Milon's group watched for Leonids from Kitt Peak the previous morning, November 16, but saw only 9 Leonids an hour per observer. That night, the full 13-man team left Tucson under cloudy skies for the 45-mile (72-kilometer) drive to Kitt Peak. From 1:30 to 2:30 local time, they saw a modest shower of about 33 an hour that "gave no indication of the spectacle to come."

During the second hour, the rate rose to 192 for a single observer. Team members were trying to make observing reports on the meteors they saw. Their timekeeper "was frantically calling out serial numbers as all nine observers, it seemed, yelled at once, 'Meteor! Meteor!' and 'More observing forms!' " The meteors kept coming faster and faster. "There were yells of 'I can't do it! There are just too many!' " So they stopped estimating magnitudes and starting making counts per minute. "The count was about 30 per minute at 4:10 when a −8 or so exploded." (A fireball with a magnitude of −8 is about 30 times brighter than Venus, the brightest starlike object in the nighttime sky.) "The sky literally began to rain shooting stars," Milon said. "It was obvious to us that this type of shower would terrify the ignorant, not to mention effects upon astrologers!"

"By 4:30," said Milon, "there were several hundred per minute. At 4:45 the meteors were so intense we guessed how many were seen by a sweep of the head in one second. The fantastic rate of about 40 per second was reached at 4:54 a.m. It was indeed difficult to gauge such a rate, but this is the consensus of the observers." One of the observers noted that "looking directly at the radiant gave the effect of depth of the Earth moving through space."

"The view was so spectacular we just didn't know where to look!" Milon wrote. Sometimes we would spin around, taking in the whole sky . . . . Everyone was yelling and laughing at the incredible, dazzling sight, and at our luck in seeing it."

"By 5:40 the shower was back down to 30 per minute," Milon reported. "We continued to see Leonids in the brightening dawn sky until a colourful Arizona sunrise closed out our observing."[23]

Remembering the 1966 Leonids

JOSE OLIVAREZ, Director of Astronomy, Chabot Observatory & Science Center, Oakland, California

Jose Olivarez. Courtesy of
Jose Olivarez

On the morning of November 17, 1966, I was part of a small group of astronomy students from Pan American University who gathered at a site north of Mission, Texas to observe the Leonids from 1:25 to 6 a.m. in a very clear sky. The display started uneventfully, with a few bright meteors early on, but only eight meteors during the first hour. As time passed, the numbers increased rapidly to dozens per hour, then hundreds, until at 5:15 they were pouring down, too many to count.

The brightest Leonids were especially striking. Many left broad granular-textured trains 10 to 30 degrees long. Yet these bright Leonids were outshone by numerous green fireballs (at least 10) that appeared at the height of the display. All the fireballs exploded with a white light and three of them cast shadows. I particularly remember some of these fireballs exploding in the south-southwestern part of the sky with flashes that illuminated our viewing site like lightning.

When the shower reached its climax for us about 5:45 a.m. Central time, the radiant was very high in the southeastern sky and the meteors falling in all directions from that point gave the impression of a gigantic umbrella. The meteors were waterfalling out of the head of Leo. The waterfalling effect seemed most intense in the south-southwest before the maximum, then more intense in the north-northwest around maximum.

Perhaps the oddest thing I noted at the height of the display was the peculiar appearance of the meteors emanating from the radiant. The greatly foreshortened meteor trails seen there resembled short bright strings of beads.

Watching the Leonid meteor shower on the morning of November 17, 1966 was both awesome and delightful. Its fantastic climax went on for over 30 minutes at my south Texas site and continued as the meteors began to fade out in the brightening light of dawn.

August 20, 1997

1967 and after

A year later, the Leonids had been largely forgotten, except by those who saw them. The *New York Times* mentioned that a meteor shower was expected the next morning and that it might be "unusually heavy." The usual rate, the article said, was 15 to 20 per hour. The story said that last year's shower had been "intense," but mentioned no rates, offered no descriptions. It said only that "the number of meteors seen tonight may again be above normal."[24]

Sky and Telescope explained why there was little hope for the 1967 Leonids as it paid tribute to the 1966 display: "Last year's Leonid storm was among the greatest in history, but its peak was very brief and limited to a small area in the southwestern states." The problem for 1967, it said, was twofold. First, a full moon would drown out most meteors. Second, the Earth would return to its meeting point with the Leonid stream in one year – 365¼ days. That extra quarter day meant that when the Earth reached the Leonids' crossing, it would be rotated on its axis one quarter day, so that whatever meteors there were would fall mostly over the Pacific Ocean and Far East.[25]

Nothing notable was seen.

In 1969, the Leonids threw a surprise farewell party for themselves – 140 meteors an hour.[26] Observers present had a good time, but attendance was sparse.

Gradually attention began to turn to 1998 and 1999.

NOTES

1. D[onald] W. R. McKinley: *Meteor Science and Engineering* (New York: McGraw-Hill, 1961), page 7.
2. Walter Sullivan: "Meteor Rain Awes Southwest With Celestial Fireworks Show," *New York Times*, November 18, 1966, page 45.
3. A. C. B[ernard] Lovell: *Meteor Science* (Oxford: Clarendon Press, 1954), page 338.
4. *Sky and Telescope* [no author given]: "Many Leonids Observed," volume 31, January 1966, pages 58–59.
5. L. J. R. [Leif J. Robinson]: "Observations of Three Meteor Showers," *Sky and Telescope*, volume 31, February 1966, pages 112–115.
6. *Science News* [no author given]: "Lost Comet Showers Sky," volume 89, June 11, 1966, page 459.
7. Thomas D. Nicholson, "Sky Reporter: The First Major Leonid Shower Since 1932 May Occur This Month,"
Natural History, volume 75, November 1966, pages 42–45.
8. *Sky and Telescope* [no author given]: "A Good Leonid Year?" volume 32, November 1966, page 251. This article also contains the warnings by Olivier and Marsden.
9. *Phoenix Gazette* [no author given]: "Watch Out for Leonids," November 15, 1966, page 8.
10. *Phoenix Gazette* [no author given]: "Meteors May Shine Tonight," November 16, 1966, page 62.
11. Walter Sullivan: "Meteor Shower Is Due Tonight," *New York Times*, November 16, 1966, pages 49, 54.
12. *New York Times* [no author given]: "Hoving Issues Invitation to Heavenly Happening," November 15, 1966.
13. Terence Smith: "Asia Sees Meteors; Show Here Foiled," *New York Times*, November 17, 1966, pages 1, 22.
14. Walter Sullivan: "Meteor Rain Awes Southwest With

Celestial Fireworks Show," *New York Times*, November 18, 1966, page 45.

15. *Atlanta Constitution* [no author given]: "Big Meteor Show Off to Slow Start," November 17, 1966, page 36.

16. Quotation from *Atlanta Constitution* [no author given]: "Meteor Spectacular Thrills Georgians," November 18, 1966, page 30. Information about how Robert E. Fried observed was taken from this article and from his later report and interpretation in *Sky and Telescope* [no author given]: "Great Leonid Meteor Shower of 1966," volume 33, January 1967, pages 4–10.

17. Quotation from *Atlanta Constitution* [no author given]: "Meteor Spectacular Thrills Georgians," November 18, 1966, page 30.

18. *Phoenix Gazette* [no author given]: "Early Birds See Meteors in a Brilliant Performance," November 17, 1966, page 1.

19. *Washington Post*, November 18, 1966, page A3.

20. *Science News* [no author given]: "Leonids Fulfill Promise," volume 90, November 26, 1966, page 453.

21. *Sky and Telescope* [no author given]: "Great Leonid Meteor Shower of 1966," volume 33, January 1967, pages 4–10.

22. Milon did not say where he found the rate he quoted for the 1833 Leonids, or indeed any reliable rate for them. Physicist Joseph Henry at the College of New Jersey (now Princeton University) estimated somewhat offhandedly "that 20 [meteors] might be counted almost at the same instant . . . " Joseph Henry: *The Papers of Joseph Henry*, volume 2, edited by Nathan Reingold (Washington, D.C.: Smithsonian Institution, 1975), page 116 of pages 116–121, 128–130, 133. In the

November 13, 1833 *New York Commercial Advertiser*, a reporter wrote: "Within the scope that the eye could contain, more than twenty could be seen at a time shooting (save upward) in every direction."

In his survey of scientists' and newspaper accounts of the 1833 Leonids, Denison Olmsted found only one attempt to estimate meteor rates. An unidentified writer in Boston's *Columbian Centinel* estimated that he saw about 1,000 meteors fall in 15 minutes, the equivalent of 4,000 meteors an hour. This was *not*, however, the peak rate, as the writer noted: "I stood observing the phenomenon till fifteen minutes before six, at which time, the meteors being fewer, I attempted to count a portion of them." Olmsted felt that the estimate was "considerably too low." Denison Olmsted: "Observations on the Meteors of November 13th, 1833," *American Journal of Science and Arts*, volume 25, number 2, January 1834, page 366 of pages 363–411.

23. Dennis Milon: "Observing the 1966 Leonids," *Journal of the British Astronomical Association*, volume 77, 1967, pages 89–93.

24. *New York Times* [no author given]: "Meteor Shower Expected to Reach Peak Tomorrow," November 16, 1967, page 55.

25. W. H. G. [William H. Glenn]: "November Meteors," *Sky and Telescope*, volume 34, November 1967, pages 345–346.

26. Estimated by Peter Millman; reported in Paul Roggemans, editor: *Handbook for Visual Meteor Observations* (Cambridge, Massachusetts: Sky Publishing, 1989), page 163.

CHAPTER 14

Killer comets and dis-asteroids

[After watching the Leonids in 1866:]
Here is a shower of missiles, of unknown weight and inconceivable velocity, always in motion from some unknown battery, and every human generation has to run the gauntlet. Will this atmosphere of ours always prove an absolute protection . . . ?

LONDON *TIMES*, November 15, 1866

THE LEONIDS occasionally provide a meteor storm, but no meteoroids from that family have ever been observed to smash into the Earth. The Perseids provide a meteor shower annually, but no meteoroids from that family have ever been observed to crash into the Earth. That is very reassuring.

Until we consider the matter further.

During meteor showers, we see from time to time spectacular meteors that rival or exceed Venus in brightness – and the traces of those fireballs show that they come from the same radiant as the shower in progress. The fireballs are created by larger particles – the size of marbles or tennis balls – that belong to that meteor stream as it orbits the Sun. Not all the particles in a meteor stream are the size of grains of sand or peas. The large particles are less numerous, but they are there. And they are hitting the Earth . . . but vaporizing up in the atmosphere.

We expect more small objects than large ones in nature because of the evidence around us. In our solar system, there are many more small bodies – comets and asteroids – than there are planets. Beyond our solar system, there are many more small stars than giant ones. In the universe there are more small galaxies than large ones.

But fireballs alert us to the possibility that even larger chunks of comets may be – must be – traveling in the streams of debris they shed. And the danger is not just from fragments of comets. It is from the comet itself. After all, a meteor *shower* is evidence that the Earth is crossing *very close* to where a comet passed or will pass. A meteor *storm* is evidence that a comet is still at work very near by. Just a little closer and . . .

When we look at the Moon, we see thousands of craters, evidence that it has suffered the impact of comets and asteroids throughout its history. Spacecraft have shown us extensive cratering on all the solid bodies in the solar system – planets, moons, even asteroids. Among the inner planets, the Earth has suffered most because it is a larger target and possesses more gravity to bend the passing debris in its direction. The Earth is not more pockmarked than the Moon because craters on Earth quickly vanish (over millions of years), erased by water and wind erosion, by volcanic activity, and by the movement of the tectonic plates that carry the continents and ocean floors around and force one underneath another, into the mantle, where the submerging plate becomes partially molten and its craters disappear. No craters older than 1.8 billion years are found on Earth.

Edmond Halley was the first scientist to propose that comets had hit the Earth in the past and had done great damage. Halley is better known for his other work. He proved that comets orbit the Sun. He showed that the comet that now bears his name returns to the Sun every 76 years. As he worked out the orbit of Halley's Comet and the paths of other comets whose sightings he had collected, he realized that comets could pass perilously close to Earth. In 1688 and again in 1694 he announced that comets could hit the Earth . . . and might have done so in the past, gouging out vast depressions such as the Caspian Sea and perhaps even creating the biblical flood.[1]

Halley was wrong about the biblical flood and the Caspian Sea, but he was right that the Earth had been dented and devastated by comets . . . and by the yet-to-be-discovered class of bodies called asteroids.

From time to time before and after Halley there were reports of buildings damaged by rocks falling from the sky. There were even a handful of claims that people had been killed by thunderstones.[2]

Scientists recognized, somewhat after many members of the general public, that the Earth was getting bombarded by rocks from outer space. Evidence to convince the scientific community was supplied principally by Ernst Chladni in Germany at the end of the 18th century and then Jean Baptiste Biot in France at the beginning of the 19th.

Once the idea of stones falling from space was accepted and the unearthly chemistry in those specimens was delineated, more meteorites were found. Huge iron meteorites turned up in remote areas – the Ahnighito, 34 tons, in Greenland; the Hoba West, 60 tons, in Namibia.

Meteorites that large got scientists thinking about great circular holes in the ground scattered around the Earth, like the Barringer Crater in Arizona, ¾ mile (1.2 kilometers) wide, 600 feet (180 meters) deep, and 50,000 years old. They had been supposed to be the craters of ancient volcanoes. But fragments of iron scattered for 10 miles (16 kilometers) in every direction and the way the rocks in the crater walls were compressed ultimately gave these craters away. They too were evidence of a pounding the Earth sustained long ago. About 150 meteorite impact craters have now been identified on Earth.

Barringer Meteor Crater in Arizona, ¾ mile (1.2 kilometers) wide and 600 feet (180 meters) deep. Courtesy Barringer Meteor Crater, Arizona

Attention grabbers

On June 30, 1908, an object about the size of a department store (60 yards; 60 meters across) slammed into the atmosphere over Russia and exploded about 5 miles (8 kilometers) above the Stony Tunguska River in Siberia. The heat wave from that aerial explosion incinerated 850 square miles (2,200 square kilometers) of forest. The fire lasted only long enough for the shock wave to arrive, which blew out the flames and flattened the forest outward in all directions from a point directly under the explosion. A herd of 1,500 reindeer died. When the concussion reached a trading post 40 miles (65 kilometers) away, it blew a man off the porch. Had that meteoroid exploded above a big city instead of remote countryside, the loss of life might have eclipsed every earthquake, volcanic eruption, storm, or flood in history. It was a thousand times more powerful than the first atomic bombs.

The Tunguska event was scarcely a fender-bender by comparison to other accidents the Earth has been involved in. The Earth has been hit by larger comets and asteroids before – hard and often. The most famous pounding came 65 million years ago when a rocky body about 6 miles (10 kilometers) in diameter hit the Earth on the northwest coast of Mexico's Yucatán peninsula. The impact vaporized the body and excavated from the Earth's crust a crater at least 110 miles (180 kilometers) in diameter, one of the largest known craters on Earth. The explosive recoil from that blow flung pulverized debris high into the atmosphere where it took most of a year to settle back to the ground. This debris was a mixture of vaporized and recondensed dust from the intruder and the Earth's crust and it fell everywhere around the world.

We know that the explosion came from an intruder from space because of the abnormally high amount of iridium and other rare metals found in this worldwide layer of dust. The Earth's original supply of iridium, along with its iron and nickel, mostly sank to our planet's core during the Earth's molten period just after it formed.

We know how much damage was done to life on Earth because this thin worldwide layer of dust marks the boundary between the Cretaceous and Tertiary Periods. It separates the age of reptiles with its fossils of the dinosaurs – below the dust layer, an earlier time – from the more recent age of mammals above the dust layer where there are no dinosaur fossils. The Tertiary Period, the next period forward in time, has no dinosaurs and is also missing at least 70 percent of the other species of life that were living on Earth when – until – the asteroid or comet hit. Life went on above the dust layer, but there was far less of it. Dinosaurs and many other kinds of plants and animals were gone forever. The variety of life was drastically reduced for a long time.

More where they came from

No event quite like Tunguska has jolted us for almost a century. No catastrophe of the magnitude of the Cretaceous/Tertiary extinction has occurred for 65 million years. How much at risk is the Earth and life on it?

The Earth is a very tiny target, presenting a bull's-eye less than one ten-thousandth the size of the Sun. The Earth's face takes up only one one-hundred-thousandth of its orbit. The Earth's gravity pulls objects toward it, but not very effectively unless they pass very close. Compared to the Sun or Jupiter, the Earth has little gravity to pull objects its way.

The fact that the Earth is hit at all means that there are many objects in the solar system that cross the orbit of the Earth. It is more than a sprinkle or a drizzle. It is a torrent, a deluge of interplanetary debris. The Earth is adding to its mass by about 500 tons a day as meteorites crash and micro-meteorites float to the ground.[3] Yet that influx is nothing compared to the rate at which comets and asteroids fell on the Earth during its first billion years. Over the 4½ billion years of its existence, the Earth has accumulated about 16 million million million tons of comets, asteroids, and their dust.[4] Still, this cosmic dumping has contributed less than one percent to the Earth's present mass.

Even though the Earth is putting on weight, we are unaware of the process except for the sight of an occasional shooting star or the report of a meteorite just found.

Asteroids where they shouldn't be

Even after scientists acknowledged that meteorites come from the heavens, not from the Earth, the danger of these impacts seemed quite remote and the size of the intruders seemed quite modest.

Until 1932. In that year a weird asteroid, Amor, was discovered. Its size, 0.6 mile (1 kilometer), was nothing special. What was startling was that Amor did not stay in the asteroid belt between Mars and Jupiter, where asteroids were supposed to be. It strayed beyond that junkyard of frag-ments from a failed small planet. Amor crossed the orbit of Mars to pass 0.3 AU from the orbit of Earth. No threat of collision . . . for now. But if one asteroid had broken loose, might there be others?

That same year asteroid Apollo was discovered. Apprehension began to mount. Apollo, about 1 mile (1.5 kilometers) in diameter, not only crossed the orbit of Mars, it crossed the orbit of the Earth as well. A collision was possible.

In 1976, asteroid Aten was discovered. It too crossed the orbit of the Earth, but most of its orbit was inside that of Earth. Aten had a period shorter than one year and so was crossing the path of the Earth most often of all the asteroids. At every crossing, Aten, 0.6 mile (0.9 kilometer) in diameter, charged by without signal lights flashing, without a crossing gate down, and with no whistle as a warning. Someone could get hurt.

By mid-1997, more than 400 near-Earth asteroids (including extinct comets) had been found: roughly 200 with orbits like Amor, 200 with orbits like Apollo, and two dozen with orbits like Aten.[5] Almost all of these aster-oids and comets are big enough to cause serious or catastrophic damage to life on Earth. And these known hazards barely hint at the frightening number of city- or civilization-destroying bodies that are flying about undetected.

Asteroid Ra-Shalom (#2100) was discovered in this photographic time exposure by Eleanor Helin on September 7, 1978 as part of the Palomar Observatory Planet-Crossing Asteroid Survey project she initiated. The 18-inch (0.46-meter) Schmidt telescope tracked the stars moving east to west as the Earth turned, allowing the asteroid's motion around the Sun to show up as a streak. Ra-Shalom was the third Aten-class asteroid to be found. Aten asteroids have orbits that cross the Earth's path and lie mostly inside the orbit of Earth. Courtesy of Eleanor Helin

More errant asteroids

The near-Earth comets and asteroids provide an immediate threat. But danger lurks everywhere. More – and bigger – objects are working their way toward us. In 1977, Charles Kowal found a large asteroid where no asteroid should be lurking. It was out past Saturn, far beyond the asteroid belt between Mars and Jupiter where almost all the asteroids travel around the Sun in nearly circular orbits. Chiron was traveling a highly elliptical path around the Sun that carried it outward just beyond the orbit of Uranus and inward just inside the orbit of Saturn in a period of 50.7 years. Chiron was 120 miles (200 kilometers) in diameter – one of the two dozen largest asteroids.[6]

However, in 1989 this asteroid developed a haze of gas and dust around it and trailing outward from the Sun. A coma and a tail. Chiron was more a comet than an asteroid. Its orbit, close to the plane of the solar system, caused it to cross near the orbits of Uranus and Saturn. So Chiron is not destined to retain its present orbit for long. An eventual close encounter with one of the giant planets may wrench its orbit inward. Over the next 200,000 years, Chiron may stagger into a Jupiter-crossing and then perhaps an Earth-crossing orbit . . . with a mass ten thousand times greater than the asteroid or comet that wiped out the dinosaurs and most of life on Earth 65 million years ago.[7]

And Chiron is not the only renegade edging inward from the Kuiper Belt and the Oort Cloud to add their masses to the impact hazard of the asteroids. Astronomers conducting near-Earth asteroid patrol programs have recently discovered three more large objects like Chiron in elliptical orbits among the giant planets: Pholus (asteroid 5145), 100 miles (160 kilometers) in diameter; asteroid 1993HA$_2$, 50 miles (80 kilometers) in diameter; and Damocles (asteroid 5335), 10 miles (16 kilometers) in diameter. They are all orbiting the Sun out beyond Jupiter, scurrying along among the giant planets. How many more objects like Chiron there are, no one knows. There may be hundreds or thousands. What astrodynamicists do know, as they calculate the orbits of these bodies and map how the gravity of the heavyweight planets distorts those orbits, is that these comets were originally in the Kuiper Belt and cannot remain in their present orbits for long. Jupiter and company will either hurl them outward to the fringes of the solar system or beyond, or will hurl them inward – where the largest planetary target is the Earth.

So says the computer. But would Jupiter ever do such a thing?

In 1993, Eugene and Carolyn Shoemaker and David Levy, the principal astronomers in an Earth-crossing asteroid search program, found a comet. As a trio, they had found eight previous ones – still more individu-

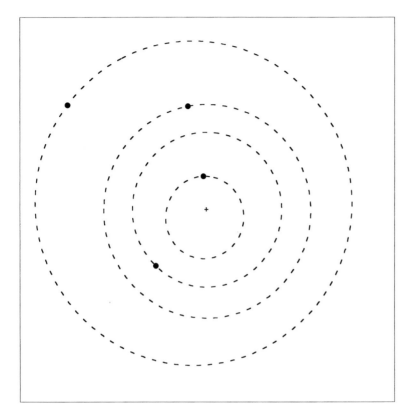

Left: orbits of the four inner planets and their positions on January 1, 1997; *Right:* orbits of 107 known near-Earth asteroids with estimated diameters of 1 kilometer (0.6 mile) or larger superimposed on the orbits of the inner planets. The orbits have been rotated into the Earth–Sun (ecliptic) plane. These 107 asteroids represent about 5% of the asteroids 1 kilometer or greater that are estimated to orbit near the Earth. Diagram by Richard P. Binzel. Courtesy of Richard P. Binzel, Massachusetts Institute of Technology

ally. But Comet Shoemaker–Levy 9 looked like no other comet. It was an elongated smear of hazy light dotted by a string of 21 bright pearls.

The discoverers reported their queer object to Brian Marsden, director of the Central Bureau for Astronomical Telegrams, a division of the International Astronomical Union that serves as a clearinghouse for reports of new comets, asteroids, solar eruptions, exploding stars, and other sudden celestial happenings that need additional astronomical observations as quickly as possible. Marsden flashed news of this bizarre comet around the world.

Eight weeks later he sent out an update on this object. Observations sent in by astronomers had allowed Marsden and Syuichi Nakano in Japan to compute the orbit of this comet. By projecting its orbit back in time, they could see that in July 1992 the comet had passed only 30,000 miles (50,000 kilometers) above the cloud tops of Jupiter, so close that the tidal effect of Jupiter's gravity had ripped the comet to pieces. The pieces were still traveling close together.

But Jupiter had done far more than that. Jupiter had torn the comet from its previous orbit around the Sun and had captured it into an orbit around itself, an orbit that would carry Shoemaker–Levy 9 back toward Jupiter, to pass even closer this next time. It would pass, Marsden wrote with astronomical whimsy, 28,000 miles (45,000 kilometers) from the center of Jupiter – pausing to remind his readers that Jupiter's cloud surface lies 44,400 miles (71,500 kilometers) above its center. In late July 1994, the face of Jupiter would stand directly in the path of a machine gun volley of fragments from a vengeful comet.

Comet Shoemaker–Levy 9 was on its last orbit . . . was already commencing its fatal dive into Jupiter. The collisions occurred as predicted over a period of four days, July 21–25, 1994, in a series of spectacular splashes. The impacts occurred just beyond the horizon on the back side of Jupiter, out of direct sight of telescopes on Earth. But the Hubble Space Telescope in Earth orbit could see within minutes plumes rising so high – 1,800 miles (3,000 kilometers) – that they were visible above the edge of the

Asteroids that come close to Earth
(including some inactive comets that look like asteroids)

Reference distances
1 astronomical unit (AU) = 93 million miles or 150 million kilometers)
Earth to Moon (mean) = 0.0026 AU; 239,000 miles; 385,000 kilometers
Earth to Venus (mean closest) = 0.277 AU; 26 million miles; 41 million kilometers
Earth to Mars (mean closest) = 0.524 AU; 49 million miles; 78 million kilometers

Asteroid (number and name)	Minimum distance from asteroid's orbit to Earth's orbit (in astronomical units [AU])
1566 Icarus	0.0357
1620 Geographos	0.0307
1862 Apollo	0.0252
1981 Midas	0.0029
2101 Adonis	0.0124
2102 Tantalus	0.0441
2201 Oljato	0.0008
3200 Phaethon	0.0221
3362 Khufu	0.0139
4015 Wilson–Harrington	0.0476
4034 1986PA	0.0187
4179 Toutatis	0.0058
4450 Pan	0.0277
4486 Mithra	0.0460
4581 Asclepius	0.0300
4660 Nereus	0.0031
4769 Castalia	0.0199
4953 1990MU	0.0290
5011 Ptah	0.0257
5189 1990UQ	0.0443
5604 1992FE	0.0339
5693 1993EQ	0.0052
—1988EG	0.0245
—1988XB	0.0068
—1989UQ	0.0140
—1991JW	0.0205
—1992SK	0.0464
—1993HP	0.0049
—1993KH	0.0014
—1997XF11	0.00638

Note: Of concern also are 7 asteroids known to pass within 1 AU of Jupiter, as they may undergo major orbital changes. These 7 asteroids are not protected from close Earth approach by resonance with Jupiter. Although these asteroids do not at present cross the orbit of Earth, their close passages by Jupiter may make them Earth-crossers within the next century. These asteroids are: 944 Hidalgo, 1921 Pala, 1922 Zulu, 3552 Don Quixote, 3688 Navajo, 5370 Taranis, and 1989SL$_5$.

Source: List from Edward Bowell and Karri Muinonen: "Earth-Crossing Asteroids and Comets: Groundbased Search Strategies," pages 149–197 in Tom Gehrels, editor: *Hazards Due to Comets and Asteroids* (Tucson: University of Arizona Press, 1994).

Comet Shoemaker–Levy 9. A close encounter with Jupiter in 1992 broke it into 21 visible pieces and put it on a course to collide with Jupiter in 1994. NASA Hubble Space Telescope

planet. As the impact sites rotated into view about 20 minutes later, observers could see black splashes of dust in Jupiter's atmosphere that were larger than Earth, caused by the plunge of comet fragments into the gas giant planet. The dark splotches remained above Jupiter's clouds for almost a year.

The hail of 100- to 200-yard (100- to 200-meter) comet fragments struck Jupiter with the force of tens of *millions* of megatons of TNT. That was more than a thousand times as powerful as the 15-megaton explosion of volcano Mount St. Helens on Earth in 1980 and more than a million times as powerful as the 15-kiloton atomic bomb that destroyed Hiroshima.[8]

In the aftermath of the Comet Shoemaker–Levy 9 spectacle, the first time that human beings had for certain observed a natural object hit another planet, astronomer Don Yeomans calculated that a close encounter with Jupiter should shatter a comet about every 1,000 years, creating additional cosmic shrapnel to wound the other planets.[9] For upsetting its tour of the planets, one comet had retaliated against Jupiter.

This time, Jupiter. Next time?

A few weeks after Comet Shoemaker–Levy 9 barreled into Jupiter, newly discovered Comet Machholz 2 split into three large fragments, all of them bound for the inner solar system, to pass 0.3 AU from Earth, about the distance of Venus at its closest. No danger – yet. But Jupiter and the other planets continue to warp the orbits of the Machholz triplets.[10]

Still more coming

As Chiron and its companions work their way inward toward possible collisions with Earth, other giant and plenty-big-enough comets are edging inward from the Kuiper Belt and the Oort Cloud to replace them and keep the impact danger high. A giant comet the size of Chiron should

be captured into the planetary realm at the rate of about one every 100,000 years.[11]

The giant comets that will replace the doomed ones that hover among the giant planets have been detected. In 1992 David Jewett and Jane Luu discovered an object orbiting the Sun beyond the outermost planets. It received the provisional minor planet designation 1992 QB$_1$ and was about 140 miles (230 kilometers) in diameter. By mid-1997, astronomers had found 44 minor planets beyond Neptune and Pluto – all between 60 and 150 miles (100 and 250 kilometers) in diameter. They are the most distant solar system objects ever seen directly by telescopes. In size and in the spectra of the light they reflect, these trans-Neptunian bodies most resemble Chiron and Pholus – giant comets caught in the gravitational resonance of the giant planets, gradually being reeled in like hooked fighting fish. Reeled in from beyond Neptune to Jupiter to the vicinity of Earth.

Near-Earth objects are interesting – and scary – because they have nowhere to go except to collide with other bodies. Left to themselves, all the near-Earth asteroids and comets would vanish in planetary impacts or

Dark plumes rising from Jupiter's atmosphere due to the impacts of fragments of Comet Shoemaker–Levy 9. For reference, Jupiter's Great Red Spot, the elliptical cloud formation in the lower left, is more than twice the size of Earth. Some of the plumes created by Comet Shoemaker–Levy 9 were larger than the Earth. NASA Hubble Space Telescope

Asteroid Ida, imaged on August 28, 1993 by NASA's spacecraft *Galileo* on its way to Jupiter. Ida is about 32 miles (52 kilometers) long. Notice all the craters where Ida has been hit by other asteroids. *Galileo* was 1,500 miles (2,400 kilometers) from Ida when this picture was taken and they passed one another at 28,000 miles per hour (12.4 kilometers per second).
Courtesy of NASA/Jet Propulsion Laboratory

disappear into rubble and dust from collisions among themselves. In 100 million years, they would be gone.[12] A hundred million years is a shade over two percent of solar system history, so the fact that the Earth still finds itself the target of hundreds of thousands of lethal bullets in a celestial shooting gallery suggests that new near-Earth objects are constantly arriving to take up the challenge of assassinating the Earth.

The true asteroids among the near-Earth objects are supplied by the asteroid belt, where collisions and repeated nudges by Jupiter's gravity drop a steady supply of asteroids inward. Once an asteroid leaves the asteroid belt for the warmer regions, it will take up an Earth-crossing orbit in about 10 million years – and become a crater on a planet in another 100,000.[13]

The comets that we see as near-Earth objects come to us from the comet reservoirs beyond the planets in the outer solar system. The Kuiper Belt and the inner Oort Cloud send us comets of all sizes, including giant comets like Chiron. The outer Oort Cloud sends us large and small comets as well, like Ikeya-Seki that skimmed by the Sun in 1965 and developed a tail more than 100 million miles (160 million kilometers) long.

Once a comet crosses inside the orbit of Jupiter, it will evolve into an Earth-crosser in only a hundred thousand to a million years. Once inside the orbit of Jupiter, exposed to the gravitational tides of the planets and the Sun, giant comets crash into something or break up in 10,000 to 100,000 years.

Because of their size and their ability to break into many destructive

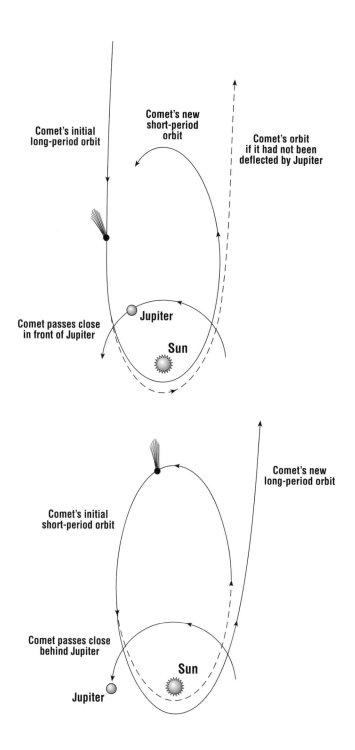

Comet's initial long-period orbit

Comet's new short-period orbit

Comet's orbit if it had not been deflected by Jupiter

Comet passes close in front of Jupiter

Jupiter

Sun

Comet's new long-period orbit

Comet's initial short-period orbit

Comet passes close behind Jupiter

Sun

Jupiter

How a planet – especially Jupiter, the most massive – can alter the orbit of a comet. *Top:* If a comet inbound to the Sun passes close in front of Jupiter, Jupiter's gravity will decelerate the comet and it will fall onto a new, shorter-period orbit. *Bottom:* If a comet inbound for the Sun passes close behind Jupiter, Jupiter's gravity will accelerate the comet and it will climb onto a new longer-period orbit. These same gravitational assist techniques are used to help interplanetary spacecraft reach their objectives without having to carry extra fuel. Diagram by Will Fontanez and Tom Wallin, University of Tennessee Cartographic Services Laboratory.

pieces, comets may bring to the inner planets as much or more material than asteroids.[14] And they deposit it on Earth in all sizes, from comets themselves – large enough to end a geological age and most of the life in it – to meteoroids that brighten the skies, to micrometeorites that float to Earth without spectacle and mingle with the clouds and oceans and soil, bringing new carbon and other chemical elements to Earth.

Assessing the danger

In 1990 the United States Congress took note of the mounting evidence that the impact of an asteroid or comet killed off the dinosaurs and most of life 65 million years ago. They realized that potentially hazardous objects still prowl near Earth. It was clear that momentous impacts had happened repeatedly in the past . . . and will happen again. Congress asked NASA to look into the threat of comet and asteroid impact on Earth and what might be done about it. Astronomers and other scientists held a series of meetings, compared their research, and in 1992 offered an assessment, *The Spaceguard Survey*.[15]

Their first objective was to understand what would happen if a near-Earth object suddenly became a hole-in-the-Earth object.

A comet or asteroid 1 kilometer (0.6 mile) in size or larger is of major concern because its impact would kill perhaps 25 percent of the world's population – 1.5 billion people – no matter where it landed. Its effect would not be confined to the initial blast or to the tsunami (tidal wave) it would create if it hit an ocean or to the molten debris the impact would launch like ballistic missiles into space to reenter the atmosphere hundreds or thousands of miles away to set fire to forests, crops, and cities. The long-term effect was equally frightening. The dust from the impact would be lifted high in the atmosphere where it would reflect sunlight back to space, thereby cooling the continents enough to disrupt crops for at least one growing season, thereby setting off a global wave of starvation that could undermine civilization itself.

Understanding the horror of such an impact led scientists to a second objective: to determine the extent of the danger. Find, catalog, and track all the Earth-crossing asteroids and comets that could destroy a sizable fraction of life on Earth.

Astronomers counted the objects with Earth-crossing orbits that were 1 kilometer or larger in diameter. At the time (1992), there were 200 known. Astronomers estimated that ten times that many lay undiscovered – 2,000 killers from space that could disrupt or exterminate human life.

How did they make such an estimate? When astronomers find an Earth-

Danger from asteroid and comet impacts on Earth

Size of object	Frequency of impact on Earth (estimated)	Explosive equivalent of impact in TNT (approximately)	Number of objects this size on Earth-crossing orbits (estimated)	Consequences
1-millionth to 1-hundredth gram (too faint for the eye to see)	1 trillion (10^{12}) a day	none		Micrometeorites
Grain of sand to size of pea				Ordinary meteor
Walnut				Fireball
Grapefruit				Dazzling fireball
Basketball	1 month			Unforgettable fireball
10 meters	1 year	10 kilotons	150 million	Most don't reach the ground; the probability of getting hit by an engine or wing falling off a jet is greater
50 meters (a 15-story office building)	100 years	10 megatons (10 million tons)	20 million	Would wipe out a city
60 meters (like 1908 Tunguska, Siberia airburst object)	300 years	10–20 megatons		Burned and flattened 2,200 square kilometers of forest
100 meters (a football field with endzones and sidelines)	1,500 years	100 megatons (more than all nuclear weapons on Earth)	300,000	Would kill 100,000 people; tsunamis caused by impact in ocean might kill 10 or 100 times that number
500 meters	40,000 years	1,000 megatons	10,000	If it fell in central Australia, it would shake all Australian cities flat
1 kilometer	200,000 years	100,000 megatons	2,000 (25% to 50% of these objects will eventually strike the Earth)	Would kill at least 25% of mankind; would lower Earth temperatures by 10 °C (18 °F)
10 kilometers	50 million years	100 million megatons	No 10-kilometer Earth-crossing asteroids *at present*; Earth-crossing short-period comets Halley and Swift-Tuttle are this size	Mass extinction like the Cretaceous–Tertiary event that wiped out the dinosaurs and 70% of all species

Notes: Information for this table gleaned from Thomas J. Ahrens and Alan W. Harris: "Deflection and Fragmentation of Near-Earth Asteroids," *Nature*, volume 360, December 3, 1992, pages 429–433; John S. Lewis: *Rain of Iron and Ice* (Reading, Massachusetts: Addison-Wesley, 1996); David Morrison: "Target: Earth!", *Astronomy*, volume 23, number 10, October 1995, pages 34–41; David Morrison: "The Spaceguard Survey: Protecting the Earth from Cosmic Impacts," *Mercury*, May/June 1992, pages 103–106, 110; Duncan Steel: *Rogue Asteroids and Doomsday Comets* (New York: John Wiley & Sons, 1995); Owen B. Toon, Richard P. Turco, Curt Covey, Kevin Zahnle, and David Morrison: "Environmental Perturbations Caused by the Impacts of Asteroids and Comets," *Reviews of Geophysics*, volume 35, February 1997; pages 41–78.

crosser 1 kilometer or larger in size, they check the records to see if that asteroid or comet is already known. Ten percent of the time they find it was discovered previously. Ninety percent of the time it is an object previously unknown. So astronomers reckon that only 10 percent of 1-kilometer-plus Earth-crossers have been spotted and 90 percent await detection. Thus the 200 Earth-crossers 1 kilometer or larger that had been discovered by 1992 constituted perhaps 10 percent of the total that exist. Therefore the total number of these menaces should be about 2,000.[16]

Astronomers are confident that all the Earth-crossing asteroids larger than about 3 miles (5 kilometers) in diameter have been discovered because, at their distance from Earth, they would be too bright to have escaped previous searches for new planets and asteroids. What troubles astronomers are long-period comets, comets arriving from the Oort Cloud that have never been seen before, because they can emerge from the darkness at any time and from any angle on a collision course with Earth ... and because new comets aren't limited to 3 miles in diameter. They may be 10, 20, 50, or even 100 miles (160 kilometers) in diameter.

Masquerade

Not many of the Earth-orbit-crossing threats are comets, gauging by how few develop comas and tails when close to the Sun. The estimates range between 2 percent and 30 percent. So comets that come close enough to Earth to create meteor showers seem like a slight to modest hazard statistically. However, astronomers suspect that many Earth-crossing asteroids are actually comets in disguise. Fifty percent or more of the Earth-crossing asteroids are probably extinct or dormant comets.[17]

A true asteroid formed between Mars and Jupiter and has orbited there for most of solar system history. It is a rocky or metallic body with an orbit lying close to the plane of the planets. It travels around the Sun in the same direction as the planets. Some of the near-Earth asteroids don't seem to fit. They have surfaces that absorb sunlight in a way suggestive of sooty ice rather than naked rock or metal. They have orbits whose tilts and elongations suggest that they have not fallen in from the asteroid belt but were captured by Jupiter as they flew close by on cometlike orbits that stretched much farther from the Sun.

In 1983, asteroid 3200 Phaethon was discovered. Fred Whipple, originator of the dirty snowball model of a comet, quickly pointed out that Phaethon had an orbit nearly identical to the Geminid meteors in December. So either an asteroid could create swarms of meteoroids – but no one could think of a way – or Phaethon was more a comet than an asteroid. Its lack of a coma and tail made it look telescopically like an asteroid. But the spectrum of the sunlight reflecting off Phaethon offered clues about the nature of its surface: an abundance of carbon compounds – more like a comet.

Phaethon is by no means the only probable masquerader in the asteroid–comet community. In 1979 an asteroid was discovered and given the provisional designation 1979VA. In 1992 Ted Bowell and Brian Marsden realized that asteroid 1979VA had been seen earlier and named Comet Wilson–Harrington because the discoverers had noticed a faint coma and tail. Apparently comets can cease outgassing activity and roam among the planets looking like asteroids.

Perhaps such comets have made so many visits close to the Sun that most of their surface ices have evaporated, leaving behind a crust of dust that insulates the ices beneath and thus chokes off outgassing... until the heat of the Sun causes the crust to crack or until a meteorite punches through the crust to expose fresh ice. Then what looked like an asteroid can become recognizably, if only briefly, a comet once more.

Will asteroid 4015 Wilson–Harrington become an active comet again?[18]

Comet versus asteroid crashes

But why worry about comets? They are merely dirty snowballs. Asteroids are rocks or metal.

Two factors make comets more fearsome hazards than asteroids, even though comets may be less dense and more crumbly. The first is size. The largest asteroid, Ceres, is 580 miles (930 kilometers) in diameter, far larger than any comet ever seen. But well-behaved Ceres stays in the asteroid belt. The largest of the Earth-crossing asteroids, Ivar (#1627), is 5 miles (8 kilometers) in diameter. Comets regularly match or exceed Ivar's size. Comet Hale–Bopp, the exceptionally bright comet of 1997, crossed inside the orbit of Earth. Its nucleus was about 25 miles (40 kilometers) in diameter. Comet Halley, oblong in shape, is 9 by 6 miles (15 by 10 kilometers) in size. Comet Swift–Tuttle, parent of the Perseid meteors, is about 5 miles (8 kilometers) across.

The second factor that makes comets especially dangerous is speed. Asteroids all travel around the Sun in the same direction as the Earth and other planets. For an asteroid to hit the Earth, it must catch up with our planet from behind, so it rear-ends the Earth at a relatively moderate speed, only about 9 miles a second (15 kilometers a second).[19] Some comets have direct orbits close to the plane of the solar system like asteroids, so they would coast into the Earth like an asteroid. Other comets,

Comet Hale–Bopp in 1997, photographed by James W. Young. Courtesy Jet Propulsion Laboratory/NASA, Table Mountain Observatory; photo by James W. Young

however, travel in more nearly polar orbits, so they can crash into the Earth like cars trying to cross an intersection, with the average speed of encounter about 26 miles (42 kilometers) per second. Still other comets travel in retrograde orbits, the opposite direction of the planets, so they will ram the Earth head on, at speeds up to 45 miles (72 kilometers) a second. It is the worst kind of accident because the energy of the crash increases with the *square* of the velocity.

The gravest threat comes from comets and asteroids 1 kilometer (0.6 mile) in diameter and larger. Against an object this size, the Earth's atmosphere protects its surface about as well as an overcoat protects a person against a bullet. A 1-kilometer object would hit the Earth with the force of 100,000 million tons of TNT. If it hit a large city, it would replace the entire metropolitan area with a crater. Millions of people would die. If the 1-kilometer object plummeted into an ocean (more likely, because 70 percent of the world is covered with water), the outlook is no better. Probably worse. The impact would generate a tsunami that would create waves at seashores a thousand miles away that would be taller than most skyscrapers. Because most of the world's population lives in cities or villages on the

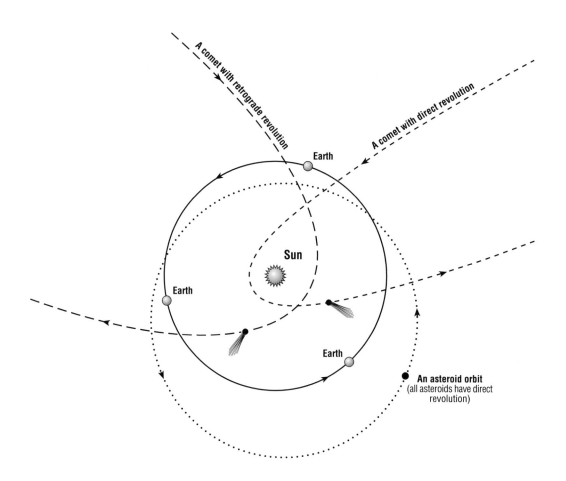

coasts, loss of life might go far into the millions. Civilization, the attempt of human beings to live together in numbers beyond a tribal group, struggling to get along with one another by ordering their lives through laws and customs, might well be scuttled.

Celestial police

On December 9, 1994, James Scotti, working with the Spacewatch near-Earth object detection patrol at the Steward Observatory in southern Arizona, looked up with his telescope just in time to see an asteroid whiz by. It was about 35 feet (10 meters) in diameter, a rock the size of a town house. It missed the Earth by only 65,000 miles (104,000 kilometers),

Comets that collide with the Earth are usually traveling much faster than asteroids. Representative asteroid and comet orbits that cross the orbit of Earth demonstrate why. All asteroids travel around the Sun in the same direction as Earth (direct orbits). In order to collide with Earth, they must overtake our planet, so their speeds of collision are modest. Comets, however, may have either direct or retrograde orbits, so they may strike the Earth from behind, from the side, or head on. Comets also tend to have much more elliptical orbits than asteroids and the more elliptical the orbit, the higher the object's velocity is at a given distance from the Sun. Diagram by Will Fontanez and Tom Wallin, University of Tennessee Cartographic Services Laboratory.

passing about four times closer than the Moon. Astronomers gave the asteroid the provisional designation 1994XM$_1$. They couldn't give it a number and name because they didn't see it long enough to calculate its orbit. They don't know when it will be back – or if it will miss the Earth next time.

As of mid-1997, 1994XM$_1$ still held the record for the nearest miss of 98 objects that pass within 4.6 million miles (7.5 million kilometers) of Earth, less than 20 times the distance of the Moon.[20] Twenty times the distance to the Moon, 1/20 the distance to the Sun, seems at first like plenty of room, until you consider that the object that almost hit you might be much larger than a townhouse. It could easily be the size of an iceberg or a mountain. One more thing to consider: it's on an orbit around the Sun. *It will be back.*

There are at present three near-Earth object detection programs in operation that are designed to find, follow, and catalog such threats. They all use telescopes that record images with charge-coupled devices (CCDs), which are far more sensitive than film. One program, initiated by Tom Gehrels at the University of Arizona, uses a 36-inch (0.91-meter) telescope at the Steward Observatory on Kitt Peak in southern Arizona.[21] A second, with Eleanor Helin of the Jet Propulsion Laboratory as the principal investigator, uses an automated camera on a 39-inch (1-meter) telescope atop the volcano Haleakala on Maui in Hawaii. A third, under Ted Bowell, uses the 18-inch (0.46-meter) Schmidt telescope of the Lowell Observatory in northern Arizona.[22]

These and previous programs,[23] despite their small size, have been responsible for a great majority of the near-Earth objects detected. And they point the way toward a search capable of finding the largest and most destructive of the Earth-crossing comets and asteroids.

The Spaceguard Survey is one proposal to find Earth's cosmic PHOs (Potentially Hazardous Objects). This detection program would equip astronomers with two 80-inch (2-meter) wide-field telescopes, upgrade two existing 40-inch (1-meter) telescopes, and obtain half the observing time on one 120- to 160-inch (3- to 4-meter) telescope. With these telescopes equipped with CCD cameras, astronomers could in about 20 years discover and plot the orbits of all or almost all asteroids and short-period comets larger than 1 kilometer in size. With those orbits, astronomers could warn the world of danger two centuries in advance.

Long-period comets, like Hale–Bopp, coming at us from the outskirts of the solar system, might furnish only a year or just a few months' warning, but they have only one shot at the Earth before they head back to the Oort Cloud or Kuiper Belt, not to return to the inner solar system for hundreds or thousands of years. Short-period comets and asteroids that are Earth-crossers, however, have periods between half a year and a few dozen years,

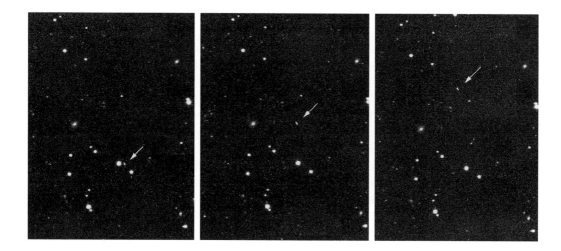

so they pass close to Earth often. Most also pass close to Jupiter and other planets that can shift their courses to increase the likelihood of collision with Earth.

NASA's 1995 final report to Congress on the danger of asteroid and comet impacts on Earth estimated that the cost of the Spaceguard Survey, this cosmic early warning system, would be $24 million for the telescopes, detectors, and computers; and $3.5 million a year for 20 years to operate them.[24] Such an investment would not rid us of the danger, but could alert us and provide the lead time for action to be taken.

Aten-class asteroid 1997 AC11 was discovered using a CCD to record the light collected by a U.S. Air Force telescope atop the volcano Haleakala on Maui on January 15, 1997. The image, made 12 minutes apart, show the motion of this Earth-orbit-crossing asteroid. This discovery is part of the Near-Earth Asteroid Tracking project, a joint venture of NASA, the Jet Propulsion Laboratory, and the U.S. Air Force, with Eleanor Helin as principal investigator. Courtesy of Eleanor Helin

Deflecting the danger

But if a minor planet were discovered to be on a collision course with Earth, what action could be taken? The knee-jerk reaction would be to blast the imminent impactor to smithereens. It is possible to break up a 1-kilometer asteroid or comet, but the only way to do it reliably is to have your vehicle bury its nuclear warhead deep in the impactor's body, which requires several times more energy than the object-destroying explosion itself. The demolition has to be done very carefully or the explosion will divide one impactor into many smaller ones that are still capable of penetrating the Earth's atmosphere and that are still on a collision course with Earth. The rifle aimed at the Earth will have been replaced with a shotgun and the many smaller impacts could wreak even more havoc than the one original body because they hit more places.

It is possible to destroy the PHO, but such violence is costly, dangerous,

Eleanor Helin. Courtesy of
Eleanor Helin

and wasteful. Reducing an asteroid or comet to powder and pebbles squanders a heavenly resource. The best approach is to use minimal force to deflect the blow, more in tune with the martial art of aikido, not like karate, which inflicts maximum damage.

The best way to do the job, Tom Ahrens and Alan Harris report, is still to use a nuclear explosion.[25] It's at least a million times more energetic than the impact of a rocket alone or even a rocket loaded with chemical explosives. It must be a preemptive attack. We cannot wait for the Death Starlet to strike first. That one blow could be fatal to mankind – and all life on Earth.

Here's what could happen. Suppose astronomers are studying a comet 1 kilometer in diameter that crosses the Earth's orbit once every 10 years. It has a mass of about a billion tons. At perihelion just inside the orbit of the Earth, it is traveling about 92,000 miles per hour (41 kilometers per second). The comet poses no threat this time around. But by projecting its orbit into the near future, astronomers can see that the comet's passage near Jupiter on its outbound journey will change its flight path so that it will hit the Earth in 50 years.

Ten years later, when the comet is returning closest to the Sun, a rocket carrying a nuclear weapon is launched from Earth. The hydrogen bomb has the explosive power of 100 to 1,000 kilotons of TNT, a modest size for a nuke, but carefully calculated to get the job done. The rocket intercepts the comet at perihelion, when the comet is closest to the Sun and traveling fastest. Altering the comet's speed even slightly here will do the most to stretch or shorten the comet's orbit and change the comet's period of revolution. The orbit of the comet will still cross the orbit of the Earth, but, at the end of 50 years, the comet will not be at that intersection when the Earth rushes by.

To accomplish this adjustment of the comet's orbit, the hydrogen bomb is detonated close to the comet, just in front or behind the comet along its orbital path. Slowing down or speeding up the comet provides better results with less energy than blasting the comet from the side to change its orbit left or right, up or down. If necessary, an orbital change that alters where the comet crosses the Earth's orbital plane, and hence sweeps the comet out of the path of the Earth for thousands of years, can be accomplished by other explosions later.

The bomb is triggered to explode at a distance above the surface of the comet equal to four-tenths the comet's radius. The nuclear explosive is a kind of hydrogen bomb called a neutron bomb, designed not for maximum shock, but to give off a maximum of neutrons and ground-penetrating x-rays. The idea is not to shatter the comet into fragments that could still hit the Earth, but to irradiate as powerfully as possible as much

Energy to divert or destroy Earth-crossing objects[a]

Asteroid or comet size	Diversion – by nuclear explosion above impactor's surface (altitude = 0.4 × impactor's radius)[b,c]	Destruction – by nuclear explosion set off deep under impactor's surface to fragment and disperse it
100 meters (~100 yards)	100 tons	3 kilotons
1 kilometer (0.6 mile)	100 kilotons	3 megatons
10 kilometers (6 miles)	100 megatons	3,000 megatons

Notes: [a]Information gleaned from Thomas J. Ahrens and Alan W. Harris: "Deflection and Fragmentation of Near-Earth Asteroids," *Nature*, volume 360, December 3, 1992, pages 429–433.
[b]If the hydrogen bomb were detonated at precisely the right altitude, the energy needed would be only one-tenth these amounts.
[c]Nuclear explosions at the surface of the impactor might also be used to divert it. Ahrens and Harris found the energy necessary for this task to be about the same but more uncertain than explosions above the impactor's surface.

of the surface of the comet as feasible with one explosion. If the comet is 1 kilometer is diameter, its radius is 500 meters, and the bomb should detonate 200 meters above the comet's surface. In this way, almost 30 percent of the comet's surface would be irradiated, the maximum dose for the maximum area. The top 8 inches (20 centimeters) of the comet's surface would be heated so suddenly and so greatly that it would rapidly expand away from the comet's body, escaping from the gravitational hold of the comet altogether. That's not hard. A comet 1 kilometer in diameter should have an escape velocity of perhaps 1 mile per hour (0.5 meter a second). The escaping vaporized surface of the comet would provide rocket thrust to slow down or speed up the comet. A change of velocity of only 1 *centimeter* a second (0.4 *inch* a second – 1/45 mile per hour) is all that is needed. That's not much compared to the speed of the comet at perihelion just inside the orbit of the Earth: 41 kilometers per second (92,000 miles per hour).

Even after we discover, track, and divert all the Earth-threatening bodies that are 1 kilometer in size or larger, we will not have saved humanity from catastrophic harm by asteroids and comets. There are estimated to be 2,000 asteroids and comets 1 kilometer or larger in Earth-crossing orbits. But what about the smaller objects, from 1 kilometer down to 100 meters in size? Objects that size are much harder to see, and they are much more numerous. The estimate is 300,000. They are smaller, but they are still dangerous. Remember the destruction caused by the airburst of an object 60 yards (60 meters) in diameter over Siberia in 1908. Now multiply the size of that object by two, by ten, by fifteen. Such a blast might not wipe out a quarter of the world's population and drop temperatures enough to starve the rest, but just one impactor 100 to 1,000 meters in size could obliterate the largest cities, killing millions.

Life on Earth would continue after an impact of that magnitude, but it seems sensible to explore whether that scale of tragedy can also be prevented.

For the most part, it can. The same Spaceguard telescopes that would scan the skies for 1-kilometer killers would be watching for the 100-meter murderers too. Because the smaller Earth-crossers are fainter and more numerous, it would take longer to spot them – not 25 years but perhaps several centuries.[26] New technology added to the search in those centuries might find nearly all of them. Once found, they too could be steered clear of Earth by nuclear nudging.

Thus, says Ahrens, hydrogen bombs used against oncoming asteroids and comets turn out to be "weapons of mass protection."[27]

Alternatives

Some scientists object to using nuclear weapons for self-defense against mountain-size rocks and icebergs in space because the technology that can send hydrogen bombs to asteroids and comets can be used to launch nuclear missiles at enemies on Earth. Several alternatives are available, if there is enough lead time. One technique is to set up a solar sail on a comet or asteroid. Imagine a sail larger than the too-near-Earth object and made of thin, highly reflective plastic like Mylar. This sail would be deployed a short distance from the comet or asteroid and tethered to it so that, despite the body's rotation, the sail could always face the Sun, using the pressure of sunlight to drag the hazard away from a collision with the Earth slowly but surely. Very slowly.

Another possibility would be to land surface-mining equipment on the homicidal comet or asteroid and set up a mass driver. Solar-powered automated mining equipment would excavate the surface of the body and load the material on a type of conveyer belt. The conveyor belt would accelerate the debris to a speed well above escape velocity and hurl it off the comet or asteroid in the proper direction so that the expelled material acts as a rocket motor to propel the body in the opposite direction, gradually changing the speed or direction of the potential impactor so that it will miss the Earth.

To divert a 1-kilometer impactor, a nuclear explosion that heats a thin surface layer of the comet or asteroid so that it bursts away, might blow off 50 tons of the threatening body. A mass driver, creating a rocket effect with lower speeds and thus lower thrust, would require more material to be flung away to create an equivalent acceleration – perhaps 10,000 tons of material to deflect a 1-kilometer asteroid.[28] Even so, that jetsam constitutes about one-thousandth of one percent of the mass of the asteroid.

Both the solar sail and the mass driver appear to be viable asteroid–comet diversion techniques, but they typically require three decades to convert what would be a fatal hit into a near miss. Mass drivers and solar sails also cost much more because of the greater mass of equipment that must be launched and then landed and assembled on the targeted body. Still, if nuclear weapons are politically unacceptable and if time permits, solar sails and mass drivers can do the job. But for objects that arrive suddenly, a nuclear weapon is the only hope. To divert oncoming impactors more than 100 meters (100 yards) in diameter, nuclear weapons are the most efficient tool. Humankind had best keep a few handy.

"For the first time since life on Earth began," says astronomer Duncan Steel, "a species has the ability to save itself and most life from extinction."[29]

A different view of near-Earth objects

The comets and asteroids that pose the highest risk of hitting Earth are the ones that come closest to Earth and therefore are the ones most accessible to the people on Earth.[30] Do near-Earth objects offer any economic benefits that could make them a greater resource than they are a threat?

When the Earth formed from planetesimals 4.6 billion years ago, those building blocks must have bequeathed to our planet more water than the Earth now possesses. But much of that water was lost to the Earth over its first half billion years when our planet was still accreting, being battered by and absorbing ever larger planetesimals that today would be called asteroids and comets. Some were so large that their impact blasted away both our water supply and the Earth's entire atmosphere as well.

In one especially nasty collision soon after the Earth coalesced, a planet the size of Mars or larger may have smashed into the Earth, creating a recoil of molten rock from the Earth so great that it coalesced in Earth orbit to become our Moon. Between devastating impacts in those early years of Earth history, perhaps oceans formed and life began, only to be wiped out – time after time.

Comets and asteroids continued to hit the Earth after the first half billion years, but not as many and none so large. They scarred the face of the planet, but they brought with them precious gifts: chemicals that made possible the development of life. There was iron. The Earth contained vast amounts of metallic iron, but not much of it was accessible. Most of the Earth's original supply of iron had sunk to its center to form the core when the Earth was molten, largely due to the heat from the impacts of planetesimals as it formed and also because of the decay of radioactive elements. Thanks especially to asteroids, the Earth gathered (and continues to gather) a new supply of iron at its surface.

The comets and asteroids brought carbon to the Earth as well, the element essential for life. The early Earth lost most of its carbon when the carbon dioxide (CO_2) or methane (CH_4) in its atmosphere was blown away by titanic impacts.

The comets that crashed (and continue to crash) into the Earth also brought a new and indispensable supply of water for the planet, the medium in which life could form. Some of Earth's water may have been liberated from rocks by volcanic heat far below ground and coughed to the surface during eruptions as clouds of steam. The rain from those clouds contributed to the formation of oceans. But most, perhaps nearly all the water on Earth was the gift of comets, dirty snowballs whose fall created conditions for life to arise.[31]

If we can remove the threat that Earth-crossing comets and asteroids pose without destroying or exiling their bodies, imagine what raw materials they will offer. And these objects even truck their goods to within our reach.

If we can nudge an asteroid or comet away from a collision course with Earth, we can also land on its surface and mine it for its treasures, to ship down to Earth in spacetruckloads or to retain in space to build orbiting habitats for humanity.

In the future, mass drivers may indeed be set up on the most threatening comets and asteroids to throw material off them into space, not so much to reduce them in size, not so much to steer them into safer orbits, but to supply iron, aluminum, and the precious and strategic metals in short supply on Earth. The cometary and asteroidal debris would be refined in space using the vast, constantly available energy of the Sun. The refined metals could then be carried to Earth by space freighters.

Even more valuable than bringing most of the excavated comet and asteroid material to Earth would be to use it to build settlements in space. Earth-threatening comets and asteroids have such low gravities that little energy is required to heave valuable ore from their surfaces and guide it to wherever space colonies are desired. Using the raw materials that space provides is far more economical than trying to lift heavy loads like iron and

A continual pelting by small comets?

Could it be that the Earth is experiencing a meteor storm every minute of every day? That's what Louis Frank of the University of Iowa thinks – and has been declaring since 1986.

Frank says that this meteor storm comes not from pinhead- and pea-size sloughoffs from traditional comets a mile or few in diameter, but from miniature comets averaging 40 feet (12 meters) across, the size of a small house. Traditional comets are icy dirtballs, perhaps half dust and rock, half water and other ices. The minicomets Frank proposes are almost entirely frozen water, perhaps protected from vaporizing in the fierce sunlight of the inner solar system by a crust of carbon compounds.

These small comets, says Frank, are raining down on Earth at the rate of about 20 a minute, 1,200 an hour – enough to qualify as a meteor storm. As these minicomets approach the Earth, they are pulled apart by the Earth's gravitational tides or radiation belts at an altitude of 15,000 down to 600 miles (25,000 to 1,000 kilometers), well before they strike the atmosphere of Earth. Because these cosmic water balloons burst into a diffuse spray before entering the Earth's atmosphere, they create few if any fireballs. Instead, a fragmented minicomet enters the atmosphere as a mist, a visually transparent cloud perhaps 30 miles (50 kilometers) across that filters down to raincloud altitude where it mingles with the droplets and ice in existing clouds and becomes rain and snow that falls to Earth.

This process, says Frank, has been going on since the Earth formed. The gentle cosmic drizzle supplied by these minicomets has filled the oceans, deposited carbon, and given the Earth the raw materials from which life could form. But, he warns, these minicomets keep coming with their water, 40 tons at a time, filling the oceans higher and higher, flooding more and more land, until, one day, millions of years from now, the Earth will be entirely covered by water.

Frank proposed this barrage of minicomets in 1986 based on satellite images of Earth that showed dark specks in the atmosphere that he interpreted as infalling water. In early 1997, NASA launched a new satellite, *Polar*. *Polar*'s cameras, developed by Frank, took ultraviolet pictures that showed cool spots in the Earth's upper atmosphere as if it were being peppered by clouds of water. Another camera took pictures in visual wavelengths that caught objects streaking toward the Earth, fluorescing in the sunlight, that Frank interpreted as the minicomets themselves. Frank is jubilant because, after 11 years, he feels that his hypothesis has been confirmed.

Trail above the Earth's atmosphere created by the fluorescence of hydroxl ions (water molecules each missing one hydrogen atom), imaged by NASA's *Polar* satellite on December 31, 1996. NASA image courtesy of Louis A. Frank, University of Iowa

Other scientists are fascinated but doubt that Frank's interpretation is correct. Comets composed almost entirely of water are previously unknown. How were they formed? Where do they come from in such vast numbers? With so many of these water comets around, how can Mars be so dry? Where are the fresh 100-yard (100-meter) craters these house-sized impactors would drill in the Moon, which has far less gravity than Earth and therefore no atmosphere to protect it? Why have none of the thousands of satellites orbiting the Earth ever recorded being hit by one of these clouds? How can almost-pure-water comets survive so near the Sun – and without ever generating comas and tails? These are just a few of the problems most scientists have with Frank's theory.

water out of the Earth's huge gravitational field, especially when those materials are badly needed on Earth. Stripping Earth-threatening comets and asteroids is far more sensible than digging up the Moon, a great research base and tourist attraction in the future. For construction and to support life in space, comets and asteroids provide every element needed, the heavier ones in far better concentrations than the Earth's crust.[32]

An iron-and-nickel asteroid like Amun could supply for space settlement construction as much iron as has been shoveled to date from all ore deposits on Earth. Larry Lebofsky of the University of Arizona's Lunar and Planetary Laboratory has studied asteroids and comets as a source of raw materials to resupply Earth and to supply propellants for future interplanetary missions, metals for solar-power satellites, and building materials for space colonies, with the leftover debris to provide radiation shielding for them. Lebofsky calculates that just one asteroid like 3554 Amun, 2 kilometers (1¼ miles) in diameter, can provide $1 trillion ($10^{12}$) worth of cobalt, $1 trillion worth of nickel, $800 billion worth of iron, and $700 billion worth of platinum and its group of rare but useful metals.[33] Don't destroy such resources. Use them. A mine is a terrible thing to waste.

Astronomer John Lewis marvels, "The total value of this single small asteroid is approximately equal to the entire national debt of the United States."[34]

Near-Earth comets and asteroids that looked at first like a dire threat – and have been throughout Earth's history, and will be again and again if no action were taken – seem on second and deeper thought to be an even greater gift, a treasure mixed to Earth's needs and delivered to its doorstep by the processes that govern the formation and operation of the solar system.

Even as we wake to the dangers from bodies near us in space, we are awakening to the opportunities those comets and asteroids provide.

NOTES

1. Edmond Halley: "An Account of Some Observations Lately Made at Nurenburg by Mr. P. Wurtzelbaur," *Philosophical Transactions of the Royal Society of London*, volume 16, 1688, pages 402–406; Edmond Halley: "Some Considerations About the Cause of the Universal Deluge, Laid Before the Royal Society, on the 12th of December 1694," *Philosophical Transactions of* the Royal Society of London, volume 33, 1726, pages 118–119.

2. John S. Lewis: *Rain of Iron and Ice: The Very Real Threat of Comet and Asteroid Bombardment* (Reading, Massachusetts: Addison-Wesley [Helix Books], 1996), pages 176–182.

3. Duncan Steel: *Rogue Asteroids and Doomsday Comets*

(New York: John Wiley & Sons, 1995), page 113. This 500-tons-a-day figure includes relatively rare large impactors that bump up the average. If meteorites over one ton are omitted, the average falls to 110 tons a day.

4. C[hristopher] F. Chyba, T[obias] C. Owen, and W[ing]-H[uen] Ip: "Impact Delivery of Volatiles and Organic Molecules to Earth," pages 9–58 in Tom Gehrels, editor: *Hazards Due to Comets and Asteroids* (Tucson: University of Arizona Press, 1994).

5. The number of known Amor, Apollo, and Aten asteroids was furnished by Brian Marsden, director of the International Astronomical Union's Central Bureau for Astronomical Telegrams, by personal communication, June 26 and 30, 1997.

6. Chiron may be as large as 231 miles (372 kilometers) in diameter. M. V. Sykes and R. G. Walker: "Constraints on the Diameter and Albedo of 2060 Chiron," *Nature*, volume 251, 1991, pages 777–780.

7. G[erhard] Hahn and M[ark] E. Bailey: "Rapid Dynamical Evolution of Giant Comet Chiron," *Nature*, volume 348, 1990, pages 132–136. See also John S. Lewis: *Rain of Iron and Ice: The Very Real Threat of Comet and Asteroid Bombardment* (Reading, Massachusetts: Addison-Wesley [Helix Books], 1996, page 146.

8. David Morrison, Clark R. Chapman, and Paul Slovic: "The Impact Hazard," pages 59–91 in Tom Gehrels, editor: *Hazards Due to Comets and Asteroids* (Tucson: University of Arizona Press, 1994).

9. Donald K. Yeomans, personal communication, October 14, 1977. Yeomans based his calculation on the nucleus of Comet Shoemaker–Levy 9 being about 3 to 6 miles (5 to 10 kilometers) in diameter before breakup. Smaller comets may be broken apart by Jupiter and hit it more often.

10. John S. Lewis: *Rain of Iron and Ice: The Very Real Threat of Comet and Asteroid Bombardment* (Reading, Massachusetts: Addison-Wesley [Helix Books], 1996, pages 148–149.

11. M. E. Bailey, S. V. M. Clube, G. Hahn, W. M. Napier, and G. B. Valsecchi: "Hazards Due to Giant Comets: Climate and Short-Term Catastrophism," pages 479–533 in Tom Gehrels: *Hazards Due to Comets and Asteroids* (Tucson: University of Arizona Press, 1994).

12. William F. Bottke, Jr., Michael C. Nolan, Richard Greenberg, and Robert A. Kolvoord: "Collisional Lifetimes and Impact Statistics of Near-Earth Asteroids," pages 337–357 in Tom Gehrels, editor: *Hazards Due to Comets and Asteroids* (Tucson: University of Arizona Press, 1994).

13. M. E. Bailey, S. V. M. Clube, G. Hahn, W. M. Napier, and G. B. Valsecchi: "Hazards Due to Giant Comets: Climate and Short-Term Catastrophism," pages 479–533 in Tom Gehrels, editor: *Hazards Due to Comets and Asteroids*

(Tucson: University of Arizona Press, 1994).

14. M. E. Bailey, S. V. M. Clube, G. Hahn, W. M. Napier, and G. B. Valsecchi: "Hazards Due to Giant Comets: Climate and Short-Term Catastrophism," pages 479–533 in Tom Gehrels, editor: *Hazards Due to Comets and Asteroids* (Tucson: University of Arizona Press, 1994).

15. David Morrison, chair and editor: *The Spaceguard Survey: Report of the NASA International Near-Earth-Object Detection Workshop*, January 10, 1992 (Pasadena: Jet Propulsion Laboratory/California Institute of Technology for NASA, 1992.)

16. The actual calculation divides the objects into size categories, with each category treated separately. Duncan Steel: *Rogue Asteroids and Doomsday Comets: The Search for the Million Megaton Menace that Threatens Life on Earth* (New York: John Wiley & Sons, 1995), pages 29–30.

17. Duncan Steel: *Rogue Asteroids and Doomsday Comets: The Search for the Million Megaton Menace that Threatens Life on Earth* (New York: John Wiley & Sons, 1995), pages 23, 27, 34.

18. Duncan Steel: *Rogue Asteroids and Doomsday Comets: The Search for the Million Megaton Menace that Threatens Life on Earth* (New York: John Wiley & Sons, 1995), page 27.

19. C[hristopher] F. Chyba, T[obias] C. Owen, and W[ing]-H[uen] Ip: "Impact Delivery of Volatiles and Organic Molecules to Earth," pages 9–58 in Tom Gehrels, editor: *Hazards Due to Comets and Asteroids* (Tucson: University of Arizona Press, 1994). Short-period comets hit the Earth with a median velocity of about 15 miles per second (25 kilometers per second). Long-period comets impact with a median velocity of about 30 miles per second (50 kilometers per second).

20. Information about 1994XM$_1$ and the number of other nearest-Earth objects was provided by Brian Marsden, personal communication, June 30, 1997.

21. Tom Gehrels: "Collisions with Comets and Asteroids," *Scientific American*, volume 274, March 1996, pages 54–59. They hope to have a 72-inch (1.8-meter) telescope in operation soon.

22. A fourth near-Earth object search program, headed by Duncan Steel and using the 48-inch (1.2-meter) Schmidt telescope at the Anglo-Australian Observatory at Siding Spring, Australia, has recently been suspended for lack of funds.

   Alain Maury hopes to begin a search program soon with a 40-inch (1-meter) telescope at the Observatoire de la Côte d'Azur in southern France.

23. The first modern near-Earth object detection program was begun by Eleanor Helin and Eugene and Carolyn Shoemaker in 1973 with the 18-inch (0.46-meter) Schmidt telescope at Palomar Mountain in their

Planet-Crossing Asteroid Survey. Helin continued PCAS after 1982, when the Shoemakers, Henry Holt, and David Levy began the Palomar Asteroid and Comet Survey. Eugene Shoemaker died and Carolyn Shoemaker was injured in an automobile accident in Australia in 1997.

24. David Morrison, chair and editor: *The Spaceguard Survey: Report of the NASA International Near-Earth-Object Detection Workshop*, January 10, 1992 (Pasadena: Jet Propulsion Laboratory/California Institute of Technology for NASA, 1992.) I have adjusted the report's 1993 fiscal year estimates by 20 percent for inflation to approximate a 1998 equivalent. New technology and the success of the existing programs might keep costs well below these levels.

25. Thomas J. Ahrens and Alan W. Harris: "Deflection and Fragmentation of Near-Earth Asteroids," *Nature*, volume 360, December 3, 1992, pages 429–433.

26. David Morrison: "The Spaceguard Survey: Protecting the Earth from Cosmic Impacts," *Mercury*, volume 21, May/June 1992, pages 103–106, 110.

27. Thomas J. Ahrens: "Weapons of Mass Protection," *The World & I*, May 1993, pages 216–221.

28. Thomas J. Ahrens and Alan W. Harris: "Deflection and Fragmentation of Near-Earth Asteroids," *Nature*, volume 360, December 3, 1992, pages 429–433.

29. Duncan Steel: *Rogue Asteroids and Doomsday Comets: The Search for the Million Megaton Menace that Threatens Life on Earth* (New York: John Wiley & Sons, 1995), page 241.

30. This phrasing is an oversimplification. Comets with retrograde or substantially inclined orbits can be reached by spacecraft only with enormous expenditures of thrust or by gravitational assists provided by planets. The most accessible comets and asteroids are those in direct, low-inclination, low-eccentricity orbits so that the speed differential between the Earth and the object is slight and hence fuel costs are low. Larry Lebofsky, personal communication, August 26, 1997.

31. Harry Y. McSween, Jr.: *Fanfare for Earth: The Origin of Our Planet and Life* (New York: St. Martin's Press, 1997), pages 171–172 and Yvonne J. Pendleton and Dale P. Cruikshank: "Life from the Stars?" *Sky & Telescope*, volume 87, March 1994, pages 36–42.

32. The platinum group of elements are enriched by factors of thousands. Thomas J. Ahrens, personal communication, September 3, 1997.

33. The platinum group of metals includes ruthenium, rhodium, palladium, osmium, iridium, and platinum itself.

   Gold and silver would also be found in the asteroid at a level per ton thousands of times greater than they are found on Earth. Thomas J. Ahrens, personal communication, September 3, 1997.

   Ahrens notes that this valuation of an asteroid is naive because the increased supply of rare metals would cause their price to drop. But it would not necessarily decline in proportion to the increased supply because new uses and hence greatly increased demand could result and because the supply of the rare metals could be rationed, as diamonds are.

34. John S. Lewis: *Rain of Iron and Ice: The Very Real Threat of Comet and Asteroid Bombardment* (Reading, Massachusetts: Addison-Wesley [Helix Books], 1996), page 222. The United States' national debt in mid-1997 was $5.3 trillion. Lewis has written extensively on the economic value of asteroids and comets. For example, see John S. Lewis and Melinda L. Hutson: "Asteroidal Resource Opportunities Suggested by Meteorite Data," pages 523–542 in John S. Lewis, Mildred S. Matthews, and Mary L. Guerrieri, editors: *Resources of Near-Earth Space* (Tucson: University of Arizona Press, 1993).

CHAPTER 15

Catch a falling star

Catch a falling star
   and put it in your pocket;
Never let it get away.

Song by PAUL VANCE and LEE TOCKRISS, 1957

Meteor Etiquette
... meet these celestial visitants out of doors, and in full dress. A constant gaze with
the neck bent backwards, for six hours or more, in a frosty night, is the kind of
etiquette they exact.   DENISON OLMSTED, 1838[1]

ARE YOU ready to observe some meteors, to join a tradition that goes
back at least 4,000 years to the first shooting stars recorded for posterity by the Chinese and Koreans? Good news: you don't need much or any astronomical equipment, or expensive gear of any kind.[2]

Just bring along the best of all meteor-observing instruments: your eyes. They are wide-field cameras that are exquisitely sensitive for detecting motion.

Even so, your eyes can allow you to watch a maximum of only one-fifth of the sky at once. Therefore, where should you look?

If you look exactly at the radiant, the meteors from that shower are coming straight at you. Therefore, they have no trails. They will be dots that appear, brighten, and vanish. Meteors a little farther from the radiant will be very short streaks. Looking at the radiant of a meteor shower is not the most spectacular spot to watch, but it is interesting to identify the position of the radiant for yourself by noticing where among the constellations (where in Leo for the Leonids) the meteors appear as trailless dots because they are coming directly at you. If the meteor shower becomes a storm, however, it is worthwhile to look back at the radiant from time to time because of an interesting illusion that arises.

If you can see five or ten meteors a second, locate the radiant and spend some time with your vision centered on it. Out of the corners of your eyes, you will catch meteors streaking past, creating the impression that you are flying through space ... *which of course you are.* Your spaceship Earth is racing around the Sun at 18.5 miles per second (29.8 kilometers per second), but almost never can you sense this motion. The one exception is during a meteor storm, as the Earth dashes through a stream of particles. Then and only then can you truly sense the Earth in motion, in high-speed flight, a little like in *Star Trek* when the *Enterprise* travels at warp speed.

Because the Earth in its orbit encounters the Leonids in their orbit nearly head-on, by looking toward the radiant, where the meteors are

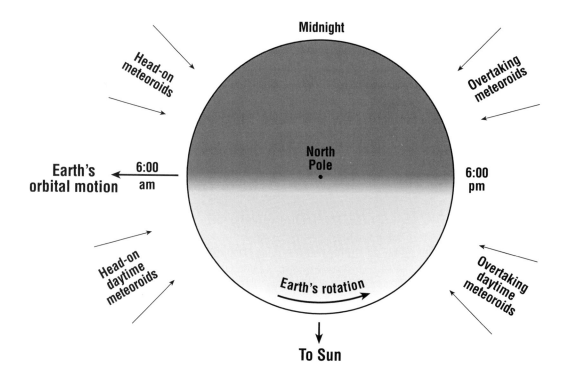

coming straight toward you, you are looking approximately in the direction that the Earth is traveling in its journey around the Sun.[3]

The farther you look away from the radiant (up to 45 degrees), the longer the meteor trails will be because you are watching them from the side, with the least front view or rear view foreshortening. It is the difference between standing next to the railroad tracks looking toward an oncoming train versus standing back from the tracks so that you see a side view of the train.

Another factor governing which direction to look to see the maximum number of meteors is the volume your eyes can take in of the layer in the atmosphere where most of the meteoroids burn up. That layer starts about 40 miles (65 kilometers) above the surface of the Earth and extends up to about 75 miles (120 kilometers). Because this meteor-burn layer wraps around the Earth, it is closest to you when you look straight up and farthest from you when you look toward the horizon. Because your eyesight gathers in a cone-shaped beam of light with your eyes at the point of the cone, the more distant the meteor layer, the greater the volume your eyes take in. The volume of the meteor layer you see varies as you scan from the zenith to the horizon.

Because of the Earth's orbital motion, meteoroids in retrograde orbits around the Sun (traveling the opposite direction of Earth) will strike our atmosphere at high relative speeds and will be seen after midnight. Meteoroids orbiting in the same direction as Earth must overtake us from behind. Therefore they will hit our atmosphere at low relative speeds and will be seen before midnight. If the Earth were a car traveling along a highway, the situation would be the same as a head-on versus a rear-end collision. Notice that daytime meteors (detected by radar) ought to be – and are – as numerous as nighttime meteors. Diagram by Will Fontanez and Tom Wallin, University of Tennessee Cartographic Services Laboratory.

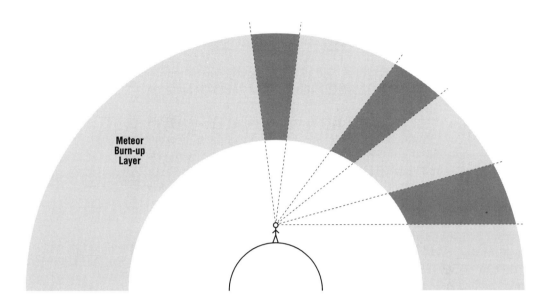

If you look just above the horizon, you are covering the largest volume of the visible sky through which the meteors pass, so you should see more meteors . . . *if* looking toward the horizon did not also mean that you are at the same time looking through the most atmosphere, and hence encountering the most haze and atmospheric light absorption, so that fewer stars – and shooting stars – are visible.

To see through the least atmosphere and hence experience the least haze and atmospheric absorption, you could look straight up. But to look toward the zenith is to watch the smallest volume of the meteor layer, and that too limits the number of meteors you will see.

The best viewing direction is a compromise. Look about halfway or a little more than halfway up the sky (45 to 50 degrees above the horizon) and about 40 degrees to one side of the radiant. Choose the side of the radiant in which the sky is clearest and darkest.

Fireballs

When observing meteors in a shower, there is always the possibility that you will see a fireball, an exceptionally bright meteor – that may or may not be part of the meteor stream which creates that shower. A fireball is defined as a meteor with an apparent magnitude brighter than -5, thus outshining Venus, the other planets, and all the nighttime stars.[4]

If the fireball is approximately 50 times brighter than that, there is a chance that some of the meteoroid survived its plunge through the atmosphere and can be found – if there are several other observations of the fireball's time, direction, brightness, and path among the constellations that can be used with yours to calculate the trajectory of the meteoroid and hence where it fell.[5]

Fireballs that will hit the ground to become meteorites are sometimes accompanied by a thunderlike rumble, the sonic boom of the meteoroid in its supersonic dive through the atmosphere of Earth. This noise travels at the speed of sound and therefore reaches the ear of the observer two or more minutes after the fireball is seen, like the sound of thunder arriving several seconds after the observer sees a flash of lightning. The late-arriving sonic boom of a fireball is an important indicator that a meteoroid has survived its flight through the atmosphere and will – *has* by the time the sound arrives – become a meteorite. These noisy fireballs are rare.[6]

Even rarer are crackling or swishing sounds that are heard at the *same time* a fireball is seen, as if the sound could travel at the speed of light. Australian astronomer Colin Keay has explored this occasionally reported phenomenon and proposes that a fireball generates not only visible light but also very-low- and extra-low-frequency radio signals. These radio waves travel at the same speed as light and are somehow converted to sound by nearby objects, such as the leaves of trees or frozen pine needles. Only fireballs destined to hit the ground generate *sustained* "electrophonic" sounds. Nothing dimmer than a fireball can generate even a momentary sound that is contemporaneous with the blaze across the sky.[7]

Photography

In a major meteor shower or storm, the radiant is a good place to point your wide-angle camera for a long exposure, to identify the position of the radiant and to catch the meteors demonstrating how they radiate from that small region. A wide-angle, long-exposure photograph can capture evidence that the meteors in a shower (except for a sprinkling of sporadic meteors) appear to come from a particular position among the stars, a realization that apparently eluded early skywatchers for thousands of years.

For the best chance of capturing bright meteors with conspicuous trails, veteran meteor observer Neil Bone recommends pointing your long-exposure camera about 20 degrees away from the radiant. Because meteors in this sector are close to the radiant, their motion across the sky appears slower because part of their motion is coming directly toward you.

Star map of the region of the constellation Leo, indicating the position of the radiant of the Leonid meteors at RA 10h08m, Dec +22°. Map by Will Fontanez and Tom Wallin, University of Tennessee Cartographic Services Laboratory.

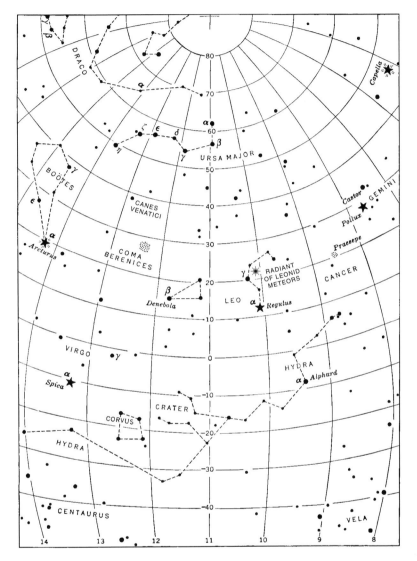

Because their apparent motion across the sky is slower than meteors 40 or 60 degrees from the radiant, these meteors spend a little longer at each position along their path and thus expose the film more, leading to pictures showing brighter meteors.

Photographing meteors

STEPHEN J. EDBERG, Author (with David Levy) of *Observing Comets, Asteroids, Meteors, and the Zodiacal Light*; Executive Director, Riverside Telescope Makers Conference

I've been photographing meteors since 1965, when I first became interested in them. My first attempts were made on super-high-speed Polaroid film in a borrowed camera, and I still have the prints on file. I've learned a lot over the years, so let me share some of that experience with you.

To capture meteors, first check your camera. Is the film high speed and is it advancing when the camera is cocked? Tighten the rewind knob and watch it turn as you cock the camera, so that at the end of a long night you don't end up with a blank roll of film or every exposure on one frame. (Yes, that's personal experience speaking.)

Bring a tripod or bean bag and a cable release. Bean bags are nice to use if a tripod isn't available. They permit the camera to be aimed with a wide variety of orientations, and they are small, portable, and cheap! A locking cable release permits the shutter to stay open as long as you want (5 minutes or more, depending on sky conditions). I prefer the type that can lock and release using only one hand.

Focus the lens at infinity, wide open, and set the shutter on B (for Bulb, allowing the shutter to stay open as long as the cable release's plunger is depressed). It's very easy to make that first time exposure last 1/60 of a second. I know!

Do you have a second camera dedicated to capturing persistent trains, loaded with very fast film and mounted on a quick-aiming tripod head? I finally captured a train during the 1992 Perseids with a 45-second exposure on ISO 3200 color film.

Watch where your camera is pointing so you know what went through its field of view, and whether you should stop the exposure. Take notes of the shutter-opening time and if you think a meteor was captured. How bright was it? What color? Shower member? Time? Keeping records now will let you figure out what will work in the future.

It is natural to think that a wide-angle lens will capture more meteors than a standard or telephoto lens. Such a thought is correct for a tele-

Your enemies

If you want to observe or photograph meteors, here are your enemies:

- A bright Moon in the sky at the same time of night that you want to watch meteors. A first quarter moon will squelch all but the brightest

photo lens, but your most efficient lens for capturing meteors is the standard 50–55 mm focal length, f/1.2–1.8 lens that probably came with your camera. The higher (slower) f/ number of most wide-angle lenses does not compensate for the wider area covered by the lens. You are more likely to capture bright meteors because of the larger sky area covered, but there are fewer of those fireballs and the slower f/ number means that you will miss the more numerous fainter meteors. I did capture a bright meteor with a homemade fisheye all-sky mirror-lens once, but you could barely see it because the image was so small.

Even with standard fast lenses and ISO/ASA 400 film, the capture rate is only a handful per night during rich showers like the Perseids and Geminids.

I like to photograph meteors using a clock-driven mount. That way I get a nice sky photograph even if no meteors are captured. If I'm aimed at the summer Milky Way, I may yet capture the annual Perseid that streaks through it.

Fireballs preferentially fall out of the camera's field of view. But not always: I captured a shadow-casting, magnitude −9 fireball once! And I just caught the starting point of it – but not the ending burst – in another camera. Good luck.

Aim 45 degrees up and 45 degrees from the radiant to cover the largest volume of meteor-making atmosphere. If there's a meteor storm, aim at the radiant, at the horizon, and everywhere in between. And don't forget to watch with your eyes!

If you are trying to capture spectra, make sure the spectrum of a meteor is dispersed approximately perpendicular to its expected direction of travel. That doesn't guarantee success. I've had the "bad" luck of getting a meteor centered so well in the field of view that its spectrum was outside the field of view. Spectra take a lot of patience, since the meteors must be very bright for their spectra to register on film.

Photographs record on film, permanently, the night's events. You will be able to go back years later and recall showers and fireballs you've seen. Enjoy, and be patient.

meteors, but it sets around midnight, leaving the period from midnight to dawn dark – just when most meteor showers are at their best.

- Clouds. If half or more of the sky is covered with clouds, meteor observing is greatly hindered. But the sky often clears as the night passes, so early evening clouds do not necessarily mean that you will not see meteors in abundance before dawn. In many places in the western United States on the evening of November 16, 1966, the skies were overcast. At midnight, the skies were still overcast. But by 3 a.m., November 17, the skies were clear and dedicated observers were treated to one of the greatest meteor storms ever seen.

Meteor observer's checklist

Here's what you will need for maximum enjoyment of any meteor shower.

A comfortable seat. Experienced meteor observers use a portable lawn chair or deck chair with an adjustable back that can provide neck support. Alternatively, they use an air mattress and a waterproof ground cloth or poncho. Remember, you will be looking about halfway up the sky to see the most meteors, and tilting your head back to that position for more than a few minutes can be a strain. You don't want meteor watching to be a pain in the neck.

Be careful about that lounge chair or air mattress though. As Stephen Edberg and David Levy warn, the reclining position assumed by meteors observers is conducive to dozing off. Falling asleep will reduce the number of meteors you see.

Warm clothes plus a blanket or sleeping bag. Meteor observing is not an aerobic sport. Clear night skies allow the day's heat to radiate back into space. The result is that you will be cooler than if you were walking or raking leaves. Your body loses heat rapidly through your head, hands, and feet. Wear layers of clothing: ski underwear, warm socks, insulated boots, a sweater, a parka, gloves, a hat, a ski face mask. Bring a blanket or a sleeping bag, or both if possible. You might also bring one-use hand and foot warmers, sold by sporting goods stores and ski shops. These small pads can be stored until needed, then activated so that they provide steady heat for about seven hours. These warmers are worn inside your gloves, between your socks and shoes, and (with care) inside a ski mask.

A red flashlight – and spare batteries. Use enough red gel to dim the light considerably or it will interfere with your night vision. The flashlight is used for moving about (as little as possible, out of consideration for other people's dark adaptation) and for making observing notes.

- Haze or smog.
- City lights. In or near a city, you can see only the brightest stars in the sky. Urban dwellers are always astonished by how many stars they can see if they leave their light-polluted cities for a dark setting. There are many more faint stars than bright ones. The same is true of meteors. If you want to see a rare meteor storm at its best, get away from city lights.
- Cold weather. The meteors don't care, but you will. Dress appropriately, and then dress a little warmer still.

Water. You might also bring coffee, tea, cocoa, fruit juices, or soft drinks – but go easy: these drinks increase the need and frequency to take time out from the meteor shower.

Snacks. To keep your energy level up and your body warm.

A **watch**. To keep track of the local time and to note times of fireballs. A stopwatch is handy for counting meteors per minute (or, if less fortunate, minutes between meteors).

Observation notebook or clipboard and paper, a star map, and some pens and pencils. To record your impressions and memorable sights.

Insect repellent (seasonal). This is your night to celebrate, not the bugs'.

First aid supplies. It's going to be dark. You could twist something (elastic bandages) or scrape against something (adhesive bandages). Your lips might crack (lip balm). Some animal might resent sharing his observing site with you (bite and pain medication). Don't forget personal medications you might need.

Toilet paper and wipes. Sorry to have to mention this, but try to use the restroom before you begin your meteor watch. Observers are frequently so anxious to begin that they find themselves in distress just when the shower is near maximum, and a distracted observer is not a happy or effective observer.

Optional: Veteran meteor watchers often bring along a **battery-powered tape recorder** so that they can note observations without averting their eyes from the sky. They also bring with them a **battery-powered radio and/or a tape or CD player** to provide background music. Caution: very cold weather can cause tape recorders and CD players to malfunction.

One more thing for all observers to bring along. **Patience**. An ample supply. There are no guarantees in meteor chasing. The hoped-for storm of meteors may not materialize, or may be seen on the opposite side of the globe. The weather may turn cloudy. But if you see a meteor storm, you will be amply rewarded for your patience.*

* Stephen J. Edberg and David H. Levy: *Observing Comets, Asteroids, Meteors, and the Zodiacal Light* (Cambridge: Cambridge University Press, 1994); Neil Bone: *Meteors* (in *Sky & Telescope* Observing Guide series) (Cambridge, Massachusetts: Sky Publishing, 1993); Paul Roggemans, editor: *Handbook for Visual Meteor Observations* (Cambridge, Massachusetts: Sky Publishing, 1989).

A night under falling stars

You don't need special observing equipment to enjoy a meteor shower or a meteor storm, but you do need a special spot. You cannot observe a great meteor display to good advantage from a city or even the suburbs. You need a dark, unobstructed spot for observing. That spot should also be away from or screened from roads where cars with night-vision-destroying headlights are frequent intruders.

Allow at least 20 minutes for your eyes to adapt to the darkness. Your night vision will continue to improve even after that.

Try to avoid returning to lighted areas, but if you must, wear sunglasses.

Watch out!

STEPHEN J. EDBERG, Author (with David Levy) of *Observing Comets, Asteroids, Meteors, and the Zodiacal Light*; Executive Director, Riverside Telescope Makers Conference

Based on over 30 years of meteor watching, I'd like to offer some hints about special things to watch for during a meteor shower (or storm!) that many observers fail to see. Oftentimes seeing something is not difficult, but remembering to look for it can be. Hopefully my hard-earned experience and a few stories will make your meteor observing experiences more fruitful.

Keep your wits about you. Meteors, fireballs, and bolides are, by their nature, surprises. Don't be astonished so long that you don't notice the nuances and subtleties of the phenomenon. Listen for anomalous (electrophonic) sounds that may be heard while a fireball is seen, and for sonic booms following, perhaps by minutes, the fireball. If you hear sonic booms, a meteorite has probably landed pretty close. Try to record the time and the altitudes and azimuths of the start and end of the fireball's path. Your report will help recovery efforts.

Stephen Edberg. Courtesy Stephen Edberg

Be prepared for the unexpected. Watch for meteors with odd colors, like 1968's scattering of ruby-red Leonids. I would have seen more of these gems that night if my girlfriend, also a member of my high school astronomy club, hadn't been such a distraction.

Watch for the rising shadow cast by a fireball as it disappears below the horizon. After a night including half a dozen Perseid fireballs, bright twilight was encouraging my friends and me to wrap up our observations. My eyes were directed down at the ground when a bright illumination and moving shadows immediately drew them skyward. A brilliant fireball was racing towards the horizon. As the fireball disappeared below the

Some very serious observers wear red goggles on such occasions.

You will catch many meteors out of the corners of your eyes – off to the side of where you are looking – because the outer regions of your retina, with a high ratio of rod cells to cone cells, are more sensitive to light than the center. The center of your retina, where the density of cone cells is greater, is more receptive to color.

Consider observing as a group, where each person views a different sector of the sky and you can talk to one another between calling off sightings. Plan to shift positions about every 15 minutes, so that no one feels deprived of the most productive sector.

Take time off about every hour to relax, nibble, sip, and refresh yourself.

hills on the western horizon, it cast the shadow of the hills on the hazy air of the Mojave Desert, and that shadow rose rapidly in the sky.

Have binoculars handy. A persistent meteor train may be visible long after the fireball that generated it. Watch for the train to expand, warp, and sweep across the sky, blown by winds tens of miles above you. During the 1980 Perseids, a fellow observer brought his new 6-inch (15-centimeter) binoculars along for general observing. Immediately after a bright fireball, I watched its persistent train through those binoculars, noting a 3-D effect and the expanding walls of a cylinder of meteoric material. I've watched other trains through telescopes and binoculars, as they twisted and turned while expanding and blowing across the sky.

Learn to distinguish between natural and artificial meteors and fireballs. Natural ones move fast and usually (but not always) are over in seconds. Reentering spacecraft and orbiting debris move slowly, about the same speed as an Earth satellite in low orbit, and can sometimes be seen for a couple of minutes moving across the sky, just like a satellite but looking more like a slow-moving firework than a moving star. In 1969, I stepped out of building in southern California just in time to watch what turned out to be the reentry of a Soviet booster rocket. It swept across about 120 degrees of sky from north to southeast in about two minutes, before fading out (and crashing into the ocean off Peru). Nearly a decade later, during an astrophotography session, I was able to stop an exposure, re-aim, and get a picture of another reentry through a telescope.

Keep your eyes up. Nothing makes seeing meteors easier than looking at the sky. The only people favored with better chances of seeing meteors are those who spend more time looking up.

It is fun to watch the immutable heavens that aren't.

Major annual meteor showers and two former showers

Shower	Dates	Maximum (varies a day due to leap year)	Solar longitude (degrees)	Radiant position at shower maximum	Velocity in miles/sec. (km/sec.)
Quadrantids (Boötids)	Jan. 1–6	Jan. 3	283.16	15h 30m, +50°	26 (42)
Lyrids	Apr. 19–25	Apr. 22	32.1	18h 16m, +34°	30 (48)
Eta Aquarids	Apr. 24–May 20	May 5	44.5	22h 27m, 00°	40 (65)
Southern Delta Aquarids	July 15–Aug. 20	July 28	125.7	22h 36m, −17°	25 (41)
Perseids	July 25–Aug. 20	Aug. 12	140.1	3h 04m, +58°	37 (60)
Draconids (Giacobinids)	Oct. 8–10	Oct. 9	196.5	17h 28m, +54°	15 (24)
Orionids	Oct. 15–Nov. 2	Oct. 22	208.7	6h 22m, +16°	41 (66)
Southern Taurids	Oct. 15–Nov. 25	Nov. 3	220.7	3h 24m, +14°	17 (28)
Northern Taurids	Oct. 15–Nov. 25	Nov. 13	230.7	3h 55m, +23°	18 (29)
Leonids	Nov. 15–20	Nov. 17	235.16	10h 08m, +22°	44 (71)
Andromedids (Bielids)	Nov. 6–Dec.1 (1971)	Nov. 27 (last storm 1885)	234 (1971) 247 (1885)	1h 40m, +44°	10 (16)
Geminids	Dec. 7–15	Dec. 14	262.3	7h 30m, +33°	22 (35)
Ursids	Dec. 19–24	Dec. 22	270.7	14h 28m, +75°	21 (34)

Sources: Compiled primarily from Neil Bone: *Meteors* (Cambridge, Massachusetts: Sky Publishing, 1993), pages 86–87, and by personal communication; Robert L. Hawkes: "Meteors, Comets, and Dust" in Roy L. Bishop, editor: *Observer's Handbook 1997* [annual] (Toronto: Royal Astronomical Society of Canada, 1996); Paul Roggemans, editor: *Handbook for Visual Meteor Observations* (Cambridge, Massachusetts: Sky Publishing, 1989); and American Meteor Society internet site.

Zenithal hourly rate	Likely observed maximum hourly rate	Parent comet (period in years)	Comments
85	40	Unknown (7)	Brief intense maximum Best after 2 a.m.
15	15	Thatcher (415)	Best after 3 a.m.
30	20	Halley (76)	Best from low northern latitudes & all southern latitudes. Seen after 3 a.m.
20	15	Unknown (3.6)	Best from low northern latitudes & all southern latitudes. Seen after 3 a.m.
100	50	Swift–Tuttle (120)	Best after 2 a.m.
		Giacobini–Zinner (6.5 before perturbations)	Storms in 1933 & 1946 when comet was near Earth
20	20	Halley (76)	Best after 2 a.m.
15	10	Encke (3.3)	Best 11 p.m. to 2 a.m.
15	10	Encke (3.3)	Best 11 p.m. to 2 a.m.
15	10	Tempel–Tuttle (33.22)	Best after 4 a.m. Periodic storms
		Biela (6.6 for 50 years after breakup)	Inactive since 1940; last storm 1885
95	50	Phaethon (asteroid or defunct comet) (1.4)	Best midnight to 3 a.m.
20	15	Tuttle (13.6)	Northern latitudes favored

Now, go take on the night

For your meteor watch, there are things you don't need. You don't need a telescope for meteor observation. A telescope allows you to see fainter meteors, but it restricts your view to a narrow region of the sky so that the number of meteors you see is likely to be reduced. Aesthetically, a telescope eliminates the panoramic view that your eyes provide, which is an essential part of the spectacle.

However, a telescope can be useful for other purposes. If you have clear, dark skies and are waiting for a meteor shower or storm to build, a telescope provides some pleasurable starwatching time.

You don't need a telescope to watch meteors, but you should bring binoculars. You won't be using the binoculars to scan for meteors, for the same reason that a telescope is not the instrument of choice. Binoculars are small telescopes, so they too narrow the field of vision for an event that is the prototype wide-field show. Instead, binoculars are useful for admiring and examining long-lasting meteor trains, the trails left behind by fireballs. With binoculars, you may be able to see those trains as expanding tubes of glowing atmospheric gases excited by the heat of friction when the meteoroid vaporizes as it enters the Earth's atmosphere. Watch for the effect of extreme high-altitude winds on these meteor trains. Notice what direction those winds are blowing.

Binoculars are also pleasant to have for admiring other astronomical objects while waiting for the meteor shower to liven up or during lulls in the shower.

You are ready to behold the big show – with (perhaps) a cast of thousands, every one a shooting star. But what do the final years of the millennium hold in store for the Leonids, greatest and most reliable of all producers of meteor storms? What will you see?

NOTES

1. Denison Olmsted: "On the Meteoric Shower of November, 1837," *American Journal of Science and Arts*, volume 33, number 2, January 1838, page 380 (footnote) of pages 379–393.

2. About 2000 B.C. seems to be the earliest date for systematic records of meteors. See Neil Bone: *Meteors*, noted below, page 47.

Excellent meteor observing tips and additional information for amateur astronomers seeking to make rigorous observations of meteors to aid scientific research may be found in: Stephen J. Edberg and David H. Levy: *Observing Comets, Asteroids, Meteors, and the Zodiacal Light* (Cambridge: Cambridge University Press, 1994); Neil Bone: *Meteors* (in *Sky & Telescope*

Observer's Guide series) (Cambridge, Massachusetts: Sky Publishing, 1993); Paul Roggemans, editor: *Handbook for Visual Meteor Observations* (Cambridge, Massachusetts: Sky Publishing, 1989).

3. Steve Edberg and David Levy urge observers to watch for one more curious phenomenon during a meteor storm. That sight is a faint glow in the direction of the actual radiant of the meteors (which may differ in position from the *apparent* radiant by many degrees) and another glow from the anti-radiant, the spot in the sky opposite the radiant (where the onslaught of shooting stars cannot be seen). Those dim glows are created by sunlight scattered by the comparatively dense stream of tiny meteoroids in their orbits beyond the Earth's atmosphere before and after their close encounters with the Earth. Stephen J. Edberg and David H. Levy: *Observing Comets, Asteroids, Meteors, and the Zodiacal Light* (Cambridge: Cambridge University Press, 1994), page 130.

4. Edberg and Levy prefer to classify meteors of magnitude −4 and brighter as fireballs.

5. Ideally, only two reports of exceptional fireballs would be necessary to determine where a meteorite fell, but a fireball always comes as a surprise and so suddenly that visual observers cannot always describe the trajectory precisely.

In the United States (and all of North America), fireball reports may be submitted to:

North American Meteor Network
c/o Mark Davis, Coordinator
1054 Anna Knapp Boulevard (Apartment 32H)
Mount Pleasant, South Carolina 29464
e-mail: MeteorObs@charleston.net
home page with reporting form:
    http://medicine.wustl.edu/~kronk/namn.html

Canadian fireball reports may be submitted to:

Meteorites and Impacts Advisory Committee (of the Canadian Space Agency)
c/o Robert Hawkes
Physics, Engineering, and Geology Department
Mt. Allison University
Sackville, New Brunswick E0A 3C0
fax: 506-364-2583
e-mail: rhawkes@mta.ca
home page with reporting form:
    http://dsaing.uqac.uquebec.ca/~mhiggins/
    MIAC/fireball.htm

Fireball reports for the United States may be submitted to:

American Meteor Society
Fireball Reporting Center
c/o Dr. David D. Meisel
Department of Physics and Astronomy
State University of New York at Geneseo
1 College Circle
Geneseo, New York 14454–1484
fax: 716-245-5282
e-mail: meisel@uno.cc.geneseo.edu
home page with reporting form:
    http://www.serve.com/meteors/

The Fireball Data Center of the International Meteor Organization serves as the clearinghouse for all fireball reports throughout the world. Reports may be sent directly to:

Fireball Data Center
c/o Andre Knoefel
Saarbrücker Strasse 8
D–40476 Düsseldorf
Germany
e-mail: fidac@imo.net
home page with reporting form:
    http://www.imo.net/fireball/index/html

Note: Reports sent to one group are shared with the others.

6. Bone applies the name bolide to a fireball that is accompanied by a sonic boom. Edberg and Levy use bolide to mean meteors that end their flights in explosions. Astronomical dictionaries and glossaries differ on this definition, although the majority seem to prefer exploding meteor.

7. See, for example, Colin S. L. Keay: "In Quest of Meteor Sounds," *Sky & Telescope*, volume 70, December 1985, pages 623–625.

In 1833, James N. Palmer reported to Denison Olmsted that he heard popping noises as he was observing the Leonid meteor storm. Was this an early report of electrophonic sounds? Doubtful. Palmer almost certainly was lying, based on the rest of his account, including the number of pops he claimed rather than an occasional sound accompanying a rare extra-bright fireball. Please see the vignette "A Fraudulent Report" in chapter 5 to judge the rest of Palmer's claims.

CHAPTER 16

Prospects for 1998, 1999, and 2000

Meteor storms are perhaps the most impressive phenomena in the night sky.
ĽUBOR KRESÁK (1993)[1]

Mother Nature delights in pulling the rug out from under astronomers who confidently predict the time of major meteor shower events.
DON YEOMANS (1997)[2]

W HAT CONDITIONS cause, foretell, or at least favor a meteor storm? Are those factors present at the close of the 20th century for the return of the Leonids?

Those factors seem to be:

Causes
- The comet must be actively shedding debris in recent visits to the Sun.
- The Earth must pass close to the comet's orbit not long before or after the comet passes.
- The comet's orbit must come close to the orbit of Earth – the closer the better.

A good omen
- The meteor stream's weak annual shower exhibits an enhanced level of activity over a period of several years before a storm is expected.

Lunar cooperation
- The phase of the Moon does not drown out the meteors with its brightness.

Factor 1: The comet

It's back. Comet Tempel–Tuttle, parent of the Leonid meteors, is alive and well. It was found on March 10, 1997 by astronomers Karen Meech, Oliver Hainaut, and James Bauer using the Keck II 10-meter (400-inch) telescope atop Mauna Kea in Hawaii. They spotted the inbound comet almost a year before it would reach perihelion, its closest point to the Sun. At that moment it was 344 million miles (554 million kilometers; 3.7 AU) from Earth and 3 million times too faint for an unaided human eye to see. The three University of Hawaii astronomers knew exactly where to look because three Jet Propulsion Laboratory astronomers had calculated a refined orbit for the comet in 1996. The calculation by Don Yeomans, Kevin

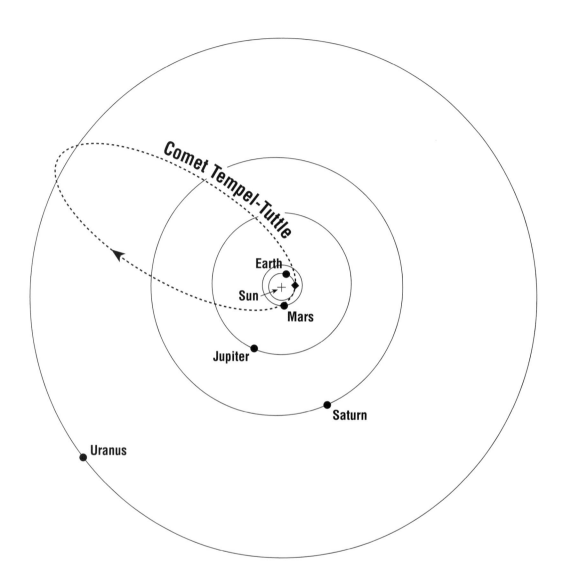

Yau, and Paul Weissman was so accurate that the predicted time of perihelion for the 33.22-year-period comet needed to be corrected by only 31 minutes.[3]

Comet Tempel–Tuttle has been passing close to the Sun frequently, losing gas and dust from its surface and developing an insulating dark, dirty-ice crust for more than a thousand years. At the distance of the asteroid belt where the comet was spotted, the temperature was about −200 degrees Fahrenheit (−130 degrees Celsius). With Tempel–Tuttle's out-

The orbit of Comet Tempel–Tuttle carries it just beyond the orbit of Uranus to just inside the orbit of Earth. This diagram shows the positions of Earth, Mars, Jupiter, Saturn, and Uranus when the comet was closest to the Sun on February 28, 1998. Diagram by Will Fontanez and Tom Wallin, University of Tennessee Cartographic Services Laboratory, based on one by Don Yeomans, Kevin Yau, and Paul Weissman, 1996.

gassing shut off by its crust and the cold, it was not surprising that the comet showed no diffuse coma or tail.

Comet Tempel–Tuttle is not a giant comet, like Comet Hale–Bopp in 1997 with a diameter of perhaps 25 miles (40 kilometers). It is not even the more average size of Halley's Comet, with an irregular shape about 9 miles long and 6 miles wide (14 by 9 kilometers). Based on its brightness and distance when recovered, Yeomans estimated that Comet Tempel–Tuttle is 2½ miles (4 kilometers) in diameter.

It is not crucial that Comet Tempel–Tuttle be bright like Comet Hale–Bopp, rivaling the brightest of the nighttime stars. To be the source of meteor storms, a comet does not even have to be bright enough to be seen with the unaided eye, as Comet Halley is. Available records indicate that Comet Tempel–Tuttle has been seen by eye alone only in 1366. Prior to 1998, in its 11 visits to the Earth and Sun since the telescope was invented, the comet has been seen by telescope only in 1699, 1865–1866, and 1965. On October 12, 901, the comet came very close to the Earth, just three times the distance of the Moon (0.008 AU), at a time near new moon, so the comet ought to have been visible in the east before dawn with a magnitude of 0, not very much dimmer than Comet Hale–Bopp (−1.4). But no record of anyone seeing the comet in 901 has been found.[4]

Yeomans estimates that Comet Tempel–Tuttle will reach an apparent total magnitude of +8.2 during the period when the comet is closest to Earth (0.36 AU) on January 17, 1998. The apparent total magnitude of a comet is calculated by integrating the brightness of its coma into a single starlike point to light, so a comet appears to the eye to be *fainter* than its apparent total magnitude. Comets with comas and tails are not visible to the eye until they reach about +3.4 apparent total magnitude,[5] whereas stars of magnitude +6.5 are visible to good eyesight under ideal conditions. So, at magnitude +8.2, Comet Tempel–Tuttle will be almost 100 times too faint for the eye to see, but within the range of most amateur telescopes.

The comet will not be visible to the eye. The comet is small. Faint, small – but adequate. "The extraordinary Leonid storm of 1966," says Yeomans, suggests that Comet Tempel–Tuttle "is still losing substantial amounts of dust," as it must if major showers and storms of the Leonids are to be seen.

Factor 2: The Earth–comet encounter

Comets near the Sun shed gas and dust along and around their paths. If their orbits carry them close to the Earth's orbit, our planet may plunge through that comet debris, lighting our skies with meteors.

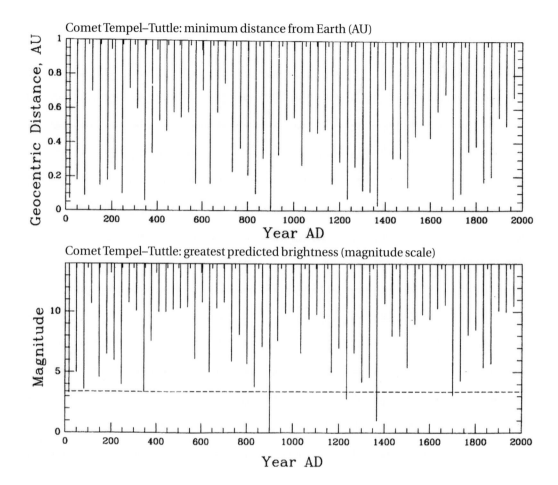

If the strewn particles have had time to spread themselves all the way around the comet's orbit rather evenly and if the Earth plows through those dust flecks and rock particles, the people on Earth will see a meteor shower every year when the Earth reaches that crossing.

A typical meteor-shower-producing comet has an orbital period of a few years, a few tens of years, or as much as a few hundred years, so many years pass between visits of that comet. Is a meteor shower likely to be enhanced if the comet that provides those meteors has just passed or is about to pass close to the Earth's orbit?

Yeomans examined that question in 1981 when he analyzed the orbit of Comet Tempel–Tuttle, parent of the Leonid meteors. He compared the return of the comet about every 33 years with the quantity of the Leonid meteors that showered the Earth. Yeomans found that, with one excep-

Comet Tempel–Tuttle: minimum distance from Earth and estimated brightness at each return from 0 to 2000 A.D. A comet is visible to the naked eye when its apparent magnitude is +3.4 or brighter (the horizontal dashed line on the magnitude plot)

Donald K. Yeomans, Kevin K. Yau, and Paul R. Weissman: "The Impending Appearance of Comet Tempel–Tuttle and the Leonid Meteors," *Icarus*, volume 124, 1996, pages 407–413

Donald K. Yeomans. Courtesy of Donald K. Yeomans

tion, all the Leonid meteor *storms* recorded in history (902 to 1966) occurred when the comet was within three years' journey of Earth.[6]

Of the 21 Leonid meteor storms[7] in the millennium between 902 and 1966, 15 occurred when the Earth passed the comet's node after the comet went by. Only 6 Leonid storms occurred when the Earth arrived ahead of the comet.

More precisely, every Leonid meteor storm but one has occurred within this range:

- The Earth crossed the orbital plane of the comet less than 1 year *ahead* of the comet's crossing through the Earth's orbital plane.
- The Earth crossed the orbital plane of the comet less than 3 years *behind* the comet's crossing through the Earth's orbital plane.

The exception was 1238, when observers reported a meteor storm 4 years after the comet passed.

Thus the meteoroids that cause almost all the Leonid storms seem to be congregated along approximately 12 percent of the length of Comet Tempel–Tuttle's orbit, with three-quarters of those particles behind the moving comet.

Clearly it is necessary for Comet Tempel–Tuttle to be on the inner part of its orbit, fairly close to Earth and sweating in the sunlight, for Leonid meteor *storms* to occur.

Factor 3: Passage close to the comet's orbit

The meteoroids that create Leonid meteor storms are concentrated along the comet's orbit within a year ahead and three years behind the comet. But what about the distribution of debris to either side of the comet's orbit, inward toward or outward from the Sun? When the Earth passes closest to the comet's orbit, does it matter whether the Earth is inside or outside the comet's orbit?

We might expect that the best position for the Earth to watch the silent celestial fireworks would be from slightly inside the comet's orbit. After all, the sunward side of the comet's nucleus is the face exposed to direct sunlight, hence the side that is losing material most rapidly. Most of the comet's ejected debris should therefore be sunward of the comet. Not only that, but the freed particles, escaping toward the sunward half of the sky and now on orbits of their own, have fallen closer to the Sun and have therefore picked up speed, necessary for them to stay in solar orbit. They are traveling slightly faster than the comet and therefore they gradually edge ahead of their parent body. Thus, based on how comets shed, we

The Leonids through history[a]

Year A.D.	Dates of observed maximum activity[b]	Estimated date of shower maximum[b]	Storm or outstanding shower[c]	Days before (−) or after (+) comet's nodal passage
902	Oct 12–13	902 Oct 13.2	Storm	+597.9
931	Oct 15–16	931 Oct 15.9	Shower	−1050.6
934	Oct 13–14, 14–15	934 Oct 14.2	Storm	+117.1
967	Oct 14–15	967 Oct 14.8	Shower	+44.3
1002	Oct 14–15	1002 Oct 14.8	Storm	+634.9
1035	Oct 14–15	1035 Oct 14.9	Shower	+349.4
1037	Oct 14–15	1037 Oct 14.7	Shower	+1080.2
1097	Oct 16–17?	1097 Oct 17.0?	Shower	−1810.0
1101	Oct 16–17	1101 Oct 17.2	Shower	−349.8
1199	Oct 21–22	1199 Oct 21.1	Shower	−480.8
1202	Oct 18–19	1202 Oct 19.0	Storm	+613.1
1237	Oct 18–19	1237 Oct 18.8	Shower	+1089.3
1238	Oct 18–19	1238 Oct 18.7	Storm	+1454.2
1366	Oct 21–22	1366 Oct 22.2	Storm	−7.2
1399	Oct 22–23?	1399 Oct 23.0?	Shower	−220.4
1466	Oct 22–23	1466 Oct 22.8	Shower	+71.4
1498	Oct 24–25	1498 Oct 24.9	Shower	−328.3
1532	Oct 24–25, 25–26	1532 Oct 24.9	Storm	−135.0
1533	Oct 24–25, 25–26	1533 Oct 25.0	Storm	+230.1
1535	Oct 25–26	1535 Oct 26.0	Shower	+961.1
1538	Oct 26–27	1538 Oct 26.8	Shower	+2057.9
1554	Oct 24–25	1554 Oct 24.8	Shower	−4533.1
1566	Oct 25–26, 26–27	1566 Oct 25.9	Storm	−149.0
1582	Nov 6–7	1582 Nov 7.0	Shower	+5697.1
1594	Nov 5–6	1594 Nov 5.9	Shower	−2092.7
1601	Nov 5–6	1601 Nov 5.9	Storm	+464.3
1602	Nov 6–7	1602 Nov 6.9	Storm	+830.3
1625	Nov 5–6, 6–7	1625 Nov 5.9	Shower	−2792.3
1666	Nov 6–7	1666 Nov 6.9	Storm	+145.9
1698	Nov 8–9	1698 Nov 8.9	Storm	−344.4
1766	Nov 11–12?	1766 Nov 12.0	Shower?[d]	−110.9
1799	Nov 11–12	1799 Nov 12.3	Storm	−117.1
1800	Nov 11–12?	1800 Nov 12.0?	Shower	+248.6
1831	Nov 12–13	1831 Nov 13.2	Shower	−423.3
1832	Nov 12–13	1832 Nov 13.1	Storm	−57.4
1833	Nov 12–13	1833 Nov 13.4	Storm	+307.9

Comet orbit minimum distance from Earth: inward (−)/outward (+) from Earth's orbit	Where observed
−0.0113	Southern Europe, North Africa
−0.0064	China
−0.0064	Europe, North Africa, China
−0.0064	Japan
−0.0129	China, Japan
−0.0249	Japan
−0.0249	Japan
	Europe
	Europe
	North Africa
−0.0059	Middle East, China
−0.0030	Japan
−0.0031	Japan
+0.0027	Europe, China
	Southern Europe
+0.0107	Japan
+0.0054	China
−0.0066	China, Korea
−0.0065	Europe, China, Korea, Japan
	China
−0.0065	Korea
−0.0024	Korea
−0.0024	China, Korea
	Europe
	China
+0.0102	China
+0.0102	China
+0.0025	Korea
-0.0043	China
-0.0162	Europe, Japan
	Northern South America
-0.0032	Eastern North & South America
	China
-0.0013	Southern Europe
-0.0013	Eastern Europe, Middle East
-0.0013	Eastern North America

The Leonids through history (*continued*)

Year A.D.	Dates of observed maximum activity[b]	Estimated date of shower maximum[b]	Storm or outstanding shower[c]	Days before (−) or after (+) comet's nodal passage
1834	Nov 13	1834 Nov 13.5	Shower?[e]	+673.0
1835	Nov 13	1835 Nov 13.5	Shower	+1038.0
1836	Nov 13–14	1836 Nov 14.2	Shower	+1404.7
1864	Nov 13–14	1864 Nov 14.1	Shower	−430.2
1865	Nov 13–14	1865 Nov 14.2	Shower	−65.1
1866	Nov 13–14	1866 Nov 14.1	**Storm**	+299.8
1867	Nov 13–14	1867 Nov 14.4	**Storm**	+665.1
1868	Nov 13–14	1868 Nov 14.4	**Storm**	+1031.1
1869	Nov 13–14	1869 Nov 14.2	Shower	+1395.9
1897	Nov 14–15	1897 Nov 14.8	Shower	−600.2
1898	Nov 14–15	1898 Nov 15.4	Shower	−234.6
1899	Nov 14–15	1899 Nov 14.8	Shower	+129.8
1900	Nov 15–16	1900 Nov 16.4	Shower	+496.4
1901	Nov 15	1901 Nov 15.5	Shower	+860.5
1903	Nov 15–16	1903 Nov 16.2	Shower	+1591.2
1930	Nov 16–17	1930 Nov 17.3	Shower	−608.8
1931	Nov 16–17	1931 Nov 17.4	Shower	−243.7
1932	Nov 17	1932 Nov 17.5	Shower	+122.4
1961	Nov 16–17	1961 Nov 17.3	Shower	−1265.2
1965	Nov 16–17	1965 Nov 17.2	**Storm**	+195.7
1966	Nov 17	1966 Nov 17.5	**Storm**	+561.0
1969	Nov 16–17	1969 Nov 17.4	Shower	+1656.9

Notes: [a]This table based on table 1 (pages 230–231) of John W. Mason: "The Leonid Meteors and Comet 55P/Tempel–Tuttle," *Journal of the British Astronomical Association*, volume 105, number 5, 1995, pages 219–235. Information on minimum Earth to comet orbit distances has been interpolated from Donald K. Yeomans: "Comet Tempel–Tuttle and the Leonid Meteors," *Icarus*, volume 47, 1981, pages 494–495 of pages 492–499. No Earth–comet orbit minimum distances are shown where Yeomans' list does not include the dates provided by Mason. Yeomans' list includes dates when orbital circumstances favored showers or storms even if no report of meteor activity has been found. Mason omits these dates. Please see Mason and Yeomans for the sources of the meteor activity reports. Note that some storms may actually have been outstanding showers, and some outstanding showers may actually have been less significant.

[b]Julian calendar for dates prior to October 1582. Gregorian calendar for dates after October 1582.
[c]If more than 1,000 meteors an hour are observed, the phenomenon is a meteor storm. If 150 to 1,000 meteors an hour are observed, it is an outstanding shower. The *typical* annual Leonid peak is about 10–15 meteors an hour. The typical annual Perseid peak is about 50 an hour.
[d]Mason classifies 1766 as a storm based on a mention by Humboldt, but no contemporaneous observations or other later reports have been found.
[e]Mason classifies 1834 as a storm based on a mention by Humboldt, but no American observations recorded large numbers of meteors.
[f]Mason says central North America.

Comet orbit minimum distance from Earth: inward (−)/outward (+) from Earth's orbit	Where observed
	North America
	North America
-0.0013	Europe, eastern North America
	Southern Europe
	Europe
-0.0065	Europe
-0.0066	Eastern North America
-0.0065	North America
	Europe
	China
	North America
-0.0117	China
-0.0117	Eastern North America
-0.0117	Western North America
-0.0118	Europe
	North America
	North America
-0.0062	North America
-0.0033	North America
-0.0032	Eastern Europe
-0.0031	Western North America[f]
-0.0032	Eastern North America

might expect that the best chance of a meteor storm would be a position slightly inside the comet's orbit and slightly ahead of the comet's arrival.

And we would be wrong. Yeomans' study shows only one example (1366) of a Leonid meteor storm that just barely fits this geometry of encounter.

What's wrong with the logic? Nothing. Comets do shed mostly to sunward. Cometary debris shed sunward will tend to draw ahead of the comet. The problem is that the train of logic has not yet been carried to completion, as illustrated by a comet's tail, which streams from the comet *away* from the Sun. Forces are at work that move much of the ejected comet gas and dust from their initial position on the sunward and forward side of the comet to positions farther from the Sun where the dust especially tends to lag behind the comet.

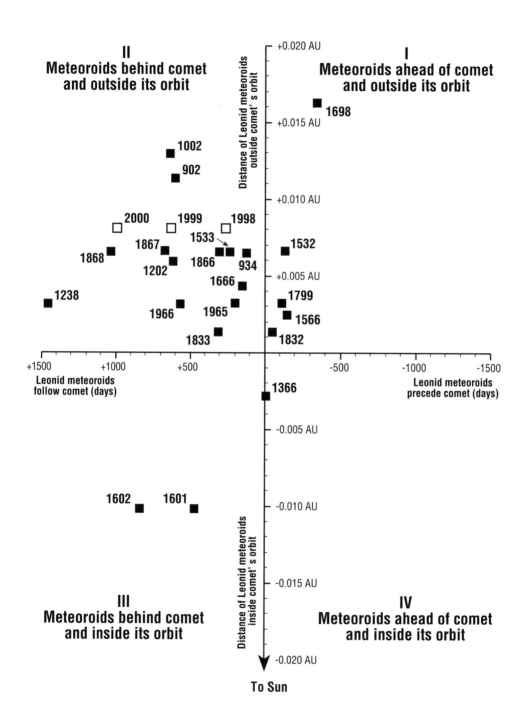

When a comet is close to the Sun, ices at the surface of the nucleus vaporize and escape from the comet, carrying away solid particles. The escaping gas and dust form a huge coma around the nucleus. Charged subatomic particles from the Sun carry the ionized gases from the comet outward directly away from the comet to form a straight ion (or gas or plasma) tail that shines bluish-white by fluorescence. The pressure of sunlight pushes dust particles away from the Sun to form the dust tail, which shines yellowish-white from the reflection of sunlight. The dust particles are harder to accelerate than the less massive gas particles, so they tend to lag behind the motion of the comet. The dust particles lost to the comet are now meteoroids that may strike the atmosphere of Earth and glow as meteors. (Not to scale.) Diagram by Will Fontanez and Tom Wallin, University of Tennessee Cartographic Services Laboratory.

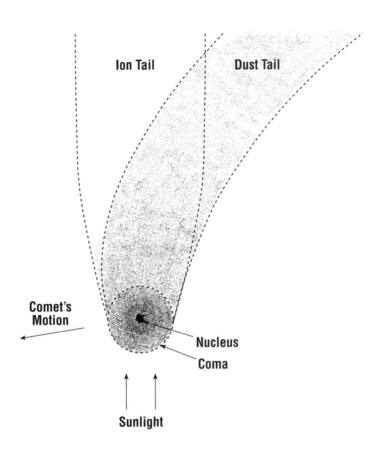

(Left) Leonid meteor storms occur most often when the Earth crosses Comet Tempel–Tuttle's orbital plane *after* (rather than before) the comet has passed (within 3 years) and close to but *farther* from the Sun than the comet's node. This diagram, based on one created by Don Yeomans in 1981, shows 21 Leonid storms and the circumstances for 1998, 1999, and 2000. Map by Will Fontanez and Tom Wallin, University of Tennessee Cartographic Services Laboratory.

Those transportation mechanisms are the subatomic particles from the Sun and the pressure of the Sun's light – forces directed outward, away from the Sun. The fast-moving subatomic charged particles from the Sun attract or repel the charged gases lost by the comet. This solar wind at the Earth's distance from the Sun is usually traveling about a million miles per hour (1.6 million kilometers per hour) and hence carries the comet gases outward with it, away from the Sun, to form the rather straight gas tail (also called an ion or plasma tail) of a comet. The gas tail shines with a bluish glow caused by one of the comet's major gases, positively charged carbon monoxide (CO^+), fluorescing when struck by the sunlight.

But gases striking the Earth's atmosphere do not provide a meteor display. Meteoroid burnup requires solid particles – those released and carried away from the comet's nucleus when the surrounding ices vaporize. These meteoroids-to-be usually have no positive or negative charge and thus the electrical charge of the solar wind's subatomic particles does not

Comet Mrkos, August 24, 1957. Notice the straight ion tail flowing directly away from the Sun and the curved dust tail that lags slightly behind the comet. Courtesy of Palomar Observatory, 48-inch (1.2-meter) Schmidt photo

affect them. Still, the solar wind's fast-moving fragments of atoms might collide with the solid cometary particles and shove them outward, but the subatomic particles have so little mass that their ability to push comet particles outward is feeble, like a gentle wind blowing against a human being.

Curiously, what herds the cometary dust and grains outward is the pressure of sunlight. Light truly does exert pressure – tiny but constant. Because the dust and grains are more massive than the lightweight charged gases of the comet, they do not accelerate as quickly, so the comet's dust tail does not extend as far into space as the gas component of the tail. And because the dust tail is moving slower than the comet, it lags behind the comet's gas tail and appears more curved. The dust tail is yellowish because it shines by sunlight reflected (and scattered) by the solid particles lost by the comet's nucleus.[8]

It is these solid particles from a comet that produce a meteor show, so the best place for the Earth to position itself is slightly *outside* the spot where the comet's orbit comes closest to the Earth's orbit.

That is exactly what scientists have found. Of the 21 Leonid meteor storms in history, 18 occurred when the Earth was outside the comet's orbit. Only one Leonid storm took place when the Earth was more than 0.013 AU outside the comet's orbit. No Leonid storms took place when the Earth was more than 0.010 AU inside the orbit of the comet.

Don Yeomans, Kevin Yau, and Paul Weissman found that no Leonid meteor showers could have been observed prior to the 8th century because the Earth passed too far inside the orbit of Comet Tempel–Tuttle.[9]

Most of the mightiest Leonid outbursts – 1799, 1833, 1966 – occurred when the Earth was outside the orbit of the comet, but very close to the node – less than 0.0032 AU (298,000 miles; 497,000 kilometers) distant, not much farther than the Moon.

The best of all possible whirls

The celestial geometry that most often creates Leonid meteor storms occurs when the Earth *follows* Comet Tempel–Tuttle's dirt-scattering passage through its descending node within two years and when the Earth is *farther* from the Sun than the comet's node, but as close to that point as possible.

When these two factors are combined for the 21 known Leonid meteor storms, some interesting tendencies emerge:

- 13 of the 21 storms occurred when the Earth, at its closest encounter with the comet's path, was *farther* from the Sun than the comet and arrived at the point *after* the comet had gone by.

Leonid meteor storms (classifications by Mason and Yeomans)

Mason[a] (23 storms)	Yeomans[b] (20 storms)
902	902
934	[934 as shower]
1002	1002
1202	1202
1238	1238
1366	1366
1532	1532
1533	1533
1566	1566
1601	1601
1602	[1602 as shower]
1666	[1666 as shower]
1698	[1698 as shower]
1766[c]	[1766 omitted]
[1798 omitted]	1798
1799	1799
1832	1832
1833	1833
1834[d]	[1834 omitted]
1866	1866
1867	1867
1868	1868
[1900 as shower]	1900
[1901 as shower]	1901
1965	1965
1966	1966

Notes: [a]John W. Mason: "The Leonid Meteors and Comet 55P/Tempel–Tuttle," *Journal of the British Astronomical Association*, volume 105, number 5, 1995, pages 219–235.
[b]D[onald] K. Yeomans: "Comet Tempel–Tuttle and the Leonid Meteors," *Icarus*, volume 47, 1981, pages 492–499.
[c]In his observation of the 1799 Leonid storm from South America, Humboldt reports that "the oldest [natives] among them remembered that the great earthquakes of 1766 were preceded by a similar phenomenon." Alexander von Humboldt: *Personal Narrative of Travels to the Equinoctial Regions of America During the Years 1799-1804 by*

Alexander von Humboldt and Aimé Bonpland, volume 1, translated and edited by Thomasina Ross (London: Henry G. Bohn, 1852), pages 352–353, 359. No month or even time of year is given. No other record of this event has been found. I believe that this one vague recollection does not justify classifying 1766 as a Leonid storm or even a major shower. It may have had no relation to the Leonids.
[d]Humboldt recorded that a sight similar to that of 1833 was seen the next year. No North American observations support more than a modest shower.

The return of the Leonids

NEIL BONE, Director, Meteor Section, British Astronomical Association; Author, *Meteors* (1993)

Neil Bone. Courtesy of
Neil Bone

To most of the current generation of active amateur meteor observers, the intense Leonid activity of the mid-1960s is the stuff of legend. Today's watchers expect their richest pickings during the Perseid, Geminid, or Quadrantid showers, whilst the Leonids through the 1970s, 1980s, and into the early 1990s produced only modest activity. British Astronomical Association observations collected from 1974 to 1991 show peak corrected zenithal hourly rates* for the Leonids about 20, on November 17, in this "quiet-time" period. Activity was far from absent during the 1974–1991 period, but levels were little above those of the contemporaneous, comparatively minor Taurid shower.

First signs of a change to higher rates, with the approach to the inner solar system of the ortho-Leonid cloud,[$] came in 1994. Observers in both Spain and the United States reported substantially elevated activity, with up to 20 Leonids an hour being seen, despite moonlight. In light of these observations, the 1995 return was keenly awaited and, in common with other groups, the BAA Meteor Section planned a major effort.

Meteor observing from maritime western Europe in the autumn is often fraught with cloud and rain, bringing disappointment all too often. In 1995, however, the preparations were rewarded with superb conditions on the critical night of Friday, November 17–18. The British Isles were immersed in a tongue of clear, frosty Arctic air extending down from Greenland, allowing the collection of hundreds of hours of watch data by scores of observers.

Around midnight, as the radiant rose out of the eastern sky, rates were comparatively low – typical counts from a dark site were 8 to 10 Leonids an hour. As the night wore on, however, activity kicked! In the early morning around 03–04 UT, observed rates approaching 30 Leonids an hour were found, giving a corrected ZHR of 40. For many, this was a first glimpse of Leonid rates comparable to those of the major annual showers.

Not only was activity much enhanced over quiet-time levels; the proportion of bright events was high (over 20% of Leonids were magnitude 0 or brighter). Observers saw many negative-magnitude Leonids cutting across the Belt of Orion. One truly spectacular event over southwest Britain at 0439 UT was comparable in brightness to the then-risen waning crescent moon, leaving an ionization train visible to the naked eye for the next five minutes.

* Zenithal hourly rates are usually higher than observed rates because they extrapolate the radiant to the top of the sky for best visibility.

[$] The ortho-Leonid meteoroids are the recently shed debris close to Comet Tempel–Tuttle. They are the particles responsible for a Leonid storm. The clino-Leonids are the "older" particles that have spread around the comet's orbit and provide a weak annual shower of Leonid meteors. G. Johnstone Stoney and Arthur M. W. Downing introduced the names ortho and clino in their predictions for the 1899 Leonid return. Please see chapter 10.

A repeat observing run planned for 1996 met with cloudy skies over most of Europe. Where the clouds did part, Leonid activity was seen at levels comparable to or even in excess of those in 1995's peak for an interval of at least 12 hours. Observers in the United States had the best views.

The onset of high Leonid activity in the morning hours of 1995 November 17–18 left few in any doubt that the shower was on the rise towards more substantial activity at the century's end. Whatever the years 1998–2000 bring, many of us already cherish memories of the Leonids breaking out of their quiet-time pattern.

- 5 of the 21 storms occurred when the Earth was *farther* from the Sun than the comet, yet reached the comet's node *ahead* of the comet.
- Only 2 in 21 recorded Leonid blizzards occurred when the Earth passed sunward of the comet's node and after the comet passed.
- In only 1 case did people see a Leonid storm when the Earth was inside the comet's orbit and ahead of the comet: in 1366 when the Earth beat Comet Tempel–Tuttle to its node by a mere 7 days.

A good omen

Another indication that a Leonid outburst is likely is the rise in Leonid activity in the years leading up to an expected storm. It's a little like a geyser beginning to burble before it erupts. In 1864 and 1865, the Leonids produced substantial meteor showers prior to meteor storms in 1866, 1867, and 1868.

On the other hand, the Leonids provided goodly showers in 1897 and 1898, but then no storms in 1899, 1900, or 1901 – just more showers. The years 1930 and 1931 produced above average Leonid showers, followed by 1932 . . . with just another shower.

Periods of intense Leonid meteor activity – a few years of heavy showers or even a storm or two – are followed by more than two dozen years with very few meteors. This cycle demonstrates that the swarms of Leonid meteoroids are strongly concentrated in the vicinity of the comet. The few years before and after the comet passes closest to the Sun and Earth – accompanied by the densest portion of its meteoroid stream – offer the only periods when Leonid meteor storms or notable meteor showers are likely.

In an ordinary year on November 17, the Leonids peak at about 10 to 15 meteors an hour, scarcely noticeable against the random background of sporadic meteors. But over the last few years, the Leonids have been significantly more active:

- 1994: a maximum of 60–100 an hour (uncertain because of bright moonlight).
- 1995: a maximum of 30–40 an hour.
- 1996: a maximum of 50–60 an hour.[10]
- 1997: a maximum of 50–100 an hour (despite an almost full moon).[11]

This increase in annual Leonid activity to a rate comparable to the annual Perseid or Geminid meteor showers holds out hope that 1998 and/or 1999 may bring far more powerful displays of Leonid meteors.

Considering moonshine

No matter how fine the meteor shower – or even storm – you won't see much if the Moon is bright and up when the meteors are streaking.[12] The Moon will cooperate in 1998 and 1999 by staying largely out of the way.

November 17, 1998 falls just before new moon (age 28 days), the dark-of-the-moon time when the Moon is not visible throughout the night.

On November 18, 1999, the Moon will be between first quarter and full moon, age 9 days. This waxing gibbous phase is bright enough to hide all but the brightest meteors. However, the Moon sets about 2 a.m., before the constellation Leo is high above the horizon, before the Leonid meteors turn up the shower to maximum spray.

On November 17, 2000, the Moon will be age 20 days, near last quarter, rising just ahead of Leo and the Leonid radiant, flooding the post-midnight sky with brightness that will hide most of any meteors available. Among meteor experts, only John Mason considers it likely that the Leonids will produce an outstanding shower in 2000, even possibly a meteor storm.[13]

Prospects for the Leonids in 1998, 1999, and 2000

Far and away the best chance for a great Leonid show occurs when the Earth passes the comet's node *after* the comet has passed (but not more than three years after) and when the Earth is *farther* from the Sun than the comet when the Earth passes the comet's node.

These conditions are met in 1998, 1999, and 2000. Comet Tempel–

Tuttle's closest approach to our planet on this circuit is January 17, 1998 when it is within 0.36 AU of Earth, although the Earth won't have its first close encounter with the comet's orbit until November 17. The comet reaches perihelion on February 28, 1998 and then begins its retreat from the Sun. Outbound, on March 5, 1998, the comet plunges through the Earth's orbital plane.[14]

Ten months after the Earth's closest encounter with Comet Tempel–Tuttle on this visit, the Earth will have its closest approach to the dust-strewn orbit of the comet – the encounter that provides the meteors. On November 17, 1998, the Earth will pass closest to the Tempel–Tuttle

A second meteor storm in 1998?

The Draconids are a new and erratic meteor shower that radiates from Draco, the Dragon, in the high northern sky. It usually peaks during a 4-hour period on October 9/10. Most years the shower is unrecognizable, but occasionally it is spectacular.

The parent of the Draconid meteors is Comet Giacobini–Zinner, so these meteors are sometimes referred to as the Giacobinids. Comet Giacobini–Zinner has an extremely short period of only 6½ years, which varies because of Jupiter's perturbations.

Comet Giacobini–Zinner is in a direct orbit, traveling around the Sun in the same direction as all the planets. So the Draconid meteor stream, following in and near the path of the comet, catches up with Earth from behind. One result is that the Draconids are most numerous in the *evening*, before midnight. Another consequence is that the Draconids, using most of their orbital speed just to catch up with the Earth, enter our atmosphere at slower speeds than most meteoroids, a mere 12 miles per second (20 kilometers per second) – less than one-third the speed of the Leonids. As a result, the Draconids are cooler, more reddish in color, and streak across the sky more slowly. They typically burn up between 60 and 55 miles (98 to 90 kilometers) above the ground.

The Earth ordinarily experiences no significant meteor activity from Giacobini–Zinner unless the comet has just passed or is just about to pass and when the comet's orbit comes very close to Earth. But even then, strong meteor showers or storms do not always happen.

However, on the evening of October 9, 1933, Europe experienced a meteor storm (briefly 800 streaks a *minute*) when the Earth passed close node 257.4 days after the comet goes by. On November 18, 1999, 622.6 days. On November 18, 2000, 987.9 days.[15]

The years 1998 and 1999 are especially promising because in those years the Earth follows the comet by less than two years. The year 2000 offers less hope because the Earth arrives 2 years 9 months after the comet has passed.

There is, however, another crucial factor to consider when gauging prospects for a return of the Leonids in dazzling numbers: how far the Earth will be from the comet's node. Yeomans found *no* Leonid storms that had occurred when the Earth was more than 930,000 miles (1.5 million

to the comet's orbit (about twice the distance of the Moon) only 80 days after Giacobini–Zinner had sped by. No unusual meteor activity warned of the coming storm, so astronomers gauge that this stream of debris from the comet must still be rather compact.

In 1946, the Earth crossed the meteoroid stream 15 days behind the comet and at a distance from its orbit of scarcely more than *half* the distance of the Moon. A powerful shower was reported, mostly in North America, despite bright moonlight. Astronomers counting meteors by radar echoes from their ionized trails in the atmosphere recorded up to 10,000 meteors an hour.

In 1952 the Earth penetrated the comet's orbital plane 196 days *ahead* of the comet's arrival at the node and observers witnessed meteors at a peak rate of 200 an hour.

On October 8, 1998, the Earth will cross the comet's orbital plane 49.5 days ahead of the comet. The 1952 encounter demonstrated that there is significant debris ahead of Comet Giacobini–Zinner as well as behind it, but how much? Will the Earth plow through a dense strip of those meteoroids? Alas, the Earth will be 3.5 million miles (5.4 million kilometers) inside the orbit of the comet – both the wrong position and too far for the Earth to be likely to cross the Draconid meteor stream. No Draconid storm or significant shower on record has ever taken place under such circumstances.

Some observers hold out hope that the Draconids could provide a preview of the Leonids less than six weeks before they arrive. It's worth watching for the Draconids in the early evening hours of October 9, particularly in western Europe. Surprises can happen. But, says Don Yeomans, "If history is any guide, there will not be a significant Giacobinid shower in 1998.*

* A. C. B[ernard] Lovell: *Meteor Astronomy* (Oxford: Oxford University Press, 1954), pages 326–327; David W. Hughes: "The History of Meteors and Meteor Showers," *Vistas in Astronomy*, volume 26, 1982, pages 325–345; Neil Bone: *Meteors* (Cambridge, Massachusetts: Sky Publishing, 1993), pages 105–106; Stephen J. Edberg and David H. Levy: *Observing Comets, Asteroids, Meteors, and the Zodiacal Light* (Cambridge: Cambridge University Press, 1994), pages 128–129; Paul Roggemans, editor: *Handbook for Visual Meteor Observations* (Cambridge, Massachusetts: Sky Publishing, 1989), pages 144–149; Donald K. Yeomans and John C. Brandt: *The Comet Giacobini–Zinner Handbook* (Pasadena: Jet Propulsion Laboratory [document 400-254], March 1985); and Donald K. Yeomans, personal communication, May 26, 1997.

kilometers; 0.010 AU) sunward of the node or more than 1.2 million miles (2 million kilometers; 0.013 AU) farther from the Sun than the node.[16]

The eyepopper in 1833 occurred when the Earth passed the node at a distance of only 121,000 miles (195,000 kilometers; 0.0013 AU) – roughly half the distance to the Moon. In 1966, the distance was 288,000 miles (464,000 kilometers; 0.0031 AU) – only 20 percent farther than the Moon. In 1799, 298,000 miles (479,000 kilometers; 0.0032 AU).

In 1998, 1999, and 2000, the Earth will be 744,000 miles (1.2 million kilometers; 0.008 AU) from the node – just over three times the distance of the Moon. Our position behind the comet's passage augurs well. Our distance from the node somewhat dampens those hopes. The Earth passes nearly three times as far from the comet's orbital path as in 1966 and more than six times farther than during the storm of 1833. Those Leonid storms provided displays at least as great as 50,000 meteors an hour – and perhaps even 150,000.

Yet the Earth's distance of 744,000 miles outward from the comet's orbit in 1998–2000 is within the range observed for major Leonid outbursts. Of the Leonid storms that have occurred when the Earth followed the comet and was outside the node, 2 of those 13 have taken place when the Earth was even farther from the node than it will be in the years at the end of the 20th century.

"The 1998–1999 circumstances are most like those for the 1866–1868 and 1931–1932 returns," says Yeomans. The 1866–1868 period provided meteor storms with up to 5,000 meteors per hour. An impressive sight. The years 1931 and 1932, with geometry almost identical to 1866–1868, yielded at most 200 meteors per hour. A vigorous shower, but not a storm. The Earth's closest approach to the comet's orbit in 1998 and 1999 is about the same distance as in 902, the first reasonably certain storm of Leonid meteors.

Where on Earth . . . ?

All the factors that control or foretell a storm or extraordinary shower of Leonid meteors are in place for the return of the Leonids, especially in 1998 and 1999:

- The meteor storm was intense in 1966, indicating that within the past century Comet Tempel–Tuttle was still actively ejecting dust.
- The Earth passes the comet's node in 1998, 1999, and 2000 within three years after the comet has passed.
- The comet's path cuts through the Earth's orbital plane inside the path of the Earth, but not too far inside.

- The Leonid meteors have fallen at well above normal levels over the past 4 years.
- The phase of the Moon will not interfere with viewing in 1998 and 1999.

But where on Earth are the Leonid meteors most likely to be seen?

In 1799 the Leonid meteors provided a storm for the western Atlantic Ocean, the east coast of North America (mostly clouded out), and northern South America. In 1833, the show was in the eastern half of the United States and Canada. In 1866, western Europe had the best view. In 1966, the meteor deluge flooded the skies of the western United States. People in the rest of the world saw little or nothing, even if their skies were clear.

It's nice to know that Comet Tempel–Tuttle is likely to bring a storm of meteors or at least an enhanced shower. It's even nicer to know where on Earth to be to see the meteors at their most numerous.

That prediction is always uncertain because the Leonid meteor storms depend on the position, size, and density of a clump of meteoroids lost from the nucleus of Comet Tempel–Tuttle during one of its six most recent visits to the Sun. The meteoroid clump that actually causes a meteor storm generally stretches out, taffy-like, into a filament because each of the particles travels a slightly different orbit, a little faster or slower than the others. Over a period of about 200 years, the older particles separate widely from one another because of their different orbital speeds, because of solar radiation, and because of orbital changes imposed by the gravity of planets. Because these particles are now widely spread, this older ejecta no longer contributes to the part of the Leonid stream responsible for the Leonid storms.[17]

The storm-causing filament cannot (yet) be detected by telescopes or radar. So even with good celestial geometry and Moon phase boding for a litany of lively Leonids in 1998 and 1999, the Earth could miss the filament or filaments and disappoint a world of watchers. Or the Earth could wallow through a new or previously unknown filament in mid-November and surprise everyone.

Still, astronomers feel an obligation to provide the scientific community and the public with their best predictions on the subject so that experiments and observers can be positioned. Meteor astronomer Peter Jenniskens advises that "the Leonids at their maximum still offer the best, and perhaps only, opportunity for a dedicated study of a meteor storm."[18]

Don Yeomans noticed that Leonid meteor storms occur when the Earth is closest to the comet's node, especially when the Earth is farther from the Sun than the node. Yeomans takes this to mean that there is a principal filament of meteoroids that produces most or all of the Leonid storms and that most of that filament *follows* the comet, bracketing its orbit. So

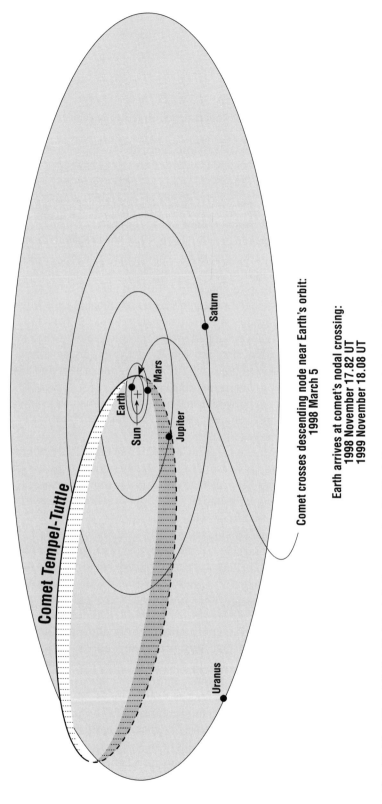

Comet crosses descending node near Earth's orbit:
1998 March 5

Earth arrives at comet's nodal crossing:
1998 November 17.82 UT
1999 November 18.08 UT

The orbit of Comet Tempel–Tuttle is tilted about 17° with respect to the orbit of Earth. The comet passes closest to the Earth's orbit when it descends through the Earth's orbital plane (ecliptic) just inside Earth's path. When the Earth each year reaches this position closest to the comet's node (ecliptic crossing), a display of Leonid meteors takes place. The time of day when the Earth reaches that position determines what continent on Earth (in the few hours before dawn) gets to see the Leonids at their best. Diagram by Will Fontanez and Tom Wallin, University of Tennessee Cartographic Services Laboratory, based on one by Don Yeomans, Kevin Yau, and Paul Weissman, 1996.

Yeomans predicts that in 1998 and 1999, the Earth is likely to have an encounter with that filament.

Yeomans predicts that a meteor storm, if it does occur in 1998 or 1999, will burst upon the skies when the Earth crosses the plane of the comet's orbit. The comet's tail extends out along and symmetrically somewhat above and below that plane, so the period just before, during, and just after the Earth crosses the plane of the comet's orbit holds the greatest likelihood of a meteor storm. Based on the orbit he has calculated for Comet Tempel–Tuttle, Yeomans believes that the hour on November 17, 1998 when the Earth will be closest to the comet's orbit and hence swimming in the meteor stream will be when Japan and Asia are positioned under the predawn skies, watching the constellation Leo rising in the east.

The Earth returns to a position closest to the comet's orbit on November 18, 1999, but this time Europe and North Africa are poised for best viewing. The geographically favored site is about 6 hours westward in consecutive years because the Earth takes about 365¼ days (365 days, 6 hours) to complete its orbit.

The Leonid peak in 2000 will return to November 17 because of leap year, when a day is added to the calendar approximately every 4 years to make up for the ¼ day ignored the past 3 years. The favored site for the Leonid meteors in 2000 should again be 6 hours west of Europe and North Africa – back in North America.

Other meteor astronomers have used Yeomans' orbit for the comet, made slightly different correlations with previous Leonid showers through the centuries, and projected the comet's orbit forward to 1998, 1999, and 2000 with slightly different results that affect where on Earth the Leonid meteors will be at their best. For comparison, those predictions are listed in the table on p. 292.

During the 33¼ years between the returns of Comet Tempel–Tuttle, Jupiter and the other planets the comet passes alter its orbit so that its period (presently 33.22 years) and path change slightly. Thus the best Leonid meteor viewing at the comet's next return is not automatically one-quarter of the Earth's circumference westward of its previous return. The inhabitants of the eastern Americas were treated to the spectacles of both 1799 and 1833.

Last call

So prospects seem quite favorable . . . but there can be no guarantees. Yeomans looked over his analysis of Comet Tempel–Tuttle and the Leonid meteors and said, "Although the conditions in 1998–1999 are optimum for a

Predictions for best observing sites for the Leonid meteors in 1998, 1999, and 2000

Caution: These experts all warn that their predictions are approximations based on Comet Tempel–Tuttle's orbit and its correlation with past Leonid activity.

Expert	November 17, 1998 Moon near new moon phase (age 28 days) (no Moon interference)	November 18, 1999 Moon past first quarter (age 9 days – sets soon after radiant rises) (little Moon interference)	November 17, 2000 Moon near last quarter (age 20 days – rises just ahead of Leo) (severe Moon interference)
Donald K. Yeomans[a]	Japan, Asia	Europe, North Africa	Eastern & central America
Peter Jenniskens[b]	China, Thailand, India (peak of 10,000 an hour)	Eastern United States (peak of 5,000 an hour)	Pacific Ocean (peak of 100 an hour)
John W. Mason[c]	Eastern & central Asia (peak of 5,000 an hour)	Eastern & central Europe (peak of 5,000 an hour)	Eastern & central America (peak of 1,000 an hour)
Ľubor Kresák[d]	Eastern North America (peak of 10,000 an hour)	Northern Pacific Ocean (peak of 100,000 an hour)	Central Asia (peak of 1,000 an hour)
Neil Bone[e]	Asia and eastern Europe	Western Europe & possibly eastern North America	
Joe Rao[f]	Asia	Europe, Africa	
Peter Brown and Jim Jones[g]	Western Pacific Ocean	Russia, China	Western Europe

Sources: [a]Donald K. Yeomans, Kevin K. Yau, and Paul Weissman: "The Impending Appearance of Comet Tempel–Tuttle and the Leonid Meteors," *Icarus*, volume 124, 1996, pages 407–413.
[b]Peter Jenniskens: "Meteor Stream Activity. III. Measurement of the First in a New Series of Leonid Outburst[s]," *Meteoritics & Planetary Science*, volume 31, 1996, pages 177–184; updated on November 18, 1997 at his web site: http://www–space.arc.nasa.gov/~leonid.
[c]John W. Mason: "The Leonid Meteors and Comet 55P/Tempel–Tuttle," *Journal of the British Astronomical Asssociation*, volume 105, number 5, 1995, pages 219–235.
[d]Ľubor Kresák: "Meteor Storms," pages 147–156 in J[an] Štohl and I[wan] P. Williams, editors: *Meteoroids and Their Parent Bodies* (Proceedings of the International Astronomical Symposium held at Smolenice, Slovakia, July 6–12, 1992) (Bratislava: Astronomical Institute of the Slovak Academy of Sciences, 1993.
[e]Neil Bone, personal communication, June 16, 1997; revised from *Meteors* (in *Sky & Telescope* Observer's Guide series) (Cambridge, Massachusetts: Sky Publishing, 1993), page 116.
[f]Joe Rao: "The Leonids: King of the Meteor Showers," *Sky & Telescope*, volume 90, number 5, November 1995, pages 24–31.
[g]P[eter] Brown and J[ames] Jones: "Evolution of the Leonid Meteor Stream," pages 57–60 in J[an] Štohl and I[wan] P. Williams, editors: *Meteoroids and Their Parent Bodies* (Proceedings of the International Astronomical Symposium held at Smolenice, Slovakia, July 6–12, 1992) (Bratislava: Astronomical Institute of the Slovak Academy of Sciences, 1993.

Weather prospects for the Leonids

JAY ANDERSON, Meteorologist, Environment Canada; President, Winnipeg Centre, Royal Astronomical Society of Canada

North America

The Leonid season brings us into the beginning of winter storm patterns, with the jet stream directing most disturbances across the middle of the continent. Cloudiness is building up from month to month as the hemisphere slides into winter. This pattern suggests the first rule for finding clear nighttime skies for a November meteor shower: *go south*.

The clearest skies for the shower in North America are found over the

Map of the United States and Canada showing the percent frequency of clear skies to scattered clouds (less than 30% sky coverage) for the hours shortly before sunrise in November. Jay Anderson plotted this map based on the average of 7 to 18 years of observations at over 700 points. Courtesy of Jay Anderson

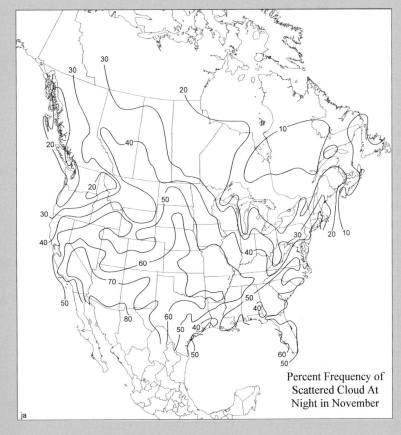

Percent Frequency of
Scattered Cloud At
Night in November

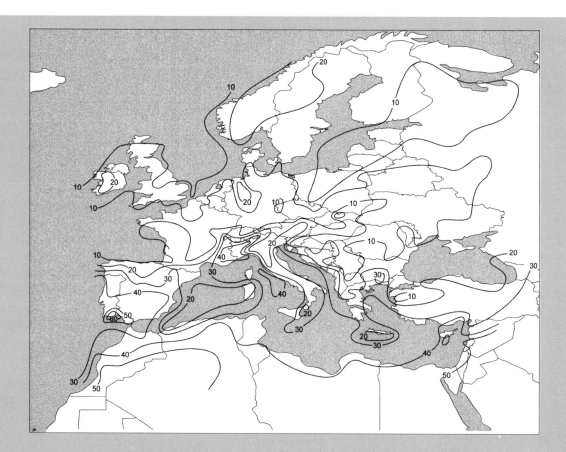

deserts of northern Mexico and the southwestern United States. There, east of California's Coastal Range, which serves as a barrier against Pacific clouds and rain, may be found the best Leonid skies for the country.

Europe

Western Europe is not protected against Atlantic storms by a range of mountains, so the north is exceptionally cloudy. Through Germany, England, Scotland, Poland, Scandinavia, France, and the Netherlands, the frequency of clear skies or scattered clouds ranges between 10% and 20%. The Baltic coasts and along the north side of the English Channel, the prospects are even more depressing, with less than 10% frequency of suitable skies.

To the south however, the Pyrenees and Alps block the Atlantic moisture. For the very best skies, Portugal, around Faro, and nearby Spain as

Map of Europe, North Africa, and the Middle East showing the percent frequency of clear skies to scattered clouds (less than 30% sky coverage) for the hours shortly before sunrise in November. Jay Anderson plotted this map based on the average of 7 to 18 years of observations at about 500 points. Courtesy of Jay Anderson

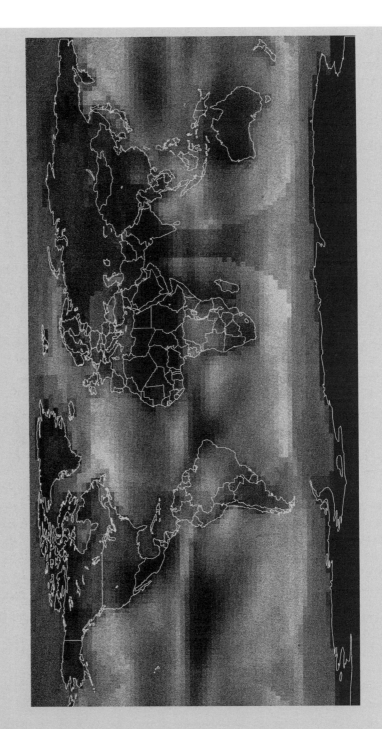

Map of the world showing cloudiness in November averaged throughout the day and night. Dark areas are clearer; white areas are cloudier. Jay Anderson constructed this map using data from the International Satellite Cloud Climatology Project, with measurements made over 8 years. Courtesy of Jay Anderson

far as Cadiz stand out clearly on the map with a 60% frequency of clear
skies or thinly scattered clouds.

Around the world

The continents tend to be less cloudy than the oceans, but not all continents are created equal. All the major cloud-suppressing mechanisms
can be seen in this map. India, at this time of year, is dry because of a high
pressure system in the Indian Ocean and the presence of the Himalayas,
which block the flow of moisture from that direction. Other locations are
blessed by their proximity to the world's persistent high pressure systems
– two in each hemisphere, straddling the equator: the Hawaiian
Anticyclone, the Peruvian Anticyclone, the Azores High, and the
Namibian Anticyclone. These highs create havens of clear nighttime
weather – and great observatories have flocked to these locations: Hawaii,
Chile, the southwestern United States, and the Canary Islands. At sea
level, low-level clouds persist in some of these regions, but a climb (by
road, of course) to higher elevations will offer a much clearer sky.

The polar regions on this map appear to have clear skies, but mostly
because synchronous weather satellites positioned over the equator have
a poor view of high and low latitudes. The "pillar" of clear skies pointing
northward through the middle of the Indian Ocean is also an artifact, due
to the lack of a geostationary satellite in this region. The poles and Indian
Ocean are actually cloudy at this time of year.

significant Leonid meteor shower, the event is not certain because the
dust particle distribution near the comet is far from uniform."[19]

The Earth might some mid-November morning run into a clump of
Leonid particles at some distance from the comet's orbit that will spangle
the night with shooting stars. It would have to be a previously unknown
filament of meteoroids some distance from the comet's orbit because the
correlations drawn by Don Yeomans, Kevin Yau, and Paul Weissman indicate that previous Leonid storms were all caused by debris left close to
Comet Tempel–Tuttle's orbit.[20]

Still, the only thing certain is that the Leonid meteors are full of surprises.

The orbit of the comet is gradually precessing, caused by the gravity of
the planets it passes. Over the next 500 years, as the comet's orbital ellipse
pivots, the comet will cross the Earth's orbital plane closer to the sun – and

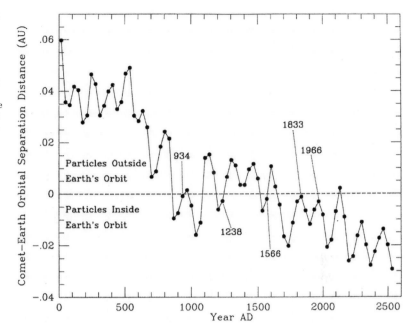

Minimum distance between Comet Tempel–Tuttle and the Earth's orbit when the comet passes through its descending node at each return from 0 to 2530 A.D. Donald K. Yeomans, Kevin K. Yau, and Paul R. Weissman: "The Impending Appearance of Comet Tempel–Tuttle and the Leonid Meteors," *Icarus*, volume 124, 1996, pages 407–413

farther from the Earth. This orbital precession of Comet Tempel–Tuttle will increase the distance between the paths of the comet and the Earth, and thus reduce the amount of material from Comet Tempel–Tuttle the Earth encounters. The spectacular bursts of meteors about three times a century will become a memory and then only a note in history.

All meteor storms, all meteor showers are evolving, never staying the same. All come to an end as the orbit of the stream of debris loses touch with the orbit of Earth.

"Because of planetary perturbations," says Yeomans, "it will be another century after the 1998–1999 events before significant Leonid meteor displays are once again likely."[21] The years 1998–1999 may be the last in a hundred years when the Leonid meteors may stage a storm. The next two returns of Comet Tempel–Tuttle will place it more than twice the distance from Earth as in 1998 and 1999 and no great Leonid display has ever occurred under those circumstances. The comet's position in 2098 is again favorable, slightly better than the end of the 20th century. In 2131 the comet passes just outside the orbit of Earth, so observers then will learn how much material accompanies Comet Tempel–Tuttle on its sunward side. The year 2164 finds the comet passing slightly farther inside the Earth's orbit than in 1998. Those are the Leonids' last gasp. Beyond 2164, there will be no more Leonid meteor storms. The precession of Comet Tempel–Tuttle's orbit will carry the meteor stream too far from Earth.[22]

Thus 1998 and 1999 are especially important. Barring a revolution in human longevity, even the youngest of today's observers will not live to see another Leonid storm. And it may be that 1998 or 1999 will produce the last Leonid meteor storms ever.

While it does not seem likely that the great storms of 1833 and 1966 will be repeated in either 1998 or 1999, the possibility cannot be ruled out, Yeomans advises. Significant displays should be looked for in both years.

Ultimately, how many Leonids are seen and where involves not only scientific data and careful interpretation, but also a sizable portion of luck. This incalculable element hinges on an unseen, undetected accident that occurred a third of a century or as much as a century or two ago when the comet trekked by. It was then (perhaps 1833 or 1866 or 1966) that the activity – or inactivity – of the Sun-approaching comet dislodged an extra – or below average – amount of particles that have remained densely clumped – or thinned out rapidly – so that the Earth could miss the pocket of greatest density and only a scattering of meteors would be seen, or the Earth could hit a filament of particles and provide a pleasant shower . . . or the heavens could blaze once again for minutes or hours as they have on unforgettable occasions in the past.

May the stars pour forth for you.

NOTES

1. Ľ[ubor] Kresák: "Meteor Storms," pages 147–156 in J[an] Štohl and I[wan] P. Williams, editors: *Meteoroids and Their Parent Bodies* (Proceedings of the International Astronomical Symposium held at Smolenice, Slovakia, July 6–12, 1992) (Bratislava: Astronomical Institute of the Slovak Academy of Sciences, 1993).

2. Donald K. Yeomans, personal communication, June 26, 1997.

3. Central Bureau for Astronomical Telegrams/International Astronomical Union Circular number 6579, March 10, 1997: "Comet 55P/1997 E1 (Tempel–Tuttle)." The University of Hawaii astronomers also had a second orbital calculation for Comet Tempel–Tuttle to crosscheck the predicted position of the comet – one by Syuichi Nakano and Ichiro Hasegawa (Minor Planet Circular 27288). Their pre-

dicted position for the comet was a little less accurate than the one by Donald K. Yeomans, Kevin K. Yau, and Paul R. Weissman: "The Impending Appearance of Comet Tempel–Tuttle and the Leonid Meteors," *Icarus*, volume 124, 1996, pages 407–413. Additional information provided by Don Yeomans, personal communication, May 1, 1997.

4. Donald K. Yeomans, Kevin K. Yau, and Paul R. Weissman: "The Impending Appearance of Comet Tempel–Tuttle and the Leonid Meteors," *Icarus*, volume 124, 1996, pages 407–413.

5. Donald K. Yeomans, Kevin K. Yau, and Paul R. Weissman: "The Impending Appearance of Comet Tempel–Tuttle and the Leonid Meteors," *Icarus*, volume 124, 1996, pages 407–413.

6. D[onald] K. Yeomans: "Comet Tempel–Tuttle and

the Leonid Meteors," *Icarus*, volume 47, 1981, pages 492–499. This article and the 1996 article by Yeomans, Yau, and Weissman are the most cited in modern Leonid meteor research.

7. Donald K. Yeomans lists 20 Leonid meteor storms in "Comet Tempel–Tuttle and the Leonid Meteors," *Icarus*, volume 47, 1981, pages 492–499. John W. Mason lists 23 Leonid storms in "The Leonid Meteors and Comet 55P/Tempel–Tuttle," *Journal of the British Astronomical Association*, volume 105, number 5, 1995, pages 219–235. Please see my table "Leonid Meteor Storms" in this chapter for a comparison.

   I will use 21 as the number of Leonid meteor storms, based on my reading of the sources. My list includes: 902, 934, 1002, 1202, 1238, 1366, 1532, 1533, 1566, 1601, 1602, 1666, 1698, 1799, 1832, 1833, 1866, 1867, 1868, 1965, 1966.

   Please remember that, because of uncertainties in the historical records, some of the events noted on any of these lists as meteor storms may technically have been showers instead or not Leonids at all.

8. The pressure of sunlight accelerates the lighter comet particles more quickly than the larger ones. So the inner portion of a comet's dust tail has a higher proportion of sand- and pea-size meteoroids than the tip of the tail. The outlying dust particles are mostly so small that they decelerate gently in the highest wisps of the Earth's atmosphere, do not burn up, and waft down to the surface as micrometeorites. It is the larger particles, situated initially closer to the comet's nucleus, that provide meteor showers and storms. Stephen J. Edberg, personal communication, June 6, 1997.

9. Donald K. Yeomans, Kevin K. Yau, and Paul R. Weissman: "The Impending Appearance of Comet Tempel–Tuttle and the Leonid Meteors," *Icarus*, volume 124, 1996, pages 407–413.

10. Preliminary International Meteor Organization data, furnished by an anonymous reviewer, December 6, 1996.

11. From internet and electronic mail reports by Robert Lunsford, Stephen Edberg, and other observers in the western United States.

12. Generally, bright moonlight seriously interferes with meteor displays. However, Steve Edberg points out that the 1946 Draconid (Giacobinid) storm was spectacular even with a full moon. Stephen J. Edberg, personal communication, June 6, 1997.

13. John W. Mason: "The Leonid Meteors and Comet 55P/Tempel–Tuttle," *Journal of the British Astronomical Association*, volume 105, number 5, 1995, pages 219–235.

14. Donald K. Yeomans, personal communication, May 1, 1997.

15. Updated values utilizing 1997 data from the recovery of Comet Tempel–Tuttle provided by Donald K. Yeomans, personal communication, June 26, 1997.

16. Yeomans lists 1698 as a shower rather than a storm, as Mason prefers. The minimum distance from the comet's orbit to Earth in 1698 was 0.0162 AU.

17. P[eter] Brown and J[im] Jones: "Evolution of the Leonid Meteor Stream," pages 57–60 in J[an] Štohl and I[wan] P. Williams, editors: *Meteoroids and Their Parent Bodies* (Proceedings of the International Astronomical Symposium held at Smolenice, Slovakia, July 6–12, 1992) (Bratislava: Astronomical Institute of the Slovak Academy of Sciences, 1993).

18. Peter Jenniskens: "Meteor Stream Activity. III. Measurement of the First in a New Series of Leonid Outburst[s]," *Meteoritics & Planetary Science*, volume 31, 1996, pages 177–184.

19. D[onald] K. Yeomans: "Comet Tempel–Tuttle and the Leonid Meteors," *Icarus*, volume 47, 1981, page 492 of pages 492–499.

20. Donald K. Yeomans, Kevin K. Yau, and Paul R. Weissman: "The Impending Appearance of Comet Tempel–Tuttle and the Leonid Meteors," *Icarus*, volume 124, 1996, pages 407–413.

21. Donald K. Yeomans, Kevin K. Yau, and Paul R. Weissman: "The Impending Appearance of Comet Tempel–Tuttle and the Leonid Meteors," *Icarus*, volume 124, 1996, pages 407–413.

22. Donald K. Yeomans, Kevin K. Yau, and Paul R. Weissman: "The Impending Appearance of Comet Tempel–Tuttle and the Leonid Meteors," *Icarus*, volume 124, 1996, pages 407–413.

CHAPTER 17

The journey of a meteoroid

[Meteors] may perhaps produce some important and necessary effects in the atmosphere surrounding this globe, for the welfare of man and its other innumerable tribes of inhabitants.   DAVID RITTENHOUSE (1786)[1]

ONE NIGHT on Earth – a mid-November night at the end of the 20th century – a sudden streak brightened the sky for an instant and vanished without a sound.

So disappeared a particle the size of a grain of sand that, with others of its kind, had survived least changed of all the material in the solar system since the Sun and planets formed 4.6 billion years ago.

If a collection device on a scientific satellite orbiting Earth had captured this cosmic particle before it burned its way to oblivion in the upper atmosphere, the particle would have told the story of its journey, something like this.

Five billion years ago, an interstellar cloud of gas and dust – a nebula – lay dark and cold in a large spiral galaxy that a family of inhabitants would one day call the Milky Way. The nebula was stirring as it absorbed the winds of matter from distant, long-ago exploded stars. The new material increased the nebula's density, in places enough that the cloud began to fragment as its densest portions fell inward on themselves by the gravity of all the atoms and molecules pulling on one another.

A supermassive star or two formed and then almost immediately – less than a million years – blew themselves apart and scattered their remnants through the nebula, adding density to other regions of the cloud and triggering a new surge in star formation. Here, a fragment of the nebula began its collapse. The passing debris had stirred it slightly and thus had given it just the barest spin. The more it collapsed, the faster it rotated, like a spinning skater pulling in her arms. Gravity pulled all the particles toward the center, but the rotation of the nebular fragment kept those particles in and near the plane of the rotation from falling inward quite so fast, so that the collapsing cloud flattened into a disk. At the center, where the density was highest, a star began to form. The star-to-be, our Sun, was only of moderate size and brightness, but it gathered in almost all the mass in the system.

Almost all, but not quite. In the disk, some matter was revolving fast enough to avoid the great infall that formed the Sun. Here and there those particles collided, clumped together, and gravitationally molded a set of planets.

Gravity pressed the particles of the forming Sun together so tightly that they generated heat and all of the solid particles turned to gas. Heat and light and accelerated particles radiated from the Sun and vaporized the ices and drove away the light gases from the inner planets. Mercury, Venus, Earth, and Mars were left as small rocky worlds, stripped of atmospheres. Only far from the scalding Sun were ices safe and gases abundant. Here too, particle by particle, bodies were forming in the plane of revolution far from the Sun.

In that place and time, far from the Sun, near the beginning of the solar system, there was a bump that no one heard because the density of the gas and dust was too low for sound to carry. Yet even if the gas were denser, the cosmic crash could not have been heard because it occurred between two tiny flecks and was so gentle that they stuck together. One was a grain composed of magnesium oxide, silicon dioxide, and iron oxide. The other was a mote made of carbon, hydrogen, oxygen, and nitrogen bound together. The two black specks clung together for the next 4½ billion years. As that grain of dust spun along its course around the newly shining Sun, it gathered other dust grains and even more ices – of water, carbon dioxide, ammonia, methane – growing. Growing, colliding, growing, fragmenting, partially vaporizing under a severe impact, refreezing, growing again.

And then in one collision, the ice-covered dust particle drifted into a far larger dusty, icy body, perhaps 3 miles (5 kilometers) in diameter, and became a coating on its surface. A later eon would call this larger body a planetesimal, a building block of planets. If, however, the planetesimal never built a planet, it might one day venture into the inner solar system and be called a comet. Other planetesimals that size and larger struck one another as they orbited. Some fragmented but others merged and grew, their own gravity holding them together and molding them into balls, the form assumed by bodies massive enough so that each particle attracts all the others sufficiently to pull them into a shape where each is as close to all the others as possible – a sphere.

Of the billions of trillions of icy planetesimals that formed in the twilight 20, 50, 100 times the Earth's distance from the Sun, many continued the process of colliding gently with one another and merging until two giant planets formed, Uranus and Neptune.

Uranus and Neptune are overgrown comets, their moons are probably captured comets, and their rings may be debris of comets that passed too close and were torn asunder by the planet's gravitational tides. These rings gradually dissipate, the particles falling into the planet's atmosphere like a steady meteoric rain, perhaps eventually to be replaced by a new ring from a newly shattered comet that ventures too deep into the gravitational rip-tides of the planet.

The planetesimals, some the size of small planets, did not know their work was done and kept right on colliding with the planets, some with such power that they tilted the axis of the planet's rotation. Uranus lies on its side as it revolves around the Sun.

Despite their wounds from the steady hail of rocky iceballs, Uranus and Neptune now ruled the outer solar system by their great gravity. And now they used that power to banish or execute the planetesimals that had not initially joined with them. Their gravitational fields deflected the remaining trillions of planetesimals. The two planets hurled dirty snowballs at one another and in all directions. Some hit Uranus and Neptune, to lose themselves in those larger worlds. Others found themselves in a cosmic pinball game. They were vectored toward the inner solar system where they would crash into the Sun or Jupiter or Earth – or be redirected into still smaller orbits by Jupiter and Saturn, themselves overgrown rock and gas adhesions, so that the infalling bodies might yet, some time later, strike the Sun or Earth or Jupiter or be hurled outward again so that they would spiral away from the Sun at speeds too great for the Sun's gravity to recall, to be lost to the solar system.

A different, more direct fate awaited the other millions of trillions of dusty iceballs that formed in the realm of Uranus and Neptune. They passed close to one of those planets and were hurled by gravity farther from the Sun. Most of them escaped from the solar system, in greater profusion than the droplets of water from a wet dog shaking itself dry. But some of the planetesimals were a few miles an hour shy of escape velocity, so they did not escape. They took up residence far beyond the outermost planet, out to a distance of 100,000 AU (1.6 light years), near the limit of the Sun's gravitational hold, where the Sun gave them no warmth and scarcely any light, for at that distance the Sun did not appear to be a globe but merely the brightest star – and not by much – in an always nighttime sky.

Departure

The planetesimal on which our particle rode was flung outward from the Sun and planets. Out it flew from the birthplace of comets, the Kuiper Belt, the region where Uranus and Neptune now lie and beyond them to where the gravity of those outer planetary giants can no longer seriously meddle with a comet's orbit.

But the speed of the particle's comet was insufficient to carry it beyond the solar system, beyond the gravitational control of the Sun. So it stayed

within the Oort Cloud, a vast pasture for exiled planetesimals beyond the Kuiper Belt that widens into a roughly spherical distribution of the dirty snowballs – comets – that surround the solar system and populate its outer fringes. Comets in the trillions.

By the time the comet had reached ten thousand times or so the Earth's distance from the Sun, its outward speed had dropped to zero and it began to fall back toward the planetary realm on a million-year orbit that would carry it inward to the spot where it had experienced its fateful planetary encounter.

But now other forces were acting on the comet to change its orbit, especially the gravitational tides created by the plane of the Milky Way. The Milky Way is a spiral-shaped galaxy of 400 billion stars. Our solar system dwells within it, 30,000 light years from the nucleus. There, three-fifths of the way out from the center of the Milky Way, our solar system wends its way at 150 miles per second (240 kilometers per second), completing its circuit every 250 million years. Our solar system lies at present just above the plane of the galaxy. The concentration of millions of stars in the plane of the Milky Way creates a gravitational tug that pulls the solar system

If we could travel beyond our Milky Way Galaxy and look back at it, our star family would look very much like this spiral galaxy in Ursa Major (M 81; NGC 3031). Our Sun and solar system would lie three-fifths of the way outward from the center, on the inner edge of one of the spiral arms, just above the plane of the galaxy. Our Sun would be much too dim to be seen in a picture like this. Kitt Peak National Observatory 4-meter telescope

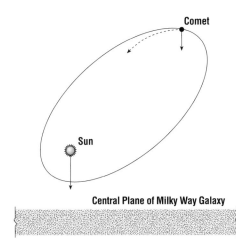

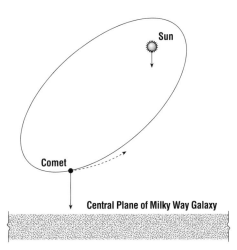

Central Plane of Milky Way Galaxy

Central Plane of Milky Way Galaxy

The tidal effect of the plane of the Milky Way Galaxy affects a comet in the Oort Cloud. Our solar system at present lies a little above the central plane of the Galaxy, so the stars concentrated in the galactic plane act together to tug the Sun and the comets in the Oort Cloud toward the plane of the Galaxy. A comet 1 light year from the Sun is tied so weakly to the Sun by gravity that the gravity of the galactic plane can modify its orbit. *Left:* The Sun and the galactic plane combine their gravitational pulls to bend a comet's orbit in closer to the Sun, decreasing the comet's perihelion distance. *Right:* The galactic plane and the Sun pull on a comet from nearly opposite directions, with the plane's tidal effect tending to bend the comet's orbit outward from the Sun, increasing the comet's perihelion distance. Diagram by Will Fontanez and Tom Wallin, University of Tennessee Cartographic Services Laboratory.

down toward it, accelerating the solar system so that it plunges through the galactic disk – a few thousand light years thick – and out the other side, where the combined gravity of the stars in the plane slow and stop the out-rushing solar system and cause it to plunge back toward the galactic plane, to overshoot once more – a cosmic, gravity-controlled version of a porpoise leaping from the surface of the water, diving back, swimming below the surface, and leaping from the water again in a continuous cycle.

Those millions of stars concentrated along the galactic plane exert by gravity a pull on the closest bodies in the Oort Cloud that is significantly different from their tug on the Sun. That tow is weak because of the distance, but the Sun's gravitational hold on the comets in the Oort Cloud is also tenuous. With the particle's comet so far from the Sun, the galactic plane either pulls harder on the comet than it does on the Sun and thereby increases the comet's distance from the Sun or the galactic plane pulls harder on the Sun than it does on the comet, in effect augmenting the Sun's pull on the comet, which causes it to fall in closer to the Sun.[2]

Over millions of years, the tidal forces of the galactic plane toyed with the orbit of the particle's comet, lifting it gradually but not steadily to a position farther out in the Oort Cloud.

In this way, through planetary encounters and then by tidal tows and tugs from the galactic plane, the particle's comet and trillions of its brethren over time fanned out to widely different distances and directions from the Sun. The comets spread into a spherical distribution, or actually an elongated sphere – shaped like a football or, more nearly, a rugby ball – with its two ends forming a line that points toward the center of the galaxy.[3] On the average, the comets were 1.4 billion miles from one another, farther apart than Saturn is from the Sun.[4] Perhaps a trillion icy planetesimals – comets – now lived in the *outer* portions of the Oort Cloud.

But by far the largest number of planetesimals left in the realm of Uranus and Neptune had not been yanked sunward or hurled starward or even heaved to the outer fringes of the Oort Cloud. Instead, they had been gently accelerated into orbits farther from the Sun by repeated alignments of their positions with the outer planets. These nudges cleared the realm of Uranus and Neptune of planetesimals, except for occasional trespassers. The small brothers of Uranus and Neptune were evicted and shoved farther from the Sun into the outer portions of the Kuiper Belt and the inner portions of the Oort Cloud.[5]

These refugees would serve as a reservoir to replenish the outer fringes of the Oort Cloud for comets lost. This reservoir in the outer Kuiper Belt and inner Oort Cloud still holds at least five trillion comets – uncountable from Earth, but estimated from the handful of comets that each year pass close to the Sun correlated with the number and size of the craters on the Moon and inner planets. In this reservoir so far from the Sun, the comets were now immune to the planetary nudges that helped to send them outward, but still some crept farther from the Sun, tugged and towed by the galactic tides, replacing the comets in the outer Oort Cloud that were lost to space or the inner solar system.

Life at the edge

Life in the Oort Cloud was slow. At that distance from the Sun, out to a light year and more, comets moved along their orbits at speeds that, at their fastest, were so meager that they scarcely exceeded those of jet airliners and at their slowest crawled along at the pace of race cars on long straight-aways.

Life was slow and dark and cold. The Sun was a bright pinpoint of light, barely outshining the brightest stars. It was always night. The surfaces of the comets were frozen near absolute zero, arresting almost all chemical activity. The comets had changed a little as they accreted near Uranus and Neptune, as they absorbed pulses of solar particles and radiation as the Sun began the nuclear fusion at its core that stabilized its size and brightness. Now, in the Oort Cloud, chemical change had almost ceased.

Exiled from the center of action and change, frozen in the near absolute zero temperatures of interstellar space, these sentinels at the outskirts of the solar system patrolled their million- or 10-million-year orbits around the distant Sun. Nothing, or almost nothing, happened for dozens or hundreds of orbits. An occasional faraway star brightened suddenly as it exploded and died. From time to time, a wisp of debris from an exploding star reached our particle's comet and embedded tiny bits of material in its surface.

Sometimes a star approached and passed the fringes of the solar system, but none barreled through the planetary realm. If it had, most of the planets would have been ripped away from the Sun. But the solar system was too small a target and the stars too widely separated for such an accident to be likely.

Meanwhile, on the third planet from the Sun life began and developed and occupied the sea and then the land. From time to time that life nearly vanished in the explosion, fire, smoke, darkness, and cold as errant comets and asteroids smashed the Earth.

While life on Earth battled to survive the blows that came from asteroids and comets, far away in the Oort Cloud the Sun fought the rest of the Milky Way Galaxy for the possession of these comets. It was a war with staggering casualties. The solar system lost at least half of the comets in the outer Oort Cloud, about a trillion over 4.6 billion years.

With the Sun's gravity holding them so weakly, these comets were easy prey for any gentle tug the solar system felt. That tug could come from a star drifting past every few million years perhaps, or it could come from the passage of an interstellar cloud of gas and dust like the Sun formed from. But most persistent was the tidal pull of the galactic plane, and therefore the galactic plane was the most effective of all in determining the future of a comet.

These gravitational weapons would either pry the comet from the solar system and set it adrift between the stars – or drop it, at first ever so slowly, but then with increasing speed, into the inner solar system once more.

Falling back into the inner solar system, back within the reach of the gravitational tentacles of the planets prowling along their orbits, the comets might escape deflection and return many times without a disturbing planetary encounter. But the planets have the gravity necessary to boost the comets' energy so that they would rush outward once again, mostly likely out of the solar system this time. Or the planets could reduce the comets' energy and thereby warp their paths so that they could not return as far as the comet cloud where they had dwelled so long.

The return begins

For one comet inbound for a few thousand years from the inner portions of the comet cloud, the gravity of Neptune perhaps shortened its orbit, lowered its closest approach to the Sun to within the orbit of Uranus, which over time reduced the orbit still more, and handed it off to Saturn, which dropped its orbit deeper into the reach of Jupiter, which could have whipped it out of the solar system but did not; instead shortened and lowered the comet's orbit still more so that it passed closer to the Sun than the Earth does.

Here, on its periodic visits to the inner solar system, closer to the Sun than it had ever been, ices boiled off the comet's surface, carrying with them flecks of embedded dust, and this comet perspiration formed a fuzzy coma surrounding the comet's solid nucleus. Particles and light streaming from the Sun pushed and carried these coma particles and gas away from the comet, outward from the Sun, to form the comet's long and graceful tail. This material could never be reclaimed by a small body with so little gravity as a comet, but each atom, molecule, particle had its own motion that carried it in its own orbit close to the orbit of the comet of which it had been a part for 4.6 billion years.

With each passage by the Sun, the comet lost more material, tens of tons a second when it was closest to the Sun.[6] It grew dimmer over the centuries, not so much because it was reduced in size but because as the volatile ices boiled away, the dirt left behind formed an insulating crust. Yet, from time to time as the comet passed the Sun, portions of that crust broke off too and fragmented, so that the comet, returning three times a century, shed new particles – ahead, behind, and to its sides.

The comet was strewing debris all around it, but for at least a thousand years, none of the material struck the Earth. When the comet passed through the plane of the Earth's orbit, it was on average several million miles outside the path of the Earth. Gradually that was changing. Over a thousand-year period, five gravitational encounters with Jupiter and one with Saturn at distances between 48 and 82 million miles (77 and 132 million kilometers) tended to bring the comet closer to the Sun until in 868 A.D. the comet for the first time passed through the Earth's orbital plane on its sunward side, allowing new and previously shed material from its coma and tail drifting outward under the pressure of sunlight to cross the orbit of Earth.[7] In the year 902, about 1,100 years ago, the Earth for the first time wallowed through that stream of debris, and the grains of grit glowed briefly as they turned to gas at a temperature they had never experienced as they plunged to extinction in the Earth's atmosphere.

Yet, then as now, not all the particles burned up as they hit the Earth's atmosphere at velocities so high that the air might as well have been a concrete wall. If the meteor particle was smaller than a shaved whisker (1/250 inch, 0.1 millimeter, 100 microns in diameter), its encounter with the faintest wisp of the Earth's atmosphere, at an altitude of about 90 miles (150 kilometers), was enough to slow it – slow it gradually so that there was too little friction to vaporize it, so these flecks and poofs and motes lost their forward momentum and floated in the upper reaches of the air, the smallest of them for months or years. Sprinkled to form a veil 50 miles (80 kilometers) above the Earth, these particles reflected sunlight, cooling the planet.

But gradually the particles wafted downward, the larger ones first, each acting as a nucleus around which moisture from the lower atmosphere could condense to form a raindrop, together a cloud, so that the cooling effect of the dust veil was balanced by the insulation and the global warming the clouds provided. By helping to form droplets, the micrometeoroids nurtured the rain and snow that returned precious water to the ground. These tiny flecks peppered the untouched snow on mountaintops and mingled with the sand on the seabottoms.

The comet gets a name

The comet bearing our particle scattered dust, dirt, and grit. This debris lit up the heavens in the most spectacular of astronomical displays to grace the skies of Earth. For centuries people were awed by what they saw, but did not know a comet had created it. That comet would be named after its 19th century telescopic discoverers, Willem Tempel and Horace Tuttle, whose observations allowed its orbit to be calculated.

Comet Tempel–Tuttle comes our way every 33 years. Thirty-three years, give or take a little, because the orbit and period of revolution of the comet are always in flux. The orbit depends on how close the comet passes to Earth, Mars, and especially Jupiter, Saturn, and Uranus, the massive orbit warpers. With a 33-year period, Comet Tempel–Tuttle never quite returns to its original homeland in the realm of Neptune and Pluto.

The comet's orbit and hence its period also change because the comet burps and sputters when it is close to the Sun. Those burps and sputters act as small rockets to nudge the comet onto a slightly different course. The solar heat vaporizes the ice on the comet's Sun-facing surface and the gases stream away, mostly sunward, from the comet's frozen nucleus. As they do, the dust and rocks trapped in the icy matrix are set free and they too drift away from the comet. The comet's gravity is too feeble to draw them back. An outward velocity of more than 2 miles per hour (1 meter per second) and they are gone.

And so, on one of its journeys close to the Sun, a tiny grain, the size of a grain of sand, lifted off the comet on which it had resided for most of 4½ billion years, sent aloft by gases boiling away, set loose to roam the solar system alone, in a stream of billions of comet refugees.[8]

Our meteoroid, loosed from the comet, survived several circuits in its own orbit, accompanied by billions of its brethren in similar but not identical orbits. Because the orbits were not quite identical to one another or to that of Comet Tempel–Tuttle, their orbital speeds were slightly different and thus they were separating one from another, thinning out, distributing themselves along the comet's orbit, all the way around, to make what someday could be a continuous and fairly even stream that the Earth could cross once each year to create an annual meteor shower of the magnitude and beauty of the Perseids in August or the Geminids in December.

But not just yet for the Leonids. A thousand years or so of Comet Tempel–Tuttle passing close to the Sun and Earth and actively shedding debris was time enough only for the castoffs to create a hardly noticeable *annual* shower. The particles were still in elongated clumps, most but not all behind the comet and slightly outside its orbit, still within two or three years' journey of the comet.

All glory is fleeting

A meteor shower is just as transient as its parent comet. The comet's fate lies in the positions of the planets whose gravity continually reshapes its orbit. One step too close to a planet and the comet is booted from the solar system or detours to become a splash and brief great dark spot in the atmosphere of Jupiter or a crater on the Moon or a momentary sizzle and puff of steam in the atmosphere of the Sun. Or the comet shatters, each fragment to live out its demise on a smaller scale. Or the comet simply wears out: the Sun boils off most of the frozen gases close to its surface, leaving the rest cloaked with a brittle black rocklike crust of dust particles welded to one another by pressure or ice. The comet stops outgassing, stops forming a puffy coma and filamentary tail, stops shining by gases spread across millions of miles fluorescing in the sunlight and by dust spread across millions of miles reflecting sunlight. The comet becomes a dark, inactive asteroid like body that no longer contributes particles for a meteor shower, though the dead comet may continue on its Earth-crossing orbit and, camouflaged in black against the interplanetary sky, stalk the Earth at murderous speed.

When a comet ceases to supply meteoroids, a meteor shower is doomed. But even if the comet lived, shedding particles for ages to come, replenishing the meteor stream, the meteor shower would still vanish from earthly skies.

Just as the comet bobs and weaves through the gauntlet of planets, its course never the same for long, so the meteoroids find their orbits bent by planetary perturbations, which can divert the stream so that it no longer crosses the path of Earth. And even if the planets leave the meteoroid stream unwarped for a while, the Sun does not. Just as the pressure of light from the Sun shoves the dust particles in a comet's tail outward, so the individual particles in the meteor stream feel and respond to the sunlight, shoved away by its tiny but incessant pressure, so that the stream itself migrates outward, away from the orbit of the Earth.

Even the Earth conspires to bring a meteor shower to an end. Year after year, the Earth hits the meteor stream as it barrels along its celestial highway, hits and runs, accumulating the particles like bugs on a car windshield. Bugs reproduce; these particles do not. A meteoroid performs once only. Year after year, there are fewer particles.

With the forces of the planets, Sun, and Earth all arrayed against it, a meteor shower cannot last for long. In 10,000 years or less, the particles that were once a comet and then a meteor stream notable for its display in the skies of Earth will be depleted to a scattering of particles scarcely different in number from those on either side of its orbit. Its remnants have become a weak, sporadic background of meteors. The meteor shower will have died.

A memorial – the memorial of all faded and never-were meteor showers – is provided by the dust still orbiting the Sun, mostly in the plane of the solar system. Sunlight reflects off these scattered particles to produce what can be seen on Earth as a faint pyramid-shaped glow resting on the horizon, best visible in the east on fall mornings just before twilight and in the west on spring evenings just after sunset fades. Astronomers call this memorial to exhausted comets and meteor showers the zodiacal light.[9]

The encounter

Now the year, the month, the day was at hand when Comet Tempel–Tuttle and former pieces of the comet, on their way outbound, would cross the orbital plane of the Earth close to the orbit of the Earth itself. For the particles, there would be the chance of – the danger of – a collision with the Earth.

Comet Tempel–Tuttle swept safely past perihelion in February 1998 and then through the Earth's orbital plane, sloping southward. Following well behind, the particles experienced the usual increase in speed and heat that accompanied approach to perihelion and the tapering downward of speed and heat as the particles sailed outward again, toward the oncoming Earth.

The Earth was whirling along at 18½ miles (30 kilometers) per second around the Sun. The particles, traveling 26 miles (42 kilometers) per second, found themselves on this voyage on a collision course with planet Earth, at a relative closing speed of 44 miles (71 kilometers) per second.

Four and a half billion years of existence was about to culminate. One of the meteoroids began to collide with molecules in the Earth's atmosphere at a speed of 150,000 miles per hour (240,000 kilometers per hour) and the friction of the collisions caused the particle to heat up, its surface vaporizing, the heat causing the molecules in the air to glow vividly, the tiny meteoroid melting, vaporizing as the temperature soared over 10,000 degrees Fahrenheit (5,500 degrees Celsius) and the particle, solid for more than 4.6 billion years, since before the Earth and Sun began to form, vanished into gas in a fraction of a second and left an evanescent blaze of light across the heavens.[10]

In that instant the meteoroid ceased to exist – except in the mind of a child who saw it and cried out "Look!" in a moment never to be forgotten.

NOTES

1. David Rittenhouse: "Account of a Meteor," *Transactions of the American Philosophical Society*, volume 2, 1786, pages 175–176. Rittenhouse is replying to an inquiry (pages 173–174) from John Page.
2. More precisely, the tidal effect of the galactic plane increases or decreases the comet's perihelion distance.
3. The dimensions of this oblate spheroid are approximately 200,000 AU along its longest axis and 160,000 AU along its shortest.
4. Donald K. Yeomans, personal communication, May 2, 1997, based on an Oort Cloud with 5 trillion comets and a radius of 100,000 AU.
5. These planetary resonances are the same mechanism that Jupiter used to clear gaps in the asteroid belt, as Daniel Kirkwood discovered.
6. Donald K. Yeomans, personal communication, May 2, 1997. Yeomans advises that at perihelion in 1997, Comet Hale–Bopp was losing about 20,000 gallons (60–90 tons) of water a second.
7. Donald K. Yeomans, Kevin K. Yau, and Paul R. Weissman: "The Impending Appearance of Comet Tempel–Tuttle and the Leonid Meteors," *Icarus*, volume 24, 1996, pages 407–413.
8. A particle this size would likely leave Comet Tempel–Tuttle at about 60 miles per hour (25 meters a second), well above escape velocity. Donald K.

Yeomans, personal communication, May 2, 1997.
9. Actually, zodiacal light is a band that stretches along the ecliptic across the entire sky, but the sky is seldom dark enough for the entire band of zodiacal light to be seen.

The position of the ecliptic at the horizon determines the best season to see the zodiacal light in the morning or evening sky. Because of its faintness, zodiacal light is best seen when it is most nearly perpendicular to the horizon so that it extends farthest above the horizon, rather than tipped so that it tends to lie along the horizon. Because the zodiacal light comes from dust distributed along the ecliptic, best viewing will occur when and where the ecliptic stands most nearly perpendicular to the horizon. For the northern hemisphere, this occurs in the eastern skies of morning in the fall, when the summer ecliptic lies to the west of the soon-to-rise Sun and it occurs in the western skies of evening in the spring, when the summer ecliptic lies to the east of the recently set Sun. For viewing the zodiacal light in the southern hemisphere, the situation is reversed.
10. This chapter is based primarily on the mainstream view of comet formation and evolution presented in Donald K. Yeomans: *Comets: A Chronological History of Observation, Science, Myth, and Folklore* (New York: John Wiley & Sons, 1991), pages 301–357.

GLOSSARY

ablation

The removal of material in successive layers, as by the vaporization of surface after surface of a meteoroid by friction as it passes through the atmosphere.

aerolite

An old name for a **stony meteorite**.

altitude

In astronomy, the elevation of a object in the sky measured by its angle upward from the horizon. The zenith (the point overhead) has an altitude of 90° from a level horizon. The exact position of an object in the sky can be specified using altitude and azimuth.

Amor asteroid

An asteroid that has a perihelion distance between 1.0167 and 1.3 AU, so that it comes close to but never crosses the Earth's orbit because it is always farther from the Sun.

aphelion

The point farthest from the Sun in the orbit of a body that is traveling around the Sun.

Apollo asteroid

An asteroid that has a perihelion distance less than 1.0167 AU but a mean distance from the Sun greater than 1.0 AU, so that its orbit lies mostly outside the orbit of the Earth but crosses the Earth's orbit, thus posing a risk of collision.

asteroid

A planetlike body in the solar system too small to be classified as a planet. Most asteroids orbit the Sun between Mars and Jupiter. As of mid-November 1997, 8,058 asteroids have been cataloged. Millions are assumed to exist. Asteroids range in diameter from rubble to (in the case of Ceres) about 617 miles (993 kilometers). Asteroids are also called **minor planets** and occasionally planetoids.

asteroid belt

The region between the orbits of Mars and Jupiter where the great majority of the asteroids orbit.

astronomical unit

The mean distance between the Sun and Earth – 93 million miles; 150 million kilometers. Abbreviated **AU**.

Aten asteroid

An asteroid that has a perihelion distance greater than 0.9833 AU and a mean distance from the Sun less than 1.0 AU, so that its orbit lies mostly inside the orbit of Earth, but crosses the Earth's orbit, thus posing a risk of collision.

AU

Abbreviation of *astronomical unit*. The mean distance between the Sun and Earth – 93 million miles; 150 million kilometers.

azimuth

The direction of an object in the sky measured by its angle along the horizon, usually starting from north and moving clockwise. East is azimuth 90°. The exact position of an object in the sky can be specified using altitude and azimuth.

bolide

A very bright meteor that explodes. Sometimes defined as a meteor accompanied by a (delayed) sonic boom.

burst

A sudden brief enhancement of light during a meteor's passage through the atmosphere (same as **flare**).

CCD

Abbreviation of *charge-coupled device*. A silicon chip used for low-light-level imaging.

charge-coupled device

A silicon chip used for low-light-level imaging. Abbreviated *CCD*.

clino- (meteor shower name)

Meteoroids ejected from a comet sufficiently long ago that they have spread out along the comet's orbit and can cause only light to moderate meteor showers if the Earth passes through the meteoroid stream.

coma (of a comet)

A diffuse gaseous region around a comet's nucleus formed by the comet's outgassing. Comas often exceed Jupiter in diameter. A comet's nucleus and coma together form the comet's head.

comet

A body (nucleus) made of ice and minerals that typically orbits the Sun in a highly elliptical orbit and which typically develops a coma and tail when close enough to the Sun to cause its surface ice to vaporize. Comet nuclei range from about 100 yards (100 meters) to 200 miles (300 kilometers) in diameter. The glow of a comet is created by sunlight which causes fluorescence of the gases and reflection off solid particles in the coma and tail. A comet's small nucleus can produce a coma 100,000 miles (160,000 kilometers) or more in diameter (larger than Jupiter) and a tail more than 100 million miles (160 million kilometers) long (longer than the distance of the Earth from the Sun).

cosmic rays

Charged subatomic particles (mostly protons) moving through space at very high speeds, even approaching the speed of light. Cosmic rays are particles, not light waves.

declination

The position of a celestial object north or south of the celestial equator, measured in degrees. Declination in astronomy is the equivalent of latitude in geography. The exact position of an object in the star field can be specified by its right ascension and declination.

differentiation

In an astronomical body, the separation of different chemicals from their original mixed state into different layers or regions. The high temperature of the newly formed Earth allowed the heavier materials (such as iron and nickel) to sink to the core and the lighter materials (such as silicates) to rise toward the surface. Thus the Earth differentiated into concentric layers: core, mantle, and crust.

direct revolution

The revolution of an object around the Sun in the same direction as the planets. Also called prograde or posigrade revolution.

Earth-crossing asteroid or **comet**

An asteroid or comet whose orbit crosses the orbit of the Earth or whose orbit brings it close enough to Earth to be captured by repeated encounters.

eccentricity

The shape of an elliptical orbit as it departs from circular. Eccentricity is expressed as a ratio between the difference and sum of aphelion and perihelion distances. An eccentricity of 0 is a circle; 1 is a parabola – an open orbit.

ecliptic

The plane of the Earth's orbit around the Sun. The ecliptic is thus also the apparent path of the Sun through the star field as seen from the Earth as it moves around the Sun over the course of a year.

Edgeworth–Kuiper Belt of comets

See **Kuiper Belt of comets.**

electrophonic sound

A sound (usually buzzing or hissing) caused by a fireball (bright meteor) that is heard on the ground at the same time that the fireball is visible. Electrophonic

sounds are thought to be produced when a vaporizing fireball generates very-low- and extra-low-frequency radio signals that travel at the speed of light and are converted to sound waves by natural or man-made objects near the observer.

elements of an orbit

Six quantities that fully describe an orbit (of a small body revolving around the Sun): semimajor axis, eccentricity, inclination, longitude of the ascending node, angular distance from the ascending node to perihelion (argument of perihelion), and time of perihelion passage.

ellipse

The gravitationally controlled shape of any closed orbit of one body around another. The planets, comets, asteroids, and meteoroid streams orbit the Sun in ellipses (except when disturbed by other bodies). An ellipse is a closed symmetrical plane curve generated by a point moving in such a way that the sums of its distances from two fixed points (foci) remain constant. An ellipse is the intersection of a plane passing through a circular cone at any angle except through the base or peak of the cone.

ephemeris

A listing that gives a celestial object's positions for various times.

escape velocity

The minimum velocity required to escape entirely from the gravitational field of an object.

fall

In meteoritics, a meteorite that was seen to fall.

falling star

Colloquial name for a **meteor.**

find

In meteoritics, a meteorite that was not seen to fall, but was found and recognized later.

fireball

A very bright meteor that rivals or exceeds Venus (the brightest planet) in brightness. The magnitude is variously defined as −5 or −4.

flare

See **burst.**

fluorescence

The process by which a substance emits light (or any electromagnetic radiation) after absorbing it.

hyperbola

An open symmetrical plane curve that is generated by the intersection of a plane passing through a circular cone where the plane is closer to perpendicular than the slope of the cone. Because no comet inbound to the Sun has exhibited a definite hyperbolic orbit, astronomers believe that all comets are part of our solar system and not strays from elsewhere in the universe.

hyperbolic velocity

A velocity so great that an object's orbit is a hyperbola and thus that object will never return to the gravitationally influential body it is near (like the Sun). If a body in the solar system has a hyperbolic velocity, it will leave the solar system.

inclination

The tilt of a celestial object's orbit to that of the Earth (ecliptic), expressed in degrees. An inclination of 0° indicates that the object is moving in the same orbital plane as Earth; 90° indicates that the object crosses the Earth's orbit at a right angle; above 90° indicates the tilt but also that the object has a retrograde orbit – is moving around the Sun in the opposite direction of Earth and the other planets and asteroids.

ion

An atom or molecule that has become electrically charged because of the loss or gain of one or more electrons.

ionization

The loss or gain of electrons by an atom or molecule so that it is electrically charged, rather than neutral. Meteoroids ionize themselves and the air around them when they streak through the atmosphere and vaporize. Radio signals (radar) can be bounced off these ionization trails (columns), allowing the speed, height, energy, path, and other characteristics of meteors to be determined accurately.

ionosphere

The portion of the atmosphere that is substantially ionized: the atoms and molecules in the air are momentarily and repeatedly stripped of one or more electrons so that they are electrically charged. The ionization occurs primarily because of the ultraviolet and x-ray energy received from the Sun. The Earth's ionosphere extends from about 50 miles (80 kilometers) to 350 miles (600 kilometers) above its surface. The ionosphere consists of several layers that vary in height and ionization according to the time of day, season, and solar cycle. The ionosphere reflects radio signals, allowing radio transmission to reach points on the Earth far beyond line of sight. The lowest part of the ionosphere corresponds to the **meteor layer** where most meteoroids vaporize.

iron meteorite

A meteorite composed primarily of metallic iron and nickel and thought to represent material from the core of a parent body that differentiated – went through a molten phase when heavier materials (metals) sank to the center and lighter materials (silicates) floated to the surface.

isotope

Any of two or more forms of the same chemical element in which the atoms have the same number of protons but different numbers of neutrons.

Kennelly–Heaviside Layer

An early name for the **ionosphere**, particularly the lower ionosphere that reflects radio waves, allowing radio transmission to extend far beyond the line of sight of the transmission tower without relay stations. Named after the two physicists who discovered it, Arthur Kennelly and Oliver Heaviside. The Heaviside Layer is mentioned in the musical *Cats*.

Kirkwood Gaps

Regions in the asteroid belt that have been cleared of asteroids by the perturbations of Jupiter. The Kirkwood Gaps are named for Daniel Kirkwood, the astronomer who first noticed and explained them in 1866.

Kuiper Belt of comets

A ring of icy planetesimals left over from the formation of the solar system that lies beyond Neptune and Pluto and extends outward to the Oort Cloud of comets. The Kuiper Belt is a reservoir of billions of comets from which short-period comets come, perturbed into the planetary portion of the solar system by gravitational resonances with the largest planets. The primordial Kuiper Belt included the realm of Uranus and Neptune, and those planets may be agglomerated icy planetesimals. The Kuiper Belt is also called the Edgeworth-Kuiper Belt, named for Gerard P. Kuiper who proposed it in 1951 and Kenneth E. Edgeworth who suggested it in 1943.

longitude of the ascending or **descending node**

The angular distance along the ecliptic (orbital plane of the Earth), measuring eastward from the vernal equinox in degrees, where a celestial object crosses the ecliptic (its node) going north (ascending) or south (descending).

long-period comet

A comet with an orbital period of more than 200 years.

magnitude

A scale for measuring the brightness of celestial objects. The lower the magnitude number, the brighter the object. A star with apparent magnitude +1.0 is approximately 2½ times brighter than a star with apparent magnitude +2.0. Very bright objects may have negative magnitudes, like Venus, about −4.

meteor

A streak or dot of light in the sky caused by a particle from space entering the Earth's atmosphere at high speed and vaporizing because of atmospheric friction. Sometimes also taken to mean the particle that creates the streak in the sky. Colloquially, meteors are often called shooting stars or falling stars.

meteor layer

The level in the atmosphere at which meteoroids vaporize. On Earth, the meteor layer extends from about 75 miles down to 50 miles (120 to 80 kilometers), although some fast meteoroids burn up higher and some meteoroids (particularly fireballs) penetrate to lower levels.

meteoroid

A natural solid particle in space that is smaller than a comet or asteroid but much larger than an atom or molecule. A meteoroid is or may one day be on a collision course with Earth or some body in the solar system. The typical meteoroid that produces a meteor is the sky is about the size of the head of a pin or a grain of sand and weighs 0.01 gram. A meteoroid that produces a **fireball** is typically the size of a grape and weighs 1 gram.

meteorite

A natural object from space that has survived its passage through the atmosphere and landed on the Earth or other body in the solar system.

meteor path

The line in the sky apparently traveled by a meteor.

micrometeorite

A very small particle from space that has survived collision with the Earth. It is generally less than 0.1 millimeter (100 microns; 0.004 inch) in size.

minor planet

See **asteroid**.

near-Earth object or **asteroid** or **comet**

An asteroid or comet that passes close to Earth.

node

One of the two points where an object in orbit crosses the plane of another object in orbit. That nodal crossing need not coincide with the actual orbital path of the other body. There are two nodes, **ascending node** and **descending node**, corresponding to the point in space where a body crosses the orbital plane of another going northward or southward with respect to the Earth's directions. For meteor, comet, and asteroid studies, the nodes are the two points where an orbiting object crosses the orbit of Earth (the ecliptic).

nongravitational effects

The motion of a comet that cannot be explained by the gravitational effects of other bodies. The nongravitational effects are caused by the rocket effect of the comet's outgassing.

nonshower meteor

A meteor which is not a member of a recognized shower.

nucleus (of a comet)

The solid main body of a comet. The surface ices of the nucleus vaporize when close to the Sun, releasing gases and dust which expand into a large cloud around the nucleus – a coma. The solar wind and light pressure from the Sun carry away gases and particles from the coma to form the comet's tail.

Oort Cloud of comets

A cloud of more than a trillion comets extending outward beyond the Kuiper Belt from about 2,000 to 100,000 AU. These comets, the farthermost objects in the solar system, move with random orientation and direction around the Sun. The cloud was inferred by Jan H. Oort in 1950, based on his study of the orbits of long-period comets. The tidal effect of the galactic plane and the gravity of passing stars and nebulae can perturb the orbits of these comets so that they enter the planetary portion of the solar system or are ejected from the Oort Cloud.

Öpik–Oort Cloud of comets

The **Oort Cloud**, with an effort to honor Ernst J. Öpik who proposed a similar idea but a different process in 1932.

orbital elements

Six quantities that fully describe an orbit (of a body revolving around the Sun): semimajor axis, eccentricity, inclination, longitude of the ascending node, angular distance from the ascending node to perihelion (argument of perihelion), and time of perihelion passage.

ortho- (meteor shower name)

Meteoroids ejected from a comet so recently that they have not had time to spread out much along the comet's orbit and are sufficiently concentrated (near the comet) that they can cause meteor storms or very heavy meteor showers if the Earth passes through the meteoroid swarm.

osculating orbit

An approximation of the orbit of a body found by considering only the most powerful gravitational effect on it. The gravitational effects of other bodies continually modify osculating orbits.

parabola

An open symmetrical plane curve that is generated by the intersection of a plane passing through a circular cone where the plane is parallel to the slope of the cone. The orbits of long-period comets approximate parabolas, but are actually very eccentric ellipses.

parabolic velocity
A velocity just high enough that an object's orbit is a parabola rather than an ellipse. If a comet were on a parabolic orbit, it would never return to the Sun.

parallax
An apparent change in the position of an object due to the motion of the observer or different locations of observers. Parallax can be used to calculate the distance of an object (such as a meteor) if the angle of displacement and the distance between sighting locations is recorded.

perihelion
The point nearest to the Sun in the orbit of a body that is traveling around the Sun.

periodic comet
Another (less desirable) name for a **short-period comet**, one that revolves around the Sun in less than 200 years.

perturbation
A gravitational effect on a body created by an object other than the one the body is revolving around. A meteoroid stream in orbit around the Sun may be perturbed onto a new orbit by passing close to a planet.

planetesimal
A small rocky or icy body formed in the primordial solar nebula, ranging in diameter from a yard (a meter) to more than 6 miles (10 kilometers), from which all the planets, satellites, asteroids, and comets are presumed to have formed.

plasma
A completely ionized gas, created by temperatures so high that neutral atoms cannot exist. The matter in a plasma consists of free electrons and free atomic nuclei.

Poynting–Robertson Effect
The loss of orbital energy by small particles (especially about 1 millimeter) in orbit around the Sun so that they tend to spiral in toward the Sun. The Poynting–Robertson Effect occurs because orbiting particles run into sunlight (photons) like cars run into falling rain, so the radiation is striking them to some extent head on, thus slowing them and causing them to spiral inward. This effect works especially well on meteoroid-size particles and is the method by which the dust that provides zodiacal light is steadily removed from interplanetary space by falling on the Sun. The Poynting-Robertson Effect is opposed by the outward pressure of sunlight that tends to push parti-

cles, especially the smallest ones (micrometeoroid size), away from the Sun, out of the solar system.

radar
Short-wave (microwave) radio signals that are bounced off a target and back to a receiver to analyze the distance, direction, speed, and other characteristics of the target. The name *radar* began as an acronym: *ra*dio *d*etecting *a*nd *r*anging.

radiant
The spot in the sky from which the meteors appear to come. The radiant is a small region, not a point.

retrograde revolution
The revolution of an object around the Sun in the opposite direction of the planets.

right ascension
The position of a celestial object measuring eastward along the celestial equator from the vernal equinox. Right ascension is usually measured in hours, minutes, and seconds, but can be measured in degrees. Right ascension in astronomy is the equivalent of longitude in geography. The exact position of an object in the star field can be specified by its right ascension and declination.

semimajor axis
The mean distance from the Sun of a solar system body.

shooting star
Colloquial name for a **meteor**.

short-period comet
A comet with an orbital period less than 200 years.

shower, meteor
A display of meteors that radiate from a common spot in the sky, created by the collision of the Earth with meteoroids traveling along together (a meteoroid stream).

siderite
An old name for an **iron meteorite**.

siderolite
An old name for a **stony–iron meteorite**.

solar longitude
The position of the Sun measured eastward in degrees along the ecliptic from the vernal equinox. Specifying the Sun's apparent location among the stars establishes the Earth's position in its orbit around the Sun and thus is very useful for recording and interpreting meteor data: meteoroids encountered when the Earth reached a specific position in its orbit.

solar sail
A large, light-absorbing panel that converts the pressure of sunlight into useful momentum as a means of space propulsion.

solar wind

A stream of charged particles (mostly protons, electrons, and helium nuclei) ejected from the Sun that flows by the Earth at 720,000 to 1.8 million miles per hour (320 to 800 kilometers per second). The solar wind creates the gas (or ion or plasma) tails of comets. Solar wind particles, especially when enhanced by solar flares, collide with molecules in the Earth's upper atmosphere so intensely that they cause the upper atmosphere to glow by fluorescence in displays of the aurora (the northern and southern lights).

spectroscope

An instrument for viewing a spectrum.

spectroscopy

The study of spectra.

spectrum (plural, **spectra**)

The pattern of colors or wavelengths obtained when light from an object is dispersed as it passes through a prism or diffraction grating. A rainbow is the spectrum of the Sun as its light passes through raindrop prisms. By examining a spectrum, scientists can tell the chemical composition, temperature, radial velocity, spin, and many other characteristics of a light source and the material that lies between the source and the observer.

sporadic meteor

A meteor that is not a member of a recognized shower. Over a period of a year, sporadic meteors greatly outnumber meteors that belong to well-known showers.

stony-iron meteorite

A rare type of meteorite composed of a mixture of silicates and metallic iron-nickel, thought to have originated near the core-mantle boundary of a parent body that underwent differentiation.

stony meteorite

The most common type of meteorite, composed primarily of silicates.

storm, meteor

An exceptionally heavy shower of meteors, usually taken to mean a rate of 1,000 an hour or greater. Some scientists define meteor storm to mean 1 meteor per second (3,600 an hour) or even 5,000 an hour. Unfortunately, such definitions reduce the number of true meteor storms in the historical records to a handful.

stream, meteor

A group of meteoroids with nearly identical orbits.

sublimation

In chemistry, the passage directly from a solid to a gas without passing through a liquid state. Frozen carbon dioxide (dry ice) sublimes at room temperature. Under the extremely low pressure conditions of space, water ice sublimes also.

tail (of a comet)

Gases and solid particles from a comet's coma forced outward, away from the Sun, by the pressure of sunlight (dust tail) and the solar wind (gas or ion or plasma tail).

Tears of St. Lawrence

The Perseid meteors.

trail

The projection of a meteor's path onto the celestial sphere to give its angular track. Often used loosely to mean the meteor's train.

train

The glow behind the head of a meteor. A meteor train of long duration is called a **persistent train**.

tsunami

A large sea wave produced by submarine earth movement or volcanic eruption. A tsunami is colloquially called a tidal wave, although it is not generated by tides.

twilight

The period of partial light between full nighttime darkness and sunrise and between sunset and full nighttime darkness. Thus twilight occurs in both the morning and evening.

vernal equinox

The point on the celestial sphere where the apparent motion of the Sun along the ecliptic crosses the celestial equator from south to north. (The autumnal equinox is the point where the Sun crosses the celestial equator from north to south.)

wake

A meteor train that lasts less than a second.

zenith

The point in the sky directly overhead of an observer (if he is standing vertically on level ground). The zenith will be different for observers in different locations.

zenithal hourly rate

The hypothetical rate at which meteors of a particular meteor shower would be observed under perfect conditions: by an experienced observer watching a clear sky with a limiting magnitude of 6.5 and with the radiant located in the zenith. Abbreviated **ZHR**. The lower the altitude of the radiant, the lower the observed rate. The zenithal hourly rate of a meteor shower can be approximated by dividing the observed hourly rate by the sine of the altitude angle of the radiant.

zenith attraction

The shift of a meteor's apparent radiant toward the zenith, caused by the Earth's gravity bending the meteoroid's path earthward.

ZHR

Zenithal hourly rate.

zodiac

A belt around the sky extending 9° above and below the ecliptic, so that it includes the Sun (on the ecliptic), Moon, all the naked-eye planets (Mercury through Saturn), and the star patterns across which these objects pass – the 12 constellations of the zodiac.

zodiacal light

Sunlight reflected off tiny dust particles (primarily from comets) that lie mostly near the plane of the solar system, thus along the zodiac. The reflection of sunlight from the particles is most pronounced near the Sun, so zodiacal light is usually seen as a cone or band of light either in the west after sunset or in the east before sunrise. It is very faint and therefore can be seen only under dark sky conditions.

BIBLIOGRAPHY
(Omits newspaper articles unless written by a major figure)

A [no other author identification given]. "Aus den Tagebüchern der Missionarien der evangelischen Brüdergemeine." *Annalen der Physik.* Volume 12, 1803, pages 206–223.

Adams, J[ohn] C[ouch]. "On the Orbit of the November Meteors." *Monthly Notices of the Royal Astronomical Society.* Volume 27, April 1867, pages 247–252.

Ahrens, Thomas J. "Weapons of Mass Protection." *The World & I.* May 1993, pages 216–221.

Ahrens, Thomas J., and Alan W. Harris. "Deflection and Fragmentation of Near-Earth Asteroids." Pages 897–927 in Tom Gehrels, editor. *Hazards Due to Comets and Asteroids.* Tucson: University of Arizona Press, 1994.

Ahrens, Thomas J., and Alan W. Harris. "Deflection and Fragmentation of Near-Earth Asteroids." *Nature.* Volume 360, December 3, 1992, pages 429–433.

American Journal of Science and Arts [no author given.] "Obituary [for Edward C. Herrick]." *American Journal of Science and Arts.* Volume 34, July 1862, pages 159–160.

Aristotle. *Meteorology.* Translated by F. W. Webster. Jonathan Barnes, editor. *The Complete Works of Aristotle.* Princeton, New Jersey: Princeton University Press, 1984.

Ashbrook, Joseph. *The Astronomical Scrapbook: Skywatchers, Pioneers, and Seekers in Astronomy.* Edited by Leif J. Robinson. Cambridge: Cambridge University Press and Cambridge, Massachusetts: Sky Publishing, 1984.

Aveni, Anthony F. *Skywatchers of Ancient Mexico.* Austin: University of Texas Press, 1980.

Bache, A[lexander] D[allas]. "Meteoric observations made on and about the 13th of November, 1834." *American Journal of Science and Arts.* Volume 27, number 2, January 1835, pages 335–338.

Bache, A[lexander] D[allas]. "Observations upon the facts recently presented by Prof. Olmsted, in relation to the meteors, seen on the 13th of Nov. 1834." *American Journal of Science and Arts.* Volume 29, number 2, January 1836, pages 383–388.

Bache, A[lexander] D[allas]. "Replies to a Circular in relation to the occurrence of an unusual Meteoric Display on the 13th Nov. 1834, addressed by the Secretary of War to the Military posts of the United States, with other facts relating to the same question." *American Journal of Science and Arts.* Volume 28, number 2, July 1835, pages 305–309.

Bailey, M. E., S. V. M. Clube, G. Hahn, W. M. Napier, and G. B. Valsecchi. "Hazards Due to Giant Comets: Climate and Short-Term Catastrophism." Pages 479–533 in Tom Gehrels, editor. *Hazards Due to Comets and Asteroids.* Tucson: University of Arizona Press, 1994.

Ball, Robert Stawell. *The Story of the Heavens.* London: Cassell, 1886.

Barone, G. "La grande pluie météorique de novembre 1899." *Memoires de la Societé Belge d'Astronomie.* 1900.

Bauer, L[ouis] A. "Biographical Sketch of Dr. John Locke." Unpublished; supplied by Kevin Grace, Archives and Rare Books Department, University of Cincinnati.

Bauer, Louis A. "The Early History of Terrestrial Magnetism in the United States With Special Reference to the Work of Dr. John Locke of Cincinnati." *Scientific Monthly.* Volume 18, June 1924, pages 625–627.

Beatty, J. Kelly. "Killer Crater in the Yucatán?" *Sky & Telescope.* Volume 82, July 1991, pages 38–40.

Beech, Martin. "Halley's Meteoric Hypothesis." *Astronomical Quarterly.* Volume 7, 1990, pages 3–18.

Beech, Martin. "Millet's Shooting Stars." *Journal of the Royal Astronomical Society of Canada.* Volume 82, December 1988, pages 349–358.

Beech, Martin. "A Simple Meteor Spectroscope." *Sky & Telescope.* Volume 80, November 1990, pages 554–556.

Beech, Martin. "The Stationary Radiant Debate Revisited." *Quarterly Journal of the Royal Astronomical Society.* Volume 32, 1991, pages 245–264.

Beech, Martin. "William Frederick Denning: In Quest of Meteors." *Journal of the Royal Astronomical Society of Canada.* Volume 84, number 6, December 1990, pages 383–396.

Beech, M[artin], P[eter] Brown, and J[im] Jones. "The Potential Danger to Space Platforms from Meteor Storm Activity." *Quarterly Journal of the Royal Astronomical Society.* Volume 36, 1995, pages 127–152.

Benzenberg, Johann Friedrich. *Die Sternschnuppen.* Hamburg: Perthes, Besser and Maure, 1839.

Benzenberg, J[ohann] F[riedrich], and H[einrich] W[ilhelm] Brandes. "Versuch die Entfernung, die Geschwindigkeit und die Bahn der Sternschnuppen zu Bestimmen." *Annalen der Physik.* Volume 6, 1800, pages 224–232.

Bertholon, [Pierre]. "Observation d'un globe de feu." *Journal des Sciences utiles.* Volume 4, 1791, pages 224–228.

Bessel, Friedrich. "Vorläufige Nachricht über eine die Berechnung der Sternschnupper betreffende Arbeit." *Annalen der Physik.* Volume 47, 1839, pages 525–527.

Boguslawski, G[eorg] von. *Die Sternschnuppen und ihre Beziehungen zu den Kometen.* Berlin: Carl Habel, 1874.

Bone, Neil. *Meteors.* (*Sky & Telescope* Observer's Guide series.) Cambridge, Massachusetts: Sky Publishing, 1993.

Botting, Douglas. *Humboldt and the Cosmos.* New York: Harper & Row, 1973.

Bottke, William F., Jr., Michael C. Nolan, Richard Greenberg, and Robert A. Kolvoord. "Collisional Lifetimes and Impact Statistics of Near-Earth Asteroids." Pages 337–357 in Tom Gehrels, editor. *Hazards Due to Comets and Asteroids.* Tucson: University of Arizona Press, 1994.

Bowell, Edward, and Karri Muinonen. "Earth-Crossing Asteroids and Comets: Groundbased Search Strategies." Pages 149–197 in Tom Gehrels, editor. *Hazards Due to Comets and Asteroids.* Tucson: University of Arizona Press, 1994.

Boyden, Seth. "Meteoric Phenomenon." *Mechanics' Magazine.* Volume 3, March 1834, page 185.

Brandt, John C., and Robert D. Chapman. *Introduction to Comets.* Cambridge: Cambridge University Press, 1981.

Brown, P[eter], and J[im] Jones. "Evolution of the Leonid Meteor Stream." Pages 57–60 in J[an] Štohl and I[wan] P. Williams, editors. *Meteoroids and Their Parent Bodies.* Proceedings of the International Astronomical Symposium held at Smolenice, Slovakia, July 6–12, 1992. Bratislava: Astronomical Institute of the Slovak Academy of Sciences, 1993.

Brownlee, [Donald] E., R. S. Rajan, and D. A. Tomandl. "A Chemical and Textual Comparison Between Carbonaceous Chondrites and Interplanetary Dust." Pages 137–141 in A. H. Delsemme, editor. *Comets, Asteroids, Meteorites: Interrelations, Evolution and Origins.* Toledo, Ohio: University of Toledo, 1977.

Brownlee, D[onald] E., F. Horz, D. A. Tomandl, and P[aul] W. Hodge. "Physical Properties of Interplanetary Grains." Pages 962–982 in B[ertram D.] Donn, M[ichael J.] Mumma, W. Jackson, M[ichael F.] A'Hearn, and R[obert S.] Harrington, editors. *The Study of Comets.* Washington, D.C.: National Aeronautics and Space Administration (NASA SP–393), 1976.

Browning, John. "On the Spectra of the Meteors of Nov. 13–14, 1866." *Monthly Notices of the Royal Astronomical Society.* Volume 26, December 1866, pages 77–79.

Burke, John G. *Cosmic Debris: Meteorites in History.* Berkeley: University of California Press, 1986.

Cevolani, Giordano, and Giuliano Trivellone. "Le Leonidi piccole bombe spaziali." *L'astronomia.* Number 164, April 1996, pages 38–45.

Chamberlain, Von Del. "Astronomical Content of North American Plains Indian Calendars." *Archaeoastronomy.* Number 6, pages S1–S54. Supplement to *Journal for the History of Astronomy.* Volume 15, 1984.

Chamberlain, Von Del. *When Stars Came Down to Earth: Cosmology of the Skidi Pawnee Indians of North America.* Los Altos, California: Ballena Press; and College Park, Maryland: Center for Archaeoastronomy, University of Maryland, 1982.

Chambers, George F. *A Handbook of Descriptive Astronomy.* Oxford: Clarendon Press, 1877.

Chapman, Robert D., and John C. Brandt. *The Comet Book: A Guide for the Return of Halley's Comet.* Boston: Jones and Bartlett, 1984.

Chladni, [Ernst F. F.]. "Account of a remarkable fiery Meteor seen in Gascony on the 24th of July 1790; by M. Baudin, Professor of Philosophy at Pau. With some Observations on Fire-Balls and Shooting-Stars." *Philosophical Magazine.* Volume 2, December 1798, pages 225–231. Reprinted and translated from *Magazin für das Neueste aus der Physik,* volume 11.

Chladni, [Ernst F. F.]. "Observations on a Mass of Iron found in Siberia by Professor Pallas, and on other Masses of the like Kind, with some Conjectures respecting their Connection with certain natural Phenomena." *Philosophical Magazine.* Volume 2, October 1798, pages 1–8. (A contemporary English-language summary of Chladni's 1794 book *Ueber den Ursprung der von Pallas gefundenen und anderer ihr ähnlicher Eisenmassen, und über einige damit in Berbindung stehende Naturerscheinungen.* Riga: Johann Friedrich Hartknoch, 1794.)

Chladni, [Ernst F. F.]. "Observations on Fire-Balls and Hard Bodies which have fallen from the Atmosphere." *Philosophical Magazine.* Volume 2, January 1799, pages 331–345.

Chladni, C. F. F. [actually Ernst F. F.]. "Observations on the Tones produced by an Organ-pipe in different Kinds of Gas." *Philosophical Magazine.* Volume 4, August 1799, pages 275–282.

Chladni, [Ernst F. F.]. "On a New Musical Instrument invented by Dr. Chladni; with some Experiments on the Vibrations of Sonorous Bodies." *Philosophical Magazine.* Volume 2, December 1798, pages 315–316.

Chladni, C. F. F. [actually Ernst F. F.]. "On the Invention of the Euphon, and other acoustic Discoveries." *Philosophical Magazine.* Volume 2, January 1799, pages 391–398.

Chladni, E[rnst] F[lorens] F[riedrich]. *Über den kosmischen Ursprung der Meteorite und Feuerkugeln (1794).* Commentary by Günter Hoppe. Leipzig: Geest & Portig, 1979. A reprint with biographical information and commentary on Chladni's entire *Ueber den Ursprung der von Pallas gefundenen und anderer ihr ähnlicher Eisenmassen, und über einige damit in Berbindung stehende Naturerscheinungen.* Riga: Johann Friedrich Hartknoch, 1794.

Chladni, Ernst Florens Friedrich. *Ueber Feuer-Meteore und über die mit denselben herabgefallenen Massen.* Vienna: J. G. Heubner, 1819.

Chyba, Christopher F., Tobias C. Owen, and Wing-Huen Ip. "Impact Delivery of Volatiles and Organic Molecules to Earth." Pages 9–58 in Tom Gehrels, editor. *Hazards Due to Comets and Asteroids.* Tucson: University of Arizona Press, 1994.

Clap, Thomas. *Conjectures upon the Nature and Motions of Meteors, which Are Above the Atmosophere.* Norwich, Connecticut: John Trumbull, 1781.

Clarke, W[illiam] B[ranwhite]. "On certain recent Meteoric Phenomena, Vicissitudes in the Seasons, prevalent Disorders, &c., contemporaneous, and in supposed connection, with Volcanic Emanations. No. 6." *Magazine of Natural History, and Journal of Zoology, Botany, Mineralogy, Geology, and Meteorology.* Edited by J. C. Loudon. (Often called Loudon's Magazine.) Volume 8, March 1835, page 129–161.

Clarke, W[illiam] B[ranwhite]. "On certain recent Meteoric Phenomena, Vicissitudes in the Seasons, prevalent Disorders, &c., contemporaneous, and in supposed connection, with Volcanic Emanations. No. 7." *Magazine of Natural History, and Journal of Zoology, Botany, Mineralogy, Geology, and Meteorology.* Edited by J. C. Loudon. (Often called Loudon's Magazine.) Volume 8, August 1835, pages 417–453.

Clarke, W[illiam] B[ranwhite]. "On the Meteors seen in America on the Night of Nov. 13, 1833." *Magazine of Natural History, and Journal of Zoology, Botany, Mineralogy, Geology, and Meteorology.* Edited by J. C. Loudon. (Often called Loudon's Magazine.) Volume 7, September 1834, pages 385–390.

Clarke, W. A. [actually William Branwhite]. "On the Origin of Shooting Stars." *American Journal of Science and Arts.* Volume 30, number 2, July 1836, pages 369–370.

Clegg, J. A., V. A. Hughes, and A. C. B. Lovell. "The Daylight Meteor Streams of 1947 May-August." *Monthly Notices of the Royal Astronomical Society.* Volume 107, 1947, pages 369–378.

Cobb, Nicholas H. "The Night the Stars Fell on Alabama." *Alabama Review.* Volume 22, April 1969, pages 147–157.

Condé, J. A. *History of the Dominion of the Arabs in Spain.* Translated by Mrs. Jonathan Foster. Volume 1 of 3. London: Henry G. Bohn, 1854.

Cook, Allan F. "A Working List of Meteor Streams." Pages 183–191 in Curtis L. Hemenway, Peter M. Millman, and Allan F. Cook, editors. *Evolutionary and Physical Properties of Meteoroids.* Washington, D.C.: National Aeronautics and Space Administration (SP–319), 1973.

Cronin, John F. "City Has Forgotten One Of Its Most Talented Sons." Cincinnati *Enquirer.* March 23, 1952.

Croswell, Ken. "Will the Lion Roar Again?" *Astronomy.* November 1991, pages 44–49.

Dall'olmo, Umberto. "Meteors, Meteor Showers and Meteorites in the Middle Ages: From European Medieval Sources." *Journal for the History of Astronomy.* Volume 9, 1978, pages 123–134.

d'Arrest, [Heinrich L.]. "Ueber einige merkwürdige Meteorfälle beim Durchgange der Erde durch die Bahn des *Biela*'schen Cometen." *Astronomische Nachrichten.* Number 1633, March 19, 1867, columns 7–10; article dated February 25, 1867.

Dauber, Philip M., and Richard A. Muller. *The Three Big Bangs: Comet Crashes, Exploding Stars, and the Creation of the Universe.* Reading, Massachusetts: Addison-Wesley (Helix Books), 1996.

Daubrée, A. "Bolide peint par Raphaël." *L'Astronomie: revue mensuelle d'astronomie populaire, de météorologie, de physique du globe et de photographie céleste.* Volume 10, June 1891, pages 201–206.

Denning, W[illiam] F. "The August Meteors." *Popular Science Monthly*. Volume 18, 1880, pages 178–190.

Denning, W[illiam] F. "The Claims of Meteoric Astronomy." *Journal of the Royal Astronomical Society of Canada*. Volume 9, 1915, pages 57–60.

Denning, W[illiam] F. "Falling Stars." Pages 431–448 in T. E. R. Phillips and W. H. Steavenson. *Splendour of the Heavens: A Popular Authoritative Astronomy*. Volume 1. New York: Robert M. McBride, 1925.

Denning, W[illiam] F. "Falling Stars" (poem; dated November 27, 1914). At the end of his "The Claims of Meteoric Astronomy." *Journal of the Royal Astronomical Society of Canada*. Volume 9, 1915, page 60.

Denning, W[illiam] F. "General Catalogue of the Radiant Points of Meteoric Showers and of Fireballs and Shooting Stars observed at more than one Station." *Memoirs of the Royal Astronomical Society*. Volume 53, 1899, pages 203–292 and plate 5.

Denning, W[illiam] F. *The Great Meteoric Shower of November*. London: Taylor and Francis, 1897.

Denning, W[illiam] F. "A History of the August Meteors." *Nature*. Volume 39, August 23, 1888, pages 393–395.

Denning, W[illiam] F. "The Leonid Meteoric Shower." *Nature*. Volume 63, November 8, 1900, pages 39–40.

Denning, W[illiam] F. "The Leonids. Mr. Denning's Report." *Nature*. Volume 61, November 23, 1899, pages 81–82.

Denning, William F. "The Radiant Centre of the Perseids." *Nature*. Volume 16, August 30, 1877, page 362.

Denning, W[illiam] F. "Report of the [Meteor] Section, 1899." *Memoirs of the British Astronomical Association*. Volume 9, 1900, pages 123–136 (also paginated 1–14 within the report).

Denning, William F. "The Stationary Meteor Showers." *Sidereal Messenger*. Volume 5, 1886, pages 167–174.

Denning, W[illiam] F. "Supposed Daylight Leonids." *Nature*. Volume 61, December 14, 1899, page 152.

Devens, R[ichard] M[iller]. "Sublime Meteoric Shower All Over the United States – 1833." Pages [329]–336 in *Our First Century: Being a Popular Descriptive Portraiture of the One Hundred Great and Memorable Events of Perpetual Interest in the History of Our Country*. Springfield, Massachusetts: C. A. Nichols, 1876.

di Cicco, Dennis. "New York's Cosmic Car Conker." *Sky & Telescope*. Volume 85, February 1993, pages 26.

Dodd, Robert T. *Thunderstones and Shooting Stars: The Meaning of Meteorites*. Cambridge, Massachusetts: Harvard University Press, 1986.

Dodson, R. S., Jr. *What You Should Know About Meteors and Meteorites*. Chapel Hill: Morehead Planetarium, University of North Carolina, 1960.

Dunkin, Edwin. *The Midnight Sky: Familiar Notes on the Stars and Planets*. London: Religious Tract Society, 1891.

Eastman, John Robie. "The Progress of Meteoric Astronomy in America." *Bulletin of the Philosophical Society of Washington* [D.C.]. Volume 11, July 1890, pages 275–358.

Edberg, Stephen J., and David H. Levy. *Observing Comets, Asteroids, Meteors, and the Zodiacal Light*. Cambridge: Cambridge University Press, 1994.

Edgeworth, K[enneth] E. "The Evolution of our Planetary System." *Journal of the British Astronomical Association*. Volume 53, July 1943, pages 181–188.

Edgeworth, K[enneth] E. "The Origin and Evolution of the Solar System." *Monthly*

Notices of the Royal Astronomical Society. Volume 109, number 5, October 14, 1949, pages 600–609.

E[lkin], W[illiam] L[ewis]. "November Meteors of 1899." *American Journal of Science.* Volume 8, 1899, pages 473–474.

Ellicott, Andrew. "Account of an Extraordinary Flight of Meteors (Commonly Called Shooting Stars)." *Transactions* of the American Philosophical Society. Volume 6, 1804, pages 28–29.

Ellicott, Andrew. *The Journal of Andrew Ellicott.* First published Philadelphia: 1803; reprinted Chicago: Quadrangle Books, 1962.

Ellyett, C. D., and J. G. Davies. "Velocity of Meteors Measured by Diffraction of Radio Waves from Trails during Formation." *Nature.* Volume 161, April 17, 1948, pages 596–597.

Espy, James P. "Remarks on Professor Olmsted's Theory of the Meteoric Phenomenon of November 12th [*sic*], 1833, denominated Shooting Stars, with some Queries towards forming a just Theory." *Journal of the Franklin Institute.* Volume 15, beginning in the January 1835 issue and continuing in the February, March, and April 1835 issues, pages 9–19, 85–92, 158–165, 234–236.

Flanagan, James. "Falling of the Stars. The Remarkable Phenomenon that Scared Everybody to Prayers in 1833. Danaldson's Long Trumpet, and What He Did with It. M. Fritz Proclaims Himself the Angel Gabriel." Transcribed by George F. Doyle. No publisher, no date. University of Kentucky Library: Cataloged as "Clark County, Kentucky – Meteorites" (4 typewritten pages).

Flaste, Richard, and Holcomb Noble, Walter Sullivan, and John Noble Wilford. The New York Times *Guide to the Return of Halley's Comet.* New York: Times Books, 1985.

Forster, Thomas Furley. *The Pocket Encyclopædia of Natural Phenomena.* London: J. Nichols and Son, 1827.

Forster, Thomas. *Researches About Atmospheric Phaenomena.* London: Thomas Underwood, 1813.

Fraehn, [Christian Martin]. Étoiles filantes. *L'Institut.* Number 252, October 25, 1838, pages 350–351.

Frank, Louis A., and Patrick Huyghe: *The Big Splash.* New York: Carol Publishing Group (Birch Lane Press), 1990.

Freitag, Ruth S. *Halley's Comet: A Bibliography.* Washington, D.C.: Library of Congress, 1984.

Fulchignoni, M., and Ľ. Kresák, editors. *The Evolution of the Small Bodies of the Solar System.* Proceedings of the International School of Physics "Enrico Fermi." Course 98. August 5–10, 1985. Amsterdam: North-Holland, 1987.

Fulkes, William. *A Goodly Gallerye: William Fulkes' Book of Meteors.* Edited by Theodore Hornberger. Philadelphia: American Philosophical Society, 1979.

Galle, J[ohann] G. "Ueber den muthmasslichen Zusammenhang der periodischen Sternschnuppen des 20. April mit dem ersten Cometen des Jahres 1861." *Astronomische Nachrichten.* Number 1635, April 2, 1867, columns 33–36; article dated March 11, 1867.

Ganapathy, R. "A Major Meteorite Impact on the Earth 65 Million Years Ago: Evidence from the Cretaceous-Tertiary Boundary Clay." *Science.* Volume 209, August 22, 1980, pages 921–923.

Gautier, Alfrède. "Notice sur les météores lumineux observés dan la nuit du 12 au 13 novembre 1832." *Bibliothèque universalle des sciences, belles-lettres, et arts.* Volume 51, October 2–December 2, 1832 (read at the December 6, 1832 meeting of the Society of Physics and Natural History of Geneva), pages 189–207.

Gehrels, Tom. "Collisions with Comets and Asteroids." *Scientific American.* Volume 274, March 1996, pages 54–59.

Gehrels, Tom, editor. *Hazards Due to Comets and Asteroids.* Tucson: University of Arizona Press, 1994.

Gibbs, J. Willard. "Memoir of Hubert Anson Newton, 1830–1896." *Biographical Memoirs of the National Academy of Sciences.* Volume 4, 1902, pages 99–124.

Glaisher, James, Robert P. Greg, E. W. Brayley, Alexander S. Herschel, and Charles Brooke. "Report on Observations of Luminous Meteors, 1867–1868." *Report of the 1868 Meeting of the British Association for the Advancement of Science.* 1868, pages 344–428. See especially section 3: "Papers Bearing on Meteoric Astronomy," pages 393–422, for a report on Schiaparelli's discoveries.

G[lenn], W[illiam] H. "November Meteors." *Sky and Telescope.* Volume 34, November 1967, pages 345–346.

Greg, R. P. "Observations on Meteor-showers and their Radiants." *Proceedings of the British Meteorological Society.* January 18, 1865, pages 308–315.

Grimm, Jacob. *Deutsche Mythologie.* Berlin: Ferd. Dümmlers, 1876.

Grimm, Jacob. *Teutonic Mythology.* Translated by James Steven Stallybrass. Gloucester, Massachusetts: Peter Smith, 1976.

Hahn, G[erhard], and M[ark] E. Bailey. "Rapid Dynamical Evolution of Giant Comet Chiron." *Nature.* Volume 348, 1990, pages 132–136.

Halley, Edmond. "An Account of several extraordinary Meteors or Lights in the Sky." *Philosophical Transactions* of the Royal Society of London. Volume 29, 1714, pages 159–164.

Halley, Edmond. "An Account of some Observations lately made at Nurenburg by Mr. P. Wurtzelbaur . . . " *Philosophical Transactions* of the Royal Society of London. Volume 16, 1688, pages 402–406.

Halley, Edmond. "An Account of the Extraordinary Meteor seen all over England, on the 19th of March 1718/9, With a Demonstration of the uncommon Height thereof." *Philosophical Transactions* of the Royal Society of London. Volume 30, number 360, 1719, pages 978–990.

Halley, Edmond. "A Discourse of the Rule of the decrease of the hight of the Mercury in the Barometer, according as Places are Elevated above the Surface of the Earth; with an attempt to discover the true reason of the Rising and Falling of the Mercury, upon change of Weather." *Philosophical Transactions* of the Royal Society of London. Volume 16, number 181, 1686, pages 104–116.

Halley, Edmond. "Some Considerations about the Cause of the universal Deluge, laid before the Royal Society, on the 12th of December 1694." *Philosophical Transactions* of the Royal Society of London. Volume 33, 1726, pages 118–119.

Hawkins, Gerald S. *Meteors, Comets, and Meteorites.* New York: McGraw-Hill, 1964.

Hemenway, Curtis L., Peter M. Millman, and Allan F. Cook, editors. *Evolutionary and Physical Properties of Meteoroids.* The proceedings of the International Astronomical Union's Colloquium #13, held at the State University of New York, Albany, N.Y., on June 14–17 1971. Washington, D.C.: National Aeronautics and Space Administration (SP-319), 1973.

Henry, Joseph. *The Papers of Joseph Henry.* Edited by Nathan Reingold. Volume 2. Washington, D.C.: Smithsonian Institution Press, 1975. (See especially pages 116–121; 128–130; 133 for commentary on Leonid meteors of 1833.)

Herrick, Edward C. "Account of a Meteor seen in Connecticut, December 14, 1837; with some considerations on the Meteorite which exploded near Weston, Dec. 14, 1807." *American Journal of Science and Arts.* Volume 37, number 1, October 1839, pages 130–135.

Herrick, Edward C. "Additional Account of the Shooting Stars of December 6 and 7, 1838."
    American Journal of Science and Arts. Volume 36, number 2, July 1839, pages 355–358.
Herrick, Edward C. "Additional Observations on the Shooting Stars of August 9th and
    10th, 1837." *American Journal of Science and Arts.* Volume 34, number 1, April 1838,
    pages 180–182.
Herrick, Edward C. "Contributions towards a History of the Star-Showers of Former
    Times." *American Journal of Science and Arts.* Volume 40, number 2, April 1841, pages
    349–365.
[Herrick, Edward C.] "Edward Cladius Herrick." Pages 51–52 in E. C. Herrick [and
    unnamed others], editors. *Obituary Record of Graduates of Yale College Deceased
    from July, 1859, to July, 1870.* New Haven, Connecticut: Yale University, 1871.
Herrick, Edward C. "Further proof of an annual Meteoric Shower in August, with remarks
    on Shooting Stars in general." *American Journal of Science and Arts.* Volume 33,
    number 2, January 1838, pages 354–364.
Herrick, Edward C. Notebooks and correspondence in the Yale University Library,
    Manuscripts and Archives Department.
Herrick, E[dward] C. "Observations on the Shooting Stars of August 9 and 10, 1841."
    American Journal of Science and Arts. Volume 41, number 2, September 1841, pages
    399–400.
Herrick, Edward C. "On Meteoric Showers in August; supplementary to Art. XX."
    American Journal of Science and Arts. Volume 33, number 2, January 1838, pages
    401–402.
Herrick, Edward C. "On the Meteoric Shower of April 20, 1803, with an account of
    observations made on and about the 20th April, 1839." *American Journal of Science
    and Arts.* Volume 36, number 2, July 1839, pages 358–363.
Herrick, Edward C. "On the Shooting Stars of August 9th and 10th, 1837, and on the
    Probability of the Annual Occurrence of a Meteoric Shower in August." *American
    Journal of Science and Arts.* Volume 33, number 1, October 1837, pages 176–180.
Herrick, Edward C. "Report on the Shooting Stars of August 9th and 10th, 1839, with other
    facts relating to the frequent occurrence of a meteoric display in August." *American
    Journal of Science and Arts.* Volume 37, number 2, October 1839, pages 325–338.
Herrick, Edward C. "Report on the Shooting Stars of December 7, 1838, with remarks on
    Shooting Stars in general." *American Journal of Science and Arts.* Volume 35, number
    2, January 1839, pages 361–368.
Herrick, E[dward] C. "Shooting Stars in June." *American Journal of Science and Arts.*
    Volume 42, number 1, December 1841, pages 201–202.
Herschel, A[lexander] S. "Contemporary Meteor Showers of the Leonid and Bielid
    Meteor-Periods." *Nature.* Volume 61, January 4, 1900, pages 222–226.
Herschel, A[lexander] S. "Prismatic Spectra of the August Meteors, 1866." *Intellectual
    Observer.* Volume 10, October 1866, pages 161–170.
Herschel, A[lexander] S. "Radiant Point of the November Meteors, 1866." *Monthly Notices
    of the Royal Astronomical Society.* Volume 27, number 2, December 14, 1866, pages
    17–19.
Herschel, J[ohn] F. W. "On the Meteoric Shower of 1866, Nov. 13–14." *Monthly Notices of the
    Royal Astronomical Society.* Volume 27, number 2, December 14, 1866, pages 19–21.
Hey, J[ames] S[tanley]. *The Evolution of Radio Astronomy.* New York: Neale Watson
    Academic Publications (Science History Publications), 1973.
Hey, J[ames] S[tanley], S. J[ohn] Parsons, and G[ordon] S. Stewart. "Radio Observations
    of the Giacobinid Meteor Shower, 1946." *Monthly Notices of the Royal Astronomical
    Society.* Volume 107, 1947, pages 176–183.

Hey, J[ames] S[tanley], and G[ordon] S. Stewart. "Derivation of Meteor Stream Radiants by Radio Reflexion Methods." *Nature*. Volume 158, October 5, 1946, pages 481–483.

Hey, J[ames] S[tanley], and G[ordon] S. Stewart. "Radar Observations of Meteors." *Proceedings of the Physical Society*. Volume 59, 1947, pages 858–883.

Hey, J[ames] S[tanley]. *The Radio Universe*. Oxford: Pergamon Press, 1983.

Hind, J[ohn] R[ussell]. "The Meteor-Shower of November." *The Times* (London). November 12, 1866. Reprinted as "The Meteor-Shower of This Morning." *The Scotsman* (Edinburgh). November 14, 1866.

Hindle, Brooke. *David Rittenhouse*. Princeton, New Jersey: Princeton University Press, 1964.

Hitchcock, Edward. "On the Meteors of Nov. 13th, 1833." *American Journal of Science and Arts*. Volume 25, number 2, January 1834, pages 354–363.

Hoffleit, Dorrit. *Astronomy at Yale: 1701–1968*. New Haven: Connecticut Academy of Arts and Sciences, 1992.

Hoffleit, Dorrit. "Yale Contributions to Meteoric Astronomy." *Vistas in Astronomy*. Volume 32, 1988, pages 117–143.

Hudson, Travis. "California's First Astronomers." Pages 11–41 in E. C. Krupp, editor. *Archaeoastronomy and the Roots of Science*. AAAS Selected Symposium 71. Boulder, Colorado: Westview Press for the American Association for the Advancement of Science, 1984.

Hughes, David. "Earth – A Cosmic Dustbin." *Physics Review*. Volume 2, September 1992, pages 22–26.

Hughes, David. "The History of Meteors and Meteor Showers." *Vistas of Astronomy*. Volume 26, 1982, pages 325–345.

Hughes, David. "Meteor Myths." *Astronomy Now*. Volume 3, November 1989, pages 43–46.

Hughes, David W. "A Mysterious Woodcut." *Sky & Telescope*. Volume 74, September 1987, page 252. Follow-up correspondence from John S. Kebabian and Paul Wirz in *Sky & Telescope*. Volume 75, April 1988, page 349.

Hughes, David W. "Sir John F. W. Herschel, Meteoroid Streams and the Solar Cycle." *Vistas in Astronomy*. Volume 39, 1995, pages 335–346.

Hughes, David W. "The World's Most Famous Meteor Shower Picture." *Earth, Moon, and Planets*. Volume 68, 1995, pages 311–322.

Humboldt, Alexander von. *Cosmos: A Sketch of a Physical Description of the Universe*. Volume 1. Translated by E. C. Otté. New York: Harper & Brothers, 1859.

Humboldt, Alexander von, and Aimé Bonpland. *Personal Narrative of Travels to the Equinoctial Regions of America, During the Years 1799–1804*. Translated and edited by Thomasina Ross. Volume 1 of 3. London: Henry G. Bohn, 1852. (Volume 3: 1853.)

L'Institut [no author given]: "Météorologie: Etoiles filantes." *L'Institut* (Paris). Number 218, Agust 1837, page 256.

Intellectual Observer [no author given.] "The Coming Meteor Shower. – The Spectra of Meteors." *Intellectual Observer*. Volume 10, August 1866, pages 38–40. (Diagram of Alexander S. Herschel's meteor spectroscope.)

Jenniskens, Peter. "Meteor stream activity. II. Meteor outbursts." *Astronomy and Astrophysics*. Volume 295, 1995, pages 206–235.

Jenniskens, Peter. "Meteor stream activity. III. Measurement of the first in a new series of Leonid outburst[s]." *Meteoritics & Planetary Science*. Volume 31, 1996, pages 177–184.

Kaiser, T. R. *Meteors*. Special supplement (Vol. II) to the Journal of Atmospheric and Terrestrial Physics. (Papers from a symposium on meteor physics held at the Jodrell Bank Experimental Station in July 1954.) London: Pergamon Press, 1955.

Kazimirchak-Polonskaya, E. I., and N. A. Belyaev, I. S. Astapovich, and A. K. Terent'eva.

"An Investigation into the Perturbed Motion of the Leonid Meteor Stream." *Soviet Astronomy – AJ*. Volume 11, number 3, November–December 1967, pages 490–500.

Keay, Colin S. L. "In Quest of Meteor Sounds." *Sky & Telescope*. Volume 70, December 1985, pages 623–625.

Keller, H. U. "Comets – Dirty Snowballs or Icy Dirtballs?" Pages 39–45 in *Proceedings of an International Workshop on Physics and Mechanics of Cometary Materials: 9–11 October 1989*. Paris: European Space Agency (ESA SP–302), December 1989.

Kingsmill, Tho[ma]s W. "Leonid Meteor Showers." *Nature*. Volume 61, March 22, 1900, page 491.

K[irkwood], D[aniel]. "Cometary Astronomy." *Danville Quarterly Review*. Volume 1, December 1861, pages 614–638.

Kirkwood, Daniel. *Meteoric Astronomy: A Treatise on Shooting-Stars, Fire-Balls, and Aerolites*. Appendix B: "Comets and Meteors." Philadelphia: J. B. Lippincott, 1867.

Klinkerfues, W[ilhelm]. "On the Re-discovery of Biela's Comet." *Monthly Notices of the Royal Astronomical Society*. Volume 33, number 3, January 10, 1873, pages 128–130. See also subsequent articles by various observers in this issue about sightings of the Bielid meteors.

Klinkerfues, W[ilhelm]. "Schreiben des Herrn Prof. *Klinkerfues* an den Herausgeber." *Astronomische Nachrichten*. Number 1918, columns 349–350; article dated January 2, 1873.

Kluger, Jeffrey. "The Gentle Cosmic Rain." *Time*. June 9, 1997, page 52.

Kresák, Ľubor. "The Cometary and Asteroidal Origins of Meteors." Pages 331–341 in Curtis L. Hemenway, Peter M. Millman, and Allan F. Cook, editors. *Evolutionary and Physical Properties of Meteoroids*. Washington, D.C.: National Aeronautics and Space Administration (SP–319), 1973.

Kresák, Ľ[ubor]. "Meteor storms." Pages 147–156 in J[an] Štohl and I[wan] P. Williams, editors. *Meteoroids and Their Parent Bodies*. Proceedings of the International Astronomical Symposium held at Smolenice, Slovakia, July 6–12, 1992. Bratislava: Astronomical Institute of the Slovak Academy of Sciences, 1993.

Kronk, Gary W. *Comets: A Descriptive Catalog*. Hillside, New Jersey: Enslow Press, 1984.

Kronk, Gary W. *Meteor Showers: A Descriptive Catalog*. Hillside, New Jersey: Enslow Press, 1988.

Kronk, Gary W. "Meteors and the Native Americans." http://www.maa.mhn.de/Comet/metlegends.html.

Kuiper, Gerard P. "On the Origin of the Solar System." Pages 357–424 in J. A. Hynek, editor. *Astrophysics: A Topical Symposium*. New York: McGraw-Hill, 1951.

Lancaster Brown, Peter. *Comets, Meteorites and Men*. New York: Taplinger, 1974.

LaPaz, Lincoln. "The Effects of Meteorites Upon the Earth (Including Its Inhabitants, Atmosphere, and Satellites)." Appendix 1: Meteoritical Pictographs and the Veneration and Exploitation of Meteorites (pages 329–336). Pages 217–350 in H. E. Landsberg and J. Van Mieghem, editors. *Advances in Geophysics*. Volume 4. New York: Academic Press, 1958.

Lebedinets, V. N. "Ablation in Meteors." Pages 259–269 in Curtis L. Hemenway, Peter M. Millman, and Allan F. Cook, editors. *Evolutionary and Physical Properties of Meteoroids*. Washington, D.C.: National Aeronautics and Space Administration (SP–319), 1973.

Le Verrier, [Urbain J. J]. "Sur les étoiles filantes du 13 novembre et du 10 août." *Comptes Rendus*. Volume 64, number 3, January 21, 1867, pages 94–99.

Lewis, John S. *Rain of Iron and Ice: The Very Real Threat of Comet and Asteroid Bombardment*. Reading, Massachusetts: Addison-Wesley (Helix Books), 1996.

Lewis, John S., and Melinda L. Hutson. "Asteroidal Resource Opportunities Suggested by Meteorite Data." Pages 523–542 in John S. Lewis, Mildred S. Matthews, and Mary L. Guerrieri, editors. *Resources of Near-Earth Space*. Tucson: University of Arizona Press, 1993.

Lewis, John S., David S. McKay, and Benton C. Clark. "Using Resources from Near-Earth Space." Pages 3–14 in John S. Lewis, Mildred S. Matthews, and Mary L. Guerrieri, editors. *Resources of Near-Earth Space*. Tucson: University of Arizona Press, 1993.

Lindsay, Robert Bruce. *Men of Physics: Julius Robert Mayer: Prophet of Energy*. Oxford: Pergamon Press, 1973.

Littmann, Mark, and Donald K. Yeomans. *Comet Halley: Once in a Lifetime*. Washington, D.C.: American Chemical Society, 1985.

Littmann, Mark. "The Discovery of the Perseid Meteors." *Sky & Telescope*. Volume 92, August 1996, pages 68–71.

Littmann, Mark. *Planets Beyond: Discovering the Outer Solar System*. New York: John Wiley & Sons, 1988, 1990.

Littmann, Mark, and Ken Willcox. *Totality: Eclipses of the Sun*. Honolulu: University of Hawaii Press, 1991.

Locke, John. "Meteors." *Cincinnati Daily Gazette*. August 11, 1834.

Locke, John. "Meteors No. II." *Cincinnati Daily Gazette*. August 12, 1834.

Lockyer, J. Norman. *The Meteorite Hypothesis*. London: Macmillan, 1890.

Lœwy, [Maurice]. "Paris. Academy of Sciences." *Nature*. Volume 61, November 30, 1899, page 119.

Loomis, Elias. *Elements of Astronomy*. New York: Harper & Brothers, 1869.

Loomis, [Elias]. "The Expected Meteoric Shower. Letter from Professor Loomis." *New York Evening Post*. November 10, 1866, page 2. Reprinted as "The Expected Shower of Shooting Stars." *New York Times*. November 12, 1866, page 8.

Lovell, A. C. B[ernard]. *Meteor Astronomy*. Oxford: Clarendon Press, 1954.

Lovell, Bernard. *Astronomer by Chance*. New York: Basic Books, 1990.

Lovell, Bernard, and J. A. Clegg. *Radio Astronomy*. London: Chapman & Hall, 1952.

Lovering, J[oseph]. "Meteoric Observations made at Cambridge, Mass." *American Journal of Science and Arts*. Volume 35, number 2, January 1839, pages 323–328.

Luyten, Willem J. "Flashing Show of Leonid Meteors Expected to Be at Height Tonight." *Atlanta Constitution*. November 15, 1932, page 11.

Luyten, Willem J. "Observations of the Leonid Shower at Minnesota." *Popular Astronomy*. Volume 40, December 1932, page 650.

Lyman, C. S. "Biographical Sketch of Professor Denison Olmsted." *American Journal of Science and Arts*. 2nd series. Volume 28, July 1859, pages 109–118.

Lyttleton, R[aymond] A[rthur]. *The Comets and Their Origin*. Cambridge: Cambridge University Press, 1953.

Lyttleton, R[aymond] A[rthur]. "Does a Continuous Solid Nucleus Exist in Comets?" *Astrophysics and Space Science*. Volume 15, 1972, pages 175–184.

Macpherson, Hector, Jr. *Astronomers of To-Day and Their Work*. London: Gall & Inglis, 1905.

Mallery, Garrick. *Picture-Writing of the American Indians*. Washington, D.C.: Smithsonian Institution, 1893.

Marsden, Brian G. "Comet 55P/1997 E1 (Tempel–Tuttle)." Central Bureau for Astronomical Telegrams Circular No. 6579, March 10, 1997.

Marsden, Brian G. "Comet Swift–Tuttle: Does It Threaten Earth?" *Sky & Telescope*. Volume 85, January 1993, pages 16–19.

Mason, John W. "The Leonid Meteors and Comet 55P/Tempel–Tuttle." *Journal of the British Astronomical Association.* Volume 105, number 5, 1995, pages 219–235.

McIntosh, Bruce A. "Origin and Evolution of Recent Leonid Meteor Showers." Pages 193–198 in Curtis L. Hemenway, Peter M. Millman, and Allan F. Cook, editors. *Evolutionary and Physical Properties of Meteoroids.* Washington, D.C.: National Aeronautics and Space Administration (SP-319), 1973.

McKinley, D[onald] W. R. *Meteor Science and Engineering.* New York: McGraw-Hill, 1961.

McSween, Harry Y., Jr. *Fanfare for Earth: The Origin of Our Planet and Life.* New York: St. Martin's Press, 1997.

McSween, Harry Y., Jr. *Meteorites and Their Parent Planets.* Cambridge: Cambridge University Press, 1987.

McSween, Harry Y., Jr. *Stardust to Planets: A Geological Tour of the Solar System.* New York: St. Martin's Press, 1993.

Meadows, A[rthur] J[ack]. *The High Firmament: A Survey of Astronomy in English Literature.* Leicester: Leicester University Press, 1969.

Mechanics' Magazine [no author given]. "Meteoric Phenomena." *Mechanics' Magazine.* Volume 2, November 1833, pages 287–288.

Milani, A., M. DiMartino, and A. Cellino, editors. *Asteroids, Comets, Meteors 1993.* Proceedings of the 160th Symposium of the International Astronomical Union, held in Belgirate, Italy, June 14–18, 1993. Dordrecht, The Netherlands: Kluwer, 1993.

Millman, Peter M. "Meteor News. The Leonids – 1969." *Journal of the Royal Astronomical Society of Canada.* Volume 64, 1970, pages 55–57.

Milner, Thomas. *The Gallery of Nature: A Pictorial and Descriptive Tour Through Creation, Illustrative of the Wonders of Astronomy, Physical Geography, and Geology.* London: Wm. S. Orr, 1849.

Milon, Dennis. "Observing the 1966 Leonids." *Journal of the British Astronomical Association.* Volume 77, 1967, pages 89–93.

Monastersky, R[ichard]. "Is Earth Pelted by Space Snowballs?" *Science News.* Volume 151, May 31, 1997, page 332.

Morrison, David. "The Spaceguard Survey: Protecting the Earth from Cosmic Impacts." *Mercury.* Volume 21, May/June 1992, pages 103–106, 110.

Morrison, David, chair and editor. *The Spaceguard Survey: Report of the NASA International Near-Earth-Object Detection Workshop,* January 10, 1992. Pasadena: Jet Propulsion Laboratory/California Institute of Technology for NASA, 1992.

Morrison, David. "Target: Earth!" *Astronomy.* Volume 23, number 10, October 1995, pages 34–41.

Morrison, David, Clark R. Chapman, and Paul Slovic. "The Impact Hazard." Page 59–91 in Tom Gehrels, editor. *Hazards Due to Comets and Asteroids.* Tucson: University of Arizona Press, 1994.

Musschenbroek, Petrus von. *The Elements of Natural Philosophy.* Volume 2. Translated from Latin by John Colson. London: J. Nourse, 1744.

Myers, G[eorge] W[illiam]. "The November Meteors at Urbana, Illinois." Pages 23–30. Chicago, 1897.

Nagaoka, Hantaro. "Possibility of the Radio Transmission being disturbed by Meteoric Showers." *Proceedings of the Imperial Academy of Tokyo.* Volume 5, 1929, pages 233–236.

Nashe, Thomas. *Summers Last Will and Testament. The Works of Thomas Nashe.* Edited by Ronald B. McKerrow. Volume 3. Oxford: Basil Blackwell, 1958.

Nature [no author given.] "Our Astronomical Column. The Leonid Meteors." *Nature*. Volume 63, November 22, 1900, page 92.

Nature [no author given.] "Our Astronomical Column. Leonid Meteors, November 1901." *Nature*. Volume 65, November 28, 1901, page 89.

Nature [no author given.] "The Leonid Meteors." *Nature*. Volume 128, November 28, 1931, page 912.

Nature [no author given.] "The Leonid Meteors." *Nature*. Volume 128, December 5, 1931, page 972.

Nature [no author given.] "The Leonid Meteors." *Nature*. Volume 128, December 12, 1931, page 1007.

Nature [no author given.] "The Leonid Meteors." *Nature*. Volume 130, November 26, 1932, page 817.

Nature [no author given.] "The Leonid Meteors." *Nature*. Volume 130, December 24, 1932, page 970.

Nature [no author given.] "The Leonids, 1933." *Nature*. Volume 132, December 16, 1933, page 928.

Newcomb, Simon. *Popular Astronomy*. London: Macmillan, 1883.

New England Magazine [no author given]. "The Meteoric Shower." *New England Magazine*. Volume 6, January 1834, pages 47–54.

Newton, H[ubert] A. "Abstract of a Memoir on Shooting Stars." *American Journal of Science and Arts*. Series 2. Volume 39, March 1865, pages 193–207.

N[ewton], H[ubert] A. "Astronomy." (Brief reports on recent national and international findings about meteors.) *American Journal of Science and Arts*. 2nd series. Volume 44, July 1867, pages 127–130.

Newton, Hubert A. "Evidence of the Cosmical Origin of Shooting Stars Derived from the Dates of Early Star-Showers." *American Journal of Science and Arts*. 2nd series. Volume 36, July 1863, pages 145–149.

Newton, H[ubert] A. "The Fireball in Raphael's Madonna di Foligno." *American Journal of Science*. 3rd series. Volume 41, March 1891, pages 235–238.

Newton, H[ubert] A. "Fireball of January 13th, 1893." *American Journal of Science*. Volume 46, September 1893, pages 161–172.

Newton, H[ubert] A. "The Meteorites, the Meteors, and the Shooting Stars." (Address of the retiring president in August 1866.) *Proceedings of the American Association for the Advancement of Science*. Volume 35, 1886, pages 1–18. Condensed as "Meteorites, Meteors, and Shooting Stars." *Popular Science Monthly*. Volume 29, 1886, pages 733–747.

Newton, [Hubert A.] "November Meteors. Letter from Prof. Newton, of Yale College, to Professor Henry, Secretary of the Smithsonian Institute [*sic*]." *New York Times*. November 11, 1866, page 5.

Newton, H[ubert] A. "On certain recent contributions to Astro-Meteorology." *American Journal of Science and Arts*. Volume 43, May 1867, pages 285–300.

Newton, H[ubert] A. "On Shooting Stars." *Memoirs of the National Academy of Sciences*. Volume 1, 1866, pages 291–312.

Newton, H[ubert] A. "On the effect upon the earth's velocity produced by small bodies passing near the earth." *American Journal of Science*. 3rd series. Volume 30, number 180, December 1885, pages 409–417.

Newton, H[ubert] A[nson]. "The original accounts of the displays in former times of the November Star-Shower; together with a determination of the length of its cycle, its annual period, and the probable orbit of the group of bodies around the sun."

American Journal of Science and Arts. 2nd series. Part 1: volume 37, number 111, May 1864, pages 377–389. Part 2: volume 38, number 112, July 1864, pages 53–61.

N[ewton], H[ubert] A. "Photographs of the August and December Meteors." *American Journal of Science.* Volume 47, 1894, pages 154–155.

N[ewton], H[ubert] A. "Procession and Periodicity of the November Star-Shower." *American Journal of Science.* Volume 36, September 1863, page 301.

Newton, Hubert A. "Relation of Meteorites to Comets." *Nature.* Volume 19, February 6, 1879, pages 315–317; 340–342. A reprint exists, paginated 1–12.

Newton, H[ubert] A[nson]. "Shooting Stars in November, 1866." *American Journal of Science and Arts.* Volume 43, January 1867, pages 78–88.

Newton, H[ubert] A. "Summary of observations of shooting stars during the August period, 1863." *American Journal of Science and Arts.* Volume 36, September 1863, pages 302–307.

Newton, Isaac. *Opticks or A Treatise of the Reflections, Refractions, Inflections & Colours of Light.* 4th edition. London: William Innys, 1730 (1st edition published in 1704); 4th edition reprinted New York: Dover, 1979.

Nicholson, Thomas D. "Sky Reporter: Do comets really break down to produce meteor showers? Radio astronomy's methods of observation provide valuable new data." *Natural History.* Volume 75, December 1966, pages 50–52.

Nicholson, Thomas D. "Sky Reporter: The first major Leonid shower since 1932 may occur this month." *Natural History.* Volume 75, November 1966, pages 42–45.

Niessl [von Mayendorf], G[ustav], and C[uno] Hoffmeister. "Katalog der Bestimmungsgrössen für 611 Bahnen grosser Meteore." *Denkschriften. [Österreichische] Akademie der Wissenschaften in Wien. Mathematisch-Naturwissenschaftliche Klasse.* Volume 100, 1926, pages 1–70. (Niessl's name appears as G. v. Niessl on the article.)

Olbers, W[ilhelm]. "Noch etwas über Sternschnuppen, als Nachtrag." Pages 278–282 in *Jahrbuch für 1837.* Edited by H. C. Schumacher. Stuttgart: J. G. Cotta, 1837. See also Olbers article "Die Sternschnuppen" in the same volume.

Olivier, Charles P. "Influences of Meteoric Astronomy on Evolution." *Journal of the Franklin Institute.* Volume 207, number 6, June 1929, pages 733–751.

Oliver [Olivier], Charles P. "Leonid Meteor Fall Expected to Be Heavy." *San Francisco Chronicle.* November 13, 1932, page 6.

Olivier, Charles P. *Meteors.* Baltimore: Williams & Wilkins, 1925.

Olivier, Charles P. "Will the Great Shower Return?" *Science News Letter.* Volume 22, November 5, 1932, pages 290–291, 297–298.

Olmsted, Denison. "Facts respecting the Meteoric Phenomena of November 13, 1834." *American Journal of Science and Arts.* Volume 29, number 1, October 1835 (although bound as January 1836), pages 168–170.

Olmsted, Denison. *An Introduction to Astronomy.* Revised edition. New York: Robert B. Collins, 1855.

Olmsted, Denison. *Letters on Astronomy, Addressed to a Lady: in which the Elements of the Science Are Familiarly Explained in Connexion with Its Literary History.* Boston: Marsh, Capen, Lyon, and Webb, 1841.

Olmsted, [Denison]. "The Meteors of November 13th, 1833." Pages 70–80 in [no editor given], *The American Almanac and Repository of Useful Knowledge for the Year 1835: Comprising a Calendar for the Year; Astronomical Information; Miscellaneous Directions, Hints, and Remarks; and Statistical and Other Particulars Respecting Foreign Countries and the United States.* Boston: Gray and Bowen, 1834.

Olmsted, Denison. "Observations on the Meteors of November 13th, 1833." *American Journal of Science and Arts.* Volume 25, number 2, January 1834, pages 363–411; continued under the same title in volume 26, number 1, April 1834 (although bound as July 1834), pages 132–174.

Olmsted, Denison. "On the cause of the Meteors of November 13th, 1833." *American Journal of Science and Arts.* Volume 29, number 2, January 1836, pages 376–383.

Olmsted, Denison. "On the Meteoric Shower of November, 1836." *American Journal of Science and Arts.* Volume 31, number 2, January 1837, pages 386–395.

Olmsted, Denison. "On the Meteoric Shower of November, 1837." *American Journal of Science and Arts.* Volume 33, number 2, January 1838, pages 379–393.

Olmsted, Denison. "On the Meteoric Shower of November, 1838." *American Journal of Science and Arts.* Volume 35, number 2, January 1839, pages 368–370.

Olmsted, Dension. "Remarks on Shooting Stars, in reply to Rev. W. A. Clarke; with additional observations on the present state of our Knowledge respecting the *origin* of these meteors." *American Journal of Science and Arts.* Volume 30, number 2, July 1836, pages 370–376. (W. B. Clarke's initials appeared as W. A. in his article immediately preceding.)

Olmsted, [Denison]. "Zodiacal Light." *American Journal of Science and Arts.* Volume 27, number 2, January 1835, pages 416–419.

Oort, Jan H. "The Structure of the Cloud of Comets Surrounding the Solar System and a Hypothesis Concerning Its Origin." *Bulletin of the Astronomical Institutes of the Netherlands.* Volume 11, 1950, pages 91–110.

Öpik, E[rnst Julius]. "Meteors." *Monthly Notices of the Royal Astronomical Society.* Volume 100, February 1940, pages 315–326.

Öpik, E[rnst J.]. "Note on Stellar Perturbations of Nearly Parabolic Orbits." *Daedalus* (Proceedings of the American Academy of Arts and Sciences). Volume 67, 1932, pages 169–183.

Öpik, E[rnst J.]. "A Statistical Method of Counting Shooting Stars and Its Application to the Perseid Shower of 1920." *Publications de l'Observatoire Astronomique de l'Université de Tartu (Dorpat).* Volume 25, number 1, 1922, pages 1–56 (entire issue).

Öpik, E[rnst J.]. "Teleskopische Beobachtungen der Perseiden." *Astronomische Nachrichten.* Volume 217, 1922, columns 41–46.

Oppolzer, Theodor. "Bahn-Bestimmung des Cometen II.1862." *Astronomische Nachrichten*, number 1396, December 19, 1862, columns 49–58; article dated November 25, 1862.

Oppolzer, Th[eodor]. "Bahnbestimmung des Cometen I.1866 (*Tempel*)." *Astronomische Nachrichten.* Number 1624, January 28, 1867, columns 241–250; article dated January 7, 1867.

Oppolzer, Th[eodor]. "Schreiben des Herrn Dr. *Th. Oppolzer* an den Herausgeber." *Astronomische Nachrichten.* Number 1629 (a combined issue with 1628), February 20, 1867, columns 333–334; article dated February 6, 1867.

Oppolzer, Theodor. "Ueber die Bahn des Cometen II.1862." *Astronomische Nachrichten*, number 1384, October 14, 1862, columns 249–250; article dated September 8, 1862.

Pendleton, Yvonne J., and Dale P. Cruikshank. "Life from the Stars?" *Sky & Telescope.* Volume 87, March 1994, pages 36–42.

Peters, C[hristian] F. W. "Bemerkung über den Sternschnuppenfall vom 13. November und 10. August 1866." *Astronomische Nachrichten.* Number 1626, February 4, 1867, columns 287–288; article dated January 29, 1867.

Phillips, T. E. R., and W. H. Steavenson, editors. *Splendour of the Heavens: A Popular Authoritative Astronomy.* Volume 1. New York: Robert M. McBride, 1925.

Pickering, E[dward] C[harles]. "The November Meteors of 1899." *Scientific American*. Volume 81, October 28, 1899, page 279.

Pliny the Elder. *Natural History*. Translated by H. Rackman. Cambridge, Massachusetts: Harvard University Press; and London: William Heineman, 1938.

Ponnamperuma, Cyril, editor. *Comets and the Origin of Life*. Proceedings of the Fifth College Park Colloquium on Chemical Evolution, University of Maryland, College Park, Maryland, U.S.A., October 29th to 31st, 1980. Dordrecht, Holland: D. Reidel, 1980.

Popular Astronomy [author missing]. "Leonid Observations for 1932 at Pomona College, Claremont, California." *Popular Astronomy*. Volume 41, March 1933, pages 174–176.

Porter, J[ohn] G[uy]. *Comets and Meteor Streams*. New York: John Wiley & Sons, 1952.

Prentice, J. P. M[anning]. "Meteor Section." ("Brief History.") *Memoirs of the British Astronomical Association*. Volume 36, 1948, pages 104–110.

Pringle, John. "Some Remarks upon the several Accounts of the fiery Meteor (which appeared on Sunday the 26th of November, 1758) and upon other such Bodies." *Philosophical Transactions* of the Royal Society of London. Volume 51, 1759, pages 259–274.

Proctor, Richard A. *Rough Ways Made Smooth: A Series of Familiar Essays on Scientific Subjects*. New York: R. Worthington, 1880.

Pruett, J. Hugh. "Experts Tell Why Meteors Are Expected." *Seattle Post-Intelligencer*. November 13, 1932, page 11.

Quetelet, A[dolphe]. "Catalogue des principales apparitions d'étoiles filantes." *Mémoires de l'Académie Royale des Sciences et Belles-Lettres de Bruxelles*. Volume 12, 1839, pages 1–63.

Quetelet, Adolphe. "Étoiles filantes." *Annuaire de l'Observatoire de Bruxelles pour l'An 1837*. 1836, pages 268–273.

Quetelet, Adolphe. "Étoiles filantes." *Bulletin de l'Académie Royale des Sciences et Belles-Lettres de Bruxelles*. December 3 session. Volume 3, number 11, 1836, pages 403–413.

Quetelet, Adolphe. "Étoiles filantes." *Bulletin de l'Académie Royale des Sciences et Belles-Lettres de Bruxelles*. March 4 session. Volume 4, number 3, 1837, pages 79–81.

Quetelet, Adolphe. "Étoiles filantes." *Bulletin de l'Académie Royale des Sciences et Belles-Lettres de Bruxelles*. October 7 session. Volume 4, number 9, 1837, pages 376–380.

Quetelet, A[dolphe]. "Nouveau catalogue des principales apparitions d'étoiles filantes." *Mémoires de l'Académie Royale des Sciences et Belles-Lettres de Bruxelles*. Volume 15, 1841, pages 3–60.

Quetelet, Adolphe. "On the Height, Motion, and Nature of Shooting Stars." *London and Edinburgh Philosophical Magazine and Journal of Science*. 3rd series. Volume 11, September 1837, pages 270–273. A translation of Quetelet's "Étoiles Filantes." *Annuaire de l'Observatoire de Bruxelles pour l'An 1837*. 1836, pages 268–273.

Quetelet, [Adolphe]. "On the question whether Shooting Stars are more numerous at certain times than at others." *London and Edinburgh Philosophical Magazine and Journal of Science*. 3rd series. Volume 11, September 1837, pages 268–270.

Quetelet, Adolphe. *Sciences Mathématiques et Physiques chez les Belges au commencement du XIXème siècle*. Bruxelles: Thiry-Van Buggenhoudt, 1866.

Rada, W. S., and F. R. Stephenson. "A Catalogue of Meteor Showers in Medieval Arab Chronicles." *Quarterly Journal of the Royal Astronomical Society*. Volume 33, 1992, pages 5–16.

Rao, Joe. "The Leonids: King of the Meteors Showers." *Sky & Telescope*. Volume 90, number 5, November 1995, pages 24–31.

Reddy, Francis, and Greg Walz-Chojnacki. *Celestial Delights: The Best Astronomical Events through 2001.* Berkeley, California: Celestial Arts, 1992.

Rittenhouse, David. "Account of a Meteor." (Letter dated January 16, 1780, responding to a letter dated December 4, 1779 from John Page to Rittenhouse; read to the American Philosophical Society May 2, 1783.) *Transactions of the American Philosophical Society.* Volume 2, 1786, pages 173–176 (including Page's letter).

David Rittenhouse. *The Scientific Writings of David Rittenhouse.* Edited by Brooke Hindle. New York: Arno Press, 1980.

R[obinson], L[eif] J. "Observations of Three Meteor Showers." *Sky and Telescope.* Volume 31, February 1966, pages 112–115.

Roe, George Mortimer, editor. *Cincinnati: The Queen City of the West.* Cincinnati: C. J. Brehbiel, 1895.

Roggemans, Paul, editor. *Handbook for Visual Meteor Observations.* Cambridge, Massachusetts: Sky Publishing, 1989.

Rubincam, David Parry, and Milton Rubincam II. "America's Foremost Early Astronomer." *Sky & Telescope.* Volume 89, number 5, May 1995, pages 38–41.

Sagan, Carl, and Ann Druyan. *Comet.* New York: Random House, 1985.

Sanderson, Richard. "The Night It Rained Fire." *Griffith Observer.* Volume 48, November 1984, pages 2–10.

Sauval, Jacques. "Adolphe Quetelet et les étoiles filantes." *Bulletin Astronomique de l'Observatoire Royal de Belgique.* Volume 11, number 1, September 1996, pages 67–82. (In French and German.)

Sauval, Jacques. "Quetelet and the Discovery of the First Meteor Showers." *WGN: the Journal of the International Meteor Organization.* Volume 25, number 1, 1997, pages 21–33.

Schaeffer, George C. "Notice of the Meteors of the 9th and 10th of August, 1837, and also of Nov. 12th and 13th, 1832." *American Journal of Science and Arts.* Volume 33, number 2, January 1838, pages 133–135.

Schafer, J[ohn] P[eter], and W[illiam] M[cHenry]. "Observations of Kennelly–Heaviside Layer Heights During the Leonid Meteor Shower of November, 1931." *Proceedings of the Institute of Radio Engineers.* Volume 20, December 1932, pages 1941–1945.

Schiaparelli, G[iovanni] V. *Entwurf einer astronomischen Theorie der Sternschnuppen.* Translated and edited by Georg von Boguslawski. Stettin: Nahmer, 1871. Originally published as *Note e riflessioni intorno alla teoria astronomica delle stelle cadenti.* Firenze: Stamperia reale, 1867.

Schiaparelli, Giovanni V. "Intorno al Corso ed all'Origine Probabile delle Stelle Meteoriche." *Bullettino Meteorologico dell'Osservatorio del Collegio Romano.* Five letters to Angelo Secchi. Volume 5, numbers 8, 10, 11, 12, 1866; and volume 6, number 2, 1867.

Schiaparelli, G[iovanni] V. *Le Opere di G. V. Schiaparelli.* Volume 3. Milan: Ulrico Hoepli, 1930. Reprinted New York: Johnson Reprint, 1968.

Schiaparelli, G[iovanni] V. "Sur la marche et l'origine probable des étoile météoriques." (French translation of Letter 1 from Schiaparelli to Secchi.) *Les Mondes.* 2nd series. Volume 12, December 13, 1866, pages 610–624.

Schiaparelli, J. V. [Giovanni V.]. "Sur la relation qui existe entre les comètes et les étoiles filantes." *Astronomische Nachrichten.* Number 1629 (a combined issue with 1628), February 20, 1867, columns 331–332; article dated February 2, 1867.

Schiaparelli, G[iovanni] V. "Sur le mouvement et l'origine probable des étoiles météoriques." (French translation of Letter 2 from Schiaparelli to Secchi.) *Les Mondes.* 2nd series. Volume 13, January 24, 1867, pages 147–162.

Schiaparelli, W. [Giovanni V.] "Sur les étoiles filantes, et spécialement sur l'identification des orbites des essaims d'août et de novembre avec celles des comètes de 1862 et de 1866." (Extrait d'une Lettre à M. Delaunay.) *Comptes rendus*. Volume 64, 1867, pages 598–599; article dated March 12, 1867.

Schiaparelli, G[iovanni] V. "Sur l'origine des étoiles filantes de novembre." *Les Mondes*. 2nd series. Volume 13, March 28, 1867, pages 501–507.

Schiaparelli, J. V. [Giovanni V.]. "Théorie des étoiles filantes." (French translation of Letters 4 and 5 from Schiaparelli to Secchi.) *Les Mondes*. 2nd series. Volume 13, February 21, 1867, pages 284–289.

Schubart, Joachim. "Comet Tempel–Tuttle (1866I)." Central Bureau for Astronomical Telegrams Circular No. 1907, May 21, 1965.

Schwarzschild, Bertram. "Polar Orbiter Shows Evidence of Minicomet Bombardment." *Physics Today*. Volume 50, July 1997, page 18.

Science News [no author given]. "Leonids Fulfill Promise." *Science News*. Volume 90, November 26, 1966, page 453.

Science News [no author given]. "Lost Comet Showers Sky." *Science News*. Volume 89, June 11, 1966, page 459.

Science News [no author given]. "'Venus Flytrap' Rocket To Collect Space Dust." *Science News*. Volume 90, November 12, 1966, page 401.

Science News Letter [no author given]. "Check-up Shows Leonid Meteors Numerous." *Science News Letter*. Volume 22, December 3, 1932, page 356.

Science News Letter [no author given]. "Earth Apparently Missed Main Leonid Meteor Swarm." *Science News Letter*. Volume 22, November 26, 1932, page 345.

Science News Letter [no author given]. "Height of Shooting Stars Measured With Wire Frame." *Science News Letter*. Volume 22, November 12, 1932, page 314.

Secchi, Angelo. "Sur les étoiles filantes de novembre 1866." *Les Mondes*. 2nd Series. Volume 12, December 20, 1866, pages 645–648.

Seneca, Lucius Annaeus. *Naturales Quaestiones*. Translated by Thomas H. Corcoran. Cambridge, Massachusetts: Harvard University Press; and London: William Heineman, 1971.

Senex. "The Meteoric Phenomenon on 13th November, 1833 – Keeping Diaries or Daily Journals." *Mechanics' Magazine*. Volume 3, January 1834, pages 61–62.

Serviss, Garrett P. "The Heavens in November." *Scientific American*. Volume 81, October 28, 1899.

Seventh-Day Adventist Church. *Bible Readings for the Home*. Nashville: Southern Publishing, 1942.

Seventh-Day Adventist Church. *Bible Readings for the Home*. Mountain View, California: Pacific Press, 1963.

Seventh-Day Adventist Church. *Bible Readings for the Home Circle*. Battle Creek, Michigan: Review and Herarld, 1889.

Seventh-Day Adventist Church. *Bible Readings for the Home Circle*. Mountain View, California: Pacific Press, 1914.

Shapley, Harlow, Ernst J. Öpik, and Samuel L. Boothroyd. "The Arizona Expedition for the Study of Meteors." *Proceedings of the National Academy of Sciences of the United States of America*. Volume 18, 1932, pages 16–23.

Sharpton, Virgil L., and Peter D. Ward, editors. *Global Catastrophes in Earth History; An Interdisciplinary Conference on Impacts, Volcanism, and Mass Mortality*. Boulder, Colorado: Geological Society of America (Special Paper 247), 1990.

[Silliman, Benjamin]. "Additional Observations on the Meteors of November 13 and 14,

1838." *American Journal of Science and Arts.* Volume 37, number 2, January 1839, page 372.

[Silliman, Benjamin]. "Foreign Accounts of the Meteoric Shower of November, 1836." *American Journal of Science and Arts.* Volume 32, number 1, April 1837, pages 181–183.

Sinding, Erik. "On the Systematic Changes of the Eccentricities of Nearly Parabolic Orbits." *Matematisk-Fysiske Meddelelser* udgivet af Det Kgl. Danske Videnskabernes Selskab. Volume 24, number 16, 1948, paginated 1–8.

Skellett, A[lbert] M[elvin]. "The Effect of Meteors on Radio Transmission Through the Kennelly–Heaviside Layer." *Physical Review.* Volume 37, June 15, 1931, page 1668.

Skellett, A[lbert] M[elvin]. "The Ionizing Effect of Meteors in Relation to Radio Propagation." *Proceedings of the Institute of Radio Engineers.* Volume 20, December 1932, pages 1933–1940.

Skellett, A[lbert] M[elvin]. "The Ionizing Effects of Meteors." *Proceedings of the Institute of Radio Engineers.* Volume 23, February 1935, pages 132–149.

Skellett, A[lbert] M[elvin]. "Radio Studies During the Leonid Meteor Shower of November 16, 1932." *Science.* Volume 76, November 11, 1932, page 434.

Sky and Telescope [no author given]. "A Good Leonid Year?" *Sky and Telescope.* Volume 32, November 1966, page 251.

Sky and Telescope [no author given]. "Great Leonid Meteor Shower of 1966." *Sky and Telescope.* Volume 33, January 1967, pages 4–10.

Sky and Telescope [no author given.] "Many Leonids Observed." *Sky and Telescope.* Volume 31, January 1966, pages 58–59.

Spratt, Christopher, and Sally Stephens. "Against All Odds: Meteorites That Have Struck Home." *Mercury.* Volume 21, March/April 1992, pages 50–56.

Steel, Duncan. "Meteoroid Streams." Pages 111–126 in A. Milani, M. DiMartino, and A. Cellino, editors. *Asteroids, Comets, Meteors 1993.* Dordrecht: Kluwer, 1993.

Steel, Duncan. *Rogue Asteroids and Doomsday Comets: The Search for the Million Megaton Menace that Threatens Life on Earth.* New York: John Wiley & Sons, 1995.

Štohl, J[an], and I[wan] P. Williams, editors. *Meteoroids and Their Parent Bodies.* Proceedings of the International Astronomical Symposium held at Smolenice, Slovakia, July 6–12, 1992. Bratislava: Astronomical Institute of the Slovak Academy of Sciences, 1993.

Stokes, Anson Phelps. *Memorials of Eminent Yale Men.* Volume 2: Science and Public Life. New Haven: Yale University Press, 1914.

Stoney, G[eorge] Johnstone. "Cause of the Non-appearance of the Shower." *Nature.* Volume 61, November 23, 1899, page 82.

Stoney, G[eorge] Johnstone. "The Story of the November Meteors." Lecture of February 14, 1879. Pages 160–171 in *Royal Institution Library of Science.* Barking: Elsevier, 1970.

Stoney, G[eorge] Johnstone, and A[rthur] M[atthew] W[eld] Downing. "Next Week's Leonid Shower." *Nature.* Volume 61, November 9, 1899, pages 28–29.

Stoney, G[eorge] Johnstone, and A[rthur] M[atthew] W[eld] Downing. "The Leonids: A Forecast." *Nature.* Volume 63, November 1, 1900, page 6.

Stoney, G[eorge] Johnstone, and A[rthur] M[atthew] W[eld] Downing. "Perturbations of the Leonids." *Proceedings of the Royal Society of London.* Volume 64, March 2, 1899, pages 403–409.

Strickland, [Samuel]. *Twenty-Seven Years in Canada West; or, the Experience of an Early Settler.* Edited by Agnes Strickland. London: Richard Bentley, 1853.

Strong, D. A. "On the Meteoric Phenomenon of November 13th, 1833." *Mechanics' Magazine.* Volume 3, January 1834, pages 60–61.

Sullivan, Walter. "Meteor Rain Awes Southwest With Celestial Fireworks Show." *New York Times*. November 18, 1966, page 45.

Sullivan, Walter. "Meteor Shower Is Due Tonight." *New York Times*. November 16, 1966, page 62.

Swezey, G. D. "The Present Status of Meteoric Astronomy." (Presidential address of December 1, 1899.) *Proceedings of the Nebraska Academy of Sciences*. Volume 7, November 1901, pages 79–89.

Sykes, Mark V. "Great balls of mire." *Nature*. Volume 362, April 22, 1993, pages 696–697.

Sykes, M[ark] V., and R. G. Walker. "Constraints on the Diameter and Albedo of 2060 Chiron." *Nature*. Volume 251, 1991, pages 777–780.

Thacher, T. A. "Edward C. Herrick." *New Englander*. Volume 21, October 1862, pages 820–859.

Theobald, John A. "Dubuque Counts of the 1931 Leonids." *Popular Astronomy*. Volume 40, January 1932, pages 54–56.

Theobald, John A. "Dubuque Counts of the 1932 Leonids." *Popular Astronomy*. Volume 41, January 1933, pages 56–59.

Toon, Owen B., Richard P. Turco, Curt Covey, Kevin Zahnle, and David Morrison. "Environmental Perturbations Caused by the Impacts of Asteroids and Comets." *Reviews of Geophysics*. Volume 35, February 1997, pages 41–78.

Tupman, G[eorge] L. "Results of Observations of Shooting Stars, made in the Mediterranean in the years 1869, 1870, and 1871," *Monthly Notices of the Royal Astronomical Society*. Volume 33, 1873, pages 298–312.

Turner, H[erbert] H[all]. "Observations of the Leonids of 1899, made at the University Observatory, Oxford." *Monthly Notices of the Royal Astronomical Society*. Volume 60, December 1899, pages 164–165.

Twining, Alexander C. "Investigations respecting the Meteors of Nov. 13th, 1833. – Remarks upon Prof. Olmsted's theory respecting the cause." *American Journal of Science and Arts*. Volume 26, number 2, July 1834, pages 320–352.

Twining, Alexander C. "Meteors on the morning of November 13th, 1834." *American Journal of Science and Arts*. Volume 27, number 2, January 1835, pages 339–340.

Tyerman, Thomas F. *The Origin of Meteors*. Oxford: Slatter and Rose, 1887.

Upton, Edward K. L. "The Leonids Were Dead, They Said." *Griffith Observer*. Volume 41, May 1977, pages 3–9.

U. S. Naval Observatory [no author given]. *November Meteors, 1866, as Observed at the U. S. Naval Observatory*. Washington, D.C.: Government Printing Office, 1866.

U. S. Naval Observatory [no author given]. *Observations and Discussions on the November Meteors of 1867*. Washington, D.C.: Government Printing Office, 1867.

Virgil. *Georgics. The Works of Virgil*. Translated by John Dryden. London: Oxford University Press, 1961.

Walker, Sears C. "Researches concerning the Periodical Meteors of August and November." *Transactions* of the American Philosophical Society. Volume 8, 1843, pages 87–140.

Waller, Adolph E. "Dr. John Locke, Early Ohio Scientist (1792–1856)." *Ohio State Archaeological and Historical Society Quarterly*. October–December 1946, pages 1–28.

Wasson, John T. "E. F. F. Chladni and Meteoritics." English introduction (pages iii–viii) in reprint of Ernst Floren Friedrich Chladni. *Ueber den Ursprung der von Pallas gefundenen und anderer ihr ähnlicher Eisenmassen, und über einige damit in Berbindung stehende Naturerscheinungen*. Riga: Johann Friedrich Hartknoch, 1794.

Watson, Fletcher G. *Between the Planets*. Revised edition. Garden City, New York: Doubleday (Anchor Books), 1962.

Weiss, Edmund. "Bemerkungen über den Zusammenhang zwischen Cometen und Sternschnuppen." *Astronomische Nachrichten*. Number 1632, March 9, 1867, columns 381–384; article dated February 22, 1867.

Weiss, Edmund. "Beiträge zu Kenntniss der Sternschnuppen." *Astronomische Nachrichten*. Number 1710–1711, 1868, columns 81–102; article dated July 16, 1868.

Weissman, Paul R. "The cometary impactor flux at the Earth." Pages 171–180 in Virgil L. Sharpton and Peter D. Ward, editors. *Global Catastrophes in Earth History*. Boulder, Colorado: Geological Society of America (Special Paper 247), 1900.

Weissman, Paul R., Michael F. A'Hearn, L. A. McFadden, and H. Rickman. "Evolution of Comets into Asteroids." Pages 880–920 in Richard P. Binzel, Tom Gehrels, and Mildred Shapley Matthews, editors. *Asteroids II*. Tucson: University of Arizona Press, 1989.

Wells, H. G. *War of the Worlds*. (First published in 1898.)

Wetherill, G[eorge] W. "Where Do the Apollo Objects Come From?" *Icarus*. Volume 76, 1988, pages 1–18.

Wetherill, G[eorge] W., and D. O. ReVelle. "Relationships Between Comets, Large Meteors, and Meteorites." Pages 297–319 in Laurel L. Wilkening, editor. *Comets*. Tucson: University of Arizona Press, 1982.

Whipple, Fred L. "A Comet Model. I. The Acceleration of Comet Encke." *Astrophysical Journal*. Volume 111, number 2, February 1950, pages 375–394.

Whipple, Fred L. "A Comet Model. II. Physical Relations for Comets and Meteors." *Astrophysical Journal*. Volume 113, number 3, March 1951, pages 464–474.

Whipple, Fred L. "The Harvard Photographic Meteor Program." *Sky and Telescope*. Volume 8, February 1949, reprint paginated 1–10.

Whipple, Fred L. "Photographic Meteor Studies. III. The Taurid Shower." *Proceedings of the American Philosophical Society*. Volume 83, number 5, October 1940, pages 711–745.

Whipple, Fred L., with Daniel W. E. Green. *The Mystery of Comets*. Washington, D.C.: Smithsonian Institution Press, 1985.

Williams, I[wan] P. and Zidian Wu. "The Geminid meteor stream and asteroid 3200 Phaethon." *Monthly Notices of the Royal Astronomical Society*. Volume 262, 1993, pages 231–248.

Williams, W. Mattieu. "The Fuel of the Sun." *Nature*. Volume 3, November 10, 1870, pages 26–27.

Wisconsin, State Historical Society of (Madison). Receipt from John Gardner and Lewis Benton to McCarty & Walker for merchandise, Skunk Grove, Racine County, Wisconsin; with notes on the back regarding a meteor shower and Indians' reaction to it, November 12 and 13, 1833. File 1833, November 1.

Woolsey, [Theodore Dwight]. "Discourse Commemorative of Professor Denison Olmsted, LL.D." *New Englander*. Volume 17, August 1859, pages 575–600.

Wright, M[armaduke] B[urr]. *An Address on the Life and Character of the Late Professor John Locke, Delivered at the Request of the Cincinnati Medical Society*. Cincinnati: Moore, Wilstach, Keys, 1857.

Wu Zidian and Iwan P. Williams. "Leonid meteor storms." *Monthly Notices of the Royal Astronomical Society*. Volume 280, 1996, pages 1210–1218.

Wylie, C[harles] C. "The 1931 Leonid Meteors." *Popular Astronomy*. Volume 39, December 1931, pages 609–611.

Wylie, C[harles] C. "The 1931 Leonid Meteors." *Popular Astronomy*. Volume 40, January 1932, pages 51–52.

Wylie, C[harles] C. "The Iowa Program for the 1932 Leonids." *Popular Astronomy*. Volume 40, November 1932, page 581.

Wylie, C[harles] C. "The 1932 Return of the Leonid Meteors." *Popular Astronomy*. Volume 40, February 1932, pages 97–99.

Wylie, C[harles] C. "Preliminary Report on the 1932 Leonids." *Popular Astronomy*. Volume 40, December 1932, pages 649–650.

Wylie, C[harles] C. "Preliminary Report on the 1933 Leonids." *Popular Astronomy*. Volume 41, December 1933, pages 582–583.

Yeomans, Donald K. *Comets: A Chronological History of Observation, Science, Myth, and Folklore*. New York: John Wiley & Sons, 1991.

Yeomans, D[onald] K. "Comet Tempel–Tuttle and the Leonid Meteors." *Icarus*. Volume 47, 1981, pages 492–499.

Yeomans, Donald K., and John C. Brandt. *The Comet Giacobini–Zinner Handbook*. Pasadena: Jet Propulsion Laboratory (Document 400–254), March 1985.

Yeomans, Donald K., Kevin K. Yau, and Paul R. Weissman. "The Impending Appearance of Comet Tempel–Tuttle and the Leonid Meteors." *Icarus*. Volume 124, 1996, pages 407–413.

Zhuang Tian-shan. "Ancient Chinese Records of Meteor Showers." *Chinese Astronomy*. Volume 1, 1977, pages 197–220.

INDEX

Note numbers are indicated by 'n.'